Lecture Notes in Mathematics

A collection of informal reports and seminars
Edited by A. Dold, Heidelberg and B. Eckmann, Zürich

209

Proceedings of Liverpool Singularities Symposium II

Edited by C. T. C. Wall,
University of Liverpool, Liverpool/G. B.

Springer-Verlag
Berlin · Heidelberg · New York 1971

AMS Subjekt Classifications (1970): 14D05, 14E15, 26A54, 32B10, 53C10, 53C45, 53C65, 57D30, 57D40, 57D45, 57D50, 57D70, 57D95, 58A10, 58A30, 34C20, 53A05, 53B25, 58C25, 81A15, 83C99

ISBN 3-540-05511-8 Springer-Verlag Berlin · Heidelberg · New York
ISBN 0-387-05511-8 Springer-Verlag New York · Heidelberg · Berlin

Offsetdruck: Julius Beltz, Hemsbach/Bergstr.

Introduction

The papers in this volume and its predecessor were submitted by participants of the Symposium on Singularities of smooth manifolds and maps, held at the Department of Pure Mathematics at the University of Liverpool from September 1969 - August 1970, and supported by the Science Research Council. They include also the texts of the courses of lectures given at a Summer School held in July 1970 in connection the the Symposium. I have added some further notes to bring out the relations of the different papers in an attempt to make this volume as complete an exposition of the present state of knowledge in the subject as is possible within its scope. For this reason also we have reprinted the classic notes by Levine of lectures by Thom, which had become unavailable, and which represent the source of many of the ideas further developed in these papers.

Volume 1 contains the papers closely related to the problem of classification of singularities of smooth maps and is thus fairly coherent. Volume 2 represents applications of these ideas (and others) to other branches of pure and applied mathematics, arranged in order of subject. The lack of coherence is perhaps compensated by the wealth of unsolved problems which are raised in these pages.

Thanks are due to the S.R.C. for financial support; to Mrs. Evelyn Quayle (nee Hastwell) and Mrs. Eileen Bratt for doing the typing; to members of the department for proofreading it; and of course to the participants in the Symposium for creating the body of mathematics here presented.

Liverpool, March 1971

CONTENTS

L'INTERPOLATION DES FONCTIONS DIFFÉRENTIABLES DE PLUSIEURS VARIABLES

G. Glaeser †

Introduction

La difficulté essentielle du calcul différentiel des fonctions de plusieurs variables apparaît dans le phénomène suivant :

Sous des hypothèses très larges, la limite d'une suite de cordes d'une surface est une tangente, alors qu'une limite de plans sécants n'est un plan tangent que dans des situations bien plus restrictives.

Ainsi l'approximation des dérivées partielles d'une fonction f par des expressions calculées à partir des valeurs qu'elle prend en plusieurs points pose des problèmes délicats.

Au contraire, on dispose d'une théorie satisfaisante et maniable pour étudier les fonctions d'une seule variable : c'est le calcul des différences (cf. par exemple [3] [14]). Mais lorsqu'une branche des mathématiques semble difficilement généralisable, elle n'intéresse qu'un nombre restreint de spécialistes. Le calcul des différences, abaissée au rang d'une technique, n'est guère pratiqué que par quelques analystes numériques.

Le calcul différentiel des fonctions de plusieurs variables n'a neanmoins jamais cessé de se developper, depuis Euler et Clairaut [6]. Mais faute d'un fondement "multiponctuel", il dut adopter deux artifices qui ont influencé accidentellement

† Ce travail a eté effectué pendant que l'auteur était invité par l'Université de Liverpool, au Symposium annuel sur les singularités des applications differentiables. L'auteur a bénéficié d'une bourse de la S.R.C. et du C.N.R.S.

son développement.

Le caractère artificiel de l'emploi systématique des fonctions d'une seule variable par restriction à des droites est bien reconnu. Ainsi s'introduisent les dérivées partielles.

Dans les traités didactiques, la formule de Taylor déguise un résultat relatif à la restriction à un segment de droite: c'est pour cette raison que l'on étudie généralement les fonctions différentiables sur des ensembles localement convexes (ensembles ouverts ou variétés différentielles) car on veut pouvoir joindre commodément chaque point à des points voisins. Enfin les tentatives concernant l'interpolation des fonctions de plusieurs variables apparaissent généralement en analyse numérique, comme une simple itération de l'interpolation à une seule variable avec des noeuds d'inter--polation disposés en quadrillage (cf. [14], [10], [11] etc.).

Que l'on ne vienne pas dire que c'est le seul cas "qui se rencontre" dans les applications pratiques. Lorsqu'un appareil de mesure explore une portion d'espace en effectuant des mesures en divers points, il est rare que ces résultats expérimentaux s'effectuent sur un réseau régulier! En fait, les cas qui se rencontrent sont ceux que l'on veut bien rencontrer.

Pour fonder le calcul différentiel sur des bases plus intrinsèques, grâce à la dérivée de Stolz-Fréchet, on utilise un autre artifice: l'emploi systématique des calculs biponctuels. Dans l'inégalité

$$|f(M) - T_A f(M)| \leqslant \|AM\|^r \ \omega(\|AM\|) ,$$

la géométrie intervient systématiquement par l'intermédiaire d'un couple de points (A, M). Le polynôme de Taylor $T_A f$ concentre une grande quantité d'information concernant le comportement de f en A. Mais de toute façon on ne fait intervenir dans chaque formule que des renseignements biponctuels.

La nécessité de se cantonner exclusivement dans l'emploi des raisonnements biponctuels a conduit à accorder une importance exagérée à la notion de variété différentielle fréquemment présentée comme le domaine naturel du calcul différentiel.

Il est clair aujourd'hui que l'on à constamment à manier des fonctions dérivables, non seulement sur des variétés mais aussi sur des ensembles analytiques, algébriques, semi-analytiques etc., et que la source des difficultés que l'on rencontre ici tient à l'impossibilité de couler cette situation dans le moule du calcul différentiel sur une variété.

Dans cette situation se développe actuellement sous l'influence de R. Thom et H. Whitney l'emploi d'un autre artifice: la notion d'ensemble stratifié : on se donne beaucoup de mal pour opérer une partition satisfaisante d'un ensemble en sous-variétés appelées strates, sur lesquelles on manie les fonctions différentiables. Puis intervient le supplice du recollement au cours duquel on tente de prouver que les solutions d'un problème, calculées séparément sur chaque strate représentent une solution globale.

Ce sont des tours de force de cette nature que constituent les théorèmes de Whitney, de Łojasiewicz-Hörmander, ainsi que le théorème de préparation de Malgrange [8]. Cette méthode est actuellement la seule disponible. Mais elle n'est naturelle que dans la mesure où l'on pense que l'emploi exclusif de formules biponctuelles tient à l'essence du calcul différentiel (ce que dément d'ailleurs le cas des fonctions d'une seule variable).

D'un autre côté, la théorie des jets de Ch. Ehresmann donne une représentation des fonctions de classe $\mathcal{C}^r$ grâce à une collection de renseignements ponctuels (ou encore atomiques), structurée d'une façon convenable. C'est là un outil très commode, qui n'est cependant pas tout à fait adapté à son objet. Son défaut essentiel tient à ce qu'une collection de jets variant sur le domaine de définition ne provient pas nécessairement d'une fonction différentiable; autrement dit, un champ continu de polynômes n'est taylorien que s'il satisfait à certaines inégalités biponctuelles (de Whitney).

Ce qui manque, c'est un calcul différentiel moléculaire (autrement dit multiponctuel). Une théorie de l'interpolation doit permettre d'exploiter des renseignements répartis sur divers ensembles finis, non nécessairement inclus dans une seule strate. A côté de la formule de Taylor, des formules analogues à celle de Lagrange devraient conduire à des majorations précises et maniables.

L'objet du présent article, ainsi que de [7], est de contribuer à forger les outils d'un calcul différentiel moléculaire.

Nous présentons ici deux théories de l'interpolation des fonctions de plusieurs variables.

<u>L'interpolation de Lagrange</u> n'est pas nouvelle, et présente de grands inconvénients que nous mettons en évidence; malgré ses défauts elle peut néanmoins servir dans la résolution de certains problèmes. Les appendices illustrent deux exemples d'emploi de cette méthode.

La <u>méthode des schémas d'interpolation</u> au contraire est plus prometteuse; pour interpoler une fonction en N points , on utilise un polynôme du $N^{\text{ième}}$ degré, à n variables, (qui comporte $\binom{N+n}{n} > N$ coefficients). Cette indétermination, qui a sans doute dissuadé d'entreprendre une telle étude, permet cependant l'élaboration d'une théorie satisfaisante.

Pour pouvoir envisager les cas d'interpolation osculatrices (ou confluentes), les systèmes de N points qui servent de noeuds d'interpolation sont remplacés par des <u>idéaux</u> de codimension N .

On est conduit à représenter une fonction $\mathcal{C}^r$ par ses <u>multi-jets</u>. Le théorème 4 donne une représentation bien plus satisfaisante que la représentation classique par des jets.

0. <u>Preliminaires</u> Les fonctions de classe $\mathcal{C}^r$, envisagées ici, sont définies soit sur un espace affine euclidien $\mathcal{E}$ à n dimensions, associé à un espace vectoriel réel $\mathbb{E}$, soit sur un pavé compact $\mathbb{K} \subset \mathcal{E}$.

Les polynômes définies sur $\mathcal{E}$ jouent un rôle essentiel : il en est, par conséquent, de même pour la structure affine de $\mathcal{E}$. On notera par $\mathcal{P}_N$ l'espace des polynômes à n variables de degré $\leq N$.

Si $\mathbb{B}$ est un espace de Banach, l'espace $\mathcal{C}^r(\mathbb{K}\,;\,\mathbb{B})$ des fonctions de classe $\mathcal{C}^r$, de source $\mathbb{K}$, à valeurs dans $\mathbb{B}$ sera muni d'une norme $\|\cdot\|^r_{\mathbb{K}}$ compatible avec la convergence <u>uniforme d'ordre r</u> (i.e. convergence uniforme sur $\mathbb{K}$, de chacune des derivées d'ordre $\leq r$).

Nous n'insistons pas spécialement ici sur les relations entre $\mathcal{E}^r(\mathbb{K}\,;\,\mathbb{R})$ et $\mathcal{E}^r(\mathbb{K}\,;\,\mathbb{B})$. Les principaux résultats de cet article s'etendent aux fonctions à valeurs dans un Banach, en utilisant les techniques de produit tensoriel topologique d'espaces normés ([13]): on rappelle que $\mathcal{E}^r(\mathbb{K}\,;\,\mathbb{B})$ est isomorphe a $\mathcal{E}^r(\mathbb{K}\,;\,\mathbb{R})\ \hat{\hat{\otimes}}_\epsilon\ \mathbb{B}$ ou $\hat{\hat{\otimes}}_\epsilon$ désigne le complété du produit tensoriel $\mathcal{E}^r(\mathbb{K}\,;\,\mathbb{R}) \otimes \mathbb{B}$ pour la plus petite norme de Schatten-Grothendieck. Le lecteur familier avec cette théorie, fera aisément les généralisations nécessaires.

On fera parfois allusion à l'espace $\mathcal{E}^{r+\mathrm{Lip}}(\mathbb{K}\,;\,\mathbb{R})$ des fonctions de classe $\mathcal{E}^r$ dont la dérivée $r^{\text{ième}}$ est lipschitzienne. En ce qui concerne les relations entre la théorie des distributions et la théorie des idéaux de codimension finie développée ici, les renseignements suivants seront peut-être utiles au lecteur, sans être strictement indispensables.

La transformation de Fourier établit un homomorphisme entre l'algèbre $\mathbb{R}[X_1, X_2, \ldots, X_n]$ des polynômes à n indeterminées et l'algèbre des opérateurs différentiels linéaires à coefficients réels : la "multiplication" des opérateurs différentiels est alors la convolution définie par

$$\frac{\partial^\alpha}{\partial x^\alpha} * \frac{\partial^\beta}{\partial x^\beta} = \frac{\partial^{\alpha+\beta}}{\partial x^{\alpha+\beta}} .$$

Au polynôme $P = P(X_1, X_2, \ldots, X_n)$ on associe l'opération

$$\hat{P} = P\left(\frac{\partial}{\partial x_1}, \frac{\partial}{\partial x_2}, \ldots, \frac{\partial}{\partial x_n}\right) .$$

Nous n'éprouvons pas le besoin, ici, de nous conformer à l'usage de l'analyse harmonique où la définition de $\hat{P}$ fait intervenir le nombre complexe i, pour faire apparaitre une symétrie dans la formule de Plancherel.

Ainsi, au polynôme $Q = \frac{\partial P}{\partial X_i}$, obtenu par dérivation partielle de P, on associe l'opérateur $\hat{Q} = Q\left(\frac{\partial}{\partial x_1}, \ldots, \frac{\partial}{\partial x_n}\right)$: on dit que Q se déduit de $\hat{P}$ par <u>co-dérivation partielle</u>.

Ainsi $2\frac{\partial}{\partial x_1}$ et $2\frac{\partial}{\partial x_2}$ sont les deux codérivées partielles du laplacien $\frac{\partial^2}{\partial x_1^2} + \frac{\partial^2}{\partial x_2^2}$.

Soit $Q \in \mathbb{R}[X_1, X_2, \ldots, X_n]$. A l'endomorphisme de $\mathbb{R}[X_1, X_2, \ldots, X_n]$ défini par la multiplication $P \to PQ$ correspond par transformation de Fourier un endomorphisme de co-dérivation des opérateurs différentiels.

Toutes ces considérations s'appliquent encore, au langage près, à des distributions atomiques (resp. moléculaires) $\Big($(i.e. à support ponctuel (resp. fini)$\Big)$.

La théorème suivant est trivial, mais utile (cf. [4]).

Théorème <u>Soit Δ_A un espace vectoriel de distributions d'ordre $< r$, à support ponctuel $A \in \mathcal{E}$. Pour que Δ_A soit précisement l'orthogonal d'un idéal $\mathcal{I}$ à spectre ponctuel, il faut et il suffit que Δ_A soit stable par co-dérivation.</u>

L'ensemble des distributions moléculaires orthogonales à un idéal $\mathcal{I}$ de $\mathcal{E}^r(\mathcal{E}\,;\mathbb{R})$ est somme direct des espaces Δ_A attachés aux divers points du spectre de $\mathcal{I}$.

<u>Exemple</u> Les distributions portées par l'origine, orthogonales à l'idéal principal de $\mathcal{E}^\infty(\mathbb{R}^n\,;\mathbb{R})$ engendré par le polynôme $X_1^2 + X_2^2 + \ldots + X_n^2$ sont de la forme $\hat{H} = H\left(\frac{\partial}{\partial x_1}, \frac{\partial}{\partial x_2}, \ldots, \frac{\partial}{\partial x_n}\right)$ ou $H(X_1, X_2, \ldots, X_n)$ est un polynôme harmonique.

(On utilise ici la dualité de Fourier entre $\sum X_i^2$ et le laplacien.)

Or, si H est un polynôme harmonique, il en est de même de chacune des dérivés partielles de H : il en résulte que l'orthogonal de l'idéal principal étudié est bien stable par co-dérivation.

I. <u>L'interpolation de Lagrange.</u> Soit $A = \{A_1, A_2, \ldots, A_N\}$ un ensemble fini de N points distincts de $\mathcal{E}$, que nous appellerons un <u>système de noeuds d'interpolation</u>.

Etant donnée une fonction numérique f définie sur $\mathcal{E}$, on se propose de trouver des polynômes qui prennent les mêmes valeurs que f aux noeuds $A_i \in A$: c'est toujours possible d'une infinité de manières. Mais si l'on n'accepte que des polynômes de degré $\leq r$ (r donné), les deux conditions (a) et (b) formulées plus loin assurent l'existence et l'unicité d'un tel polynôme d'interpolation. C'est le <u>polynôme de Lagrange</u> $\mathcal{L}_A f$.

Ce polynôme s'obtient en résolvant le système linéaire

$$(1) \quad \mathcal{L}_A f(A_i) = f(A_i) \qquad i \leqslant N .$$

Le nombre des coefficients à determiner est égal à $\binom{n+r}{n} = \binom{n+r}{r}$.

Pour que le système (1) soit cramérien, il faut et il suffit:

I.1 (a) que le nombre N des points distincts de A soit égal à $\binom{n+r}{r}$;

(b) que son déterminant $\Delta(A)$ ne soit pas nul.

Dans le cas classique des fonctions d'une variable $n = 1$ l'entier N peut prendre une valeur arbitraire, et le déterminant $\Delta(A)$, qui est alors vandermondien ne s'annule pas, dès lors que les points A_i sont distincts.

Mais il n'en est plus de même pour $n > 1$. Ainsi pour $n = 2$, il faut 3, 6, 10, etc. points distincts pour effectuer l'interpolation de Lagrange de degré 1, 2, 3. Le déterminant $\Delta(A)$ s'écrit en disposant en colonnes les valeurs que prennent aux points A_i, les divers polynômes constituant une base de l'espace $\mathcal{P}_r$ des polynômes de degré $\leqslant r$.

Ce déterminant s'annule, si et seulement si, les points A_i sont situés sur une même variété algébrique de degré $\leqslant r$.

Par exemple, si $n = 2$, 3 points (resp. 6 points) satisfont à la condition (b) pour $r = 1$ (resp. $r = 2$), s'ils ne sont pas alignés (resp. sont situés sur une même cônique éventuellement décomposée).

Dans le cas contraire, on dit que A est <u>unisolvant.</u> Tout système unisolvant A détermine sa base de Lagrange constituée par N polynômes $p_i\{A\}$, où $i = N$, satisfaisant à

$$p_i\{A\}(A_j) = \delta_j^i \quad \text{(symbole de Kronecker).}$$

Pour toute fonction $\vec{f}$ définie sur $\mathcal{E}$, à valeurs dans $\mathbb{B}$, on a

$$\text{I.2} \qquad \mathcal{L}_A \vec{f} = \sum_{i \leqslant N} \vec{f}(A_i)\, p_i\{A\} .$$

On conçoit que si $\mathcal{A}$ est particulièrement mal choisi $\mathcal{L}_{\mathcal{A}}\vec{f}$ traduise très mal les propriétés différentielles de $\vec{f}$. On associe à $\mathcal{A}$ (supposé à support dans $\mathbb{K}$) une norme $\|\mathcal{A}\|$: c'est la norme du projecteur $\mathcal{L}_{\mathcal{A}}$ dans $\mathcal{E}^r(\mathbb{K};\mathbb{B})$

I.3 $$\|\mathcal{L}_{\mathcal{A}}\vec{f}\|^r_{\mathbb{K}} \leq \|\mathcal{A}\|\cdot\|\vec{f}\|^r_{\mathbb{K}} .$$

Par exemple, si une suite de triangles $\{A_1, A_2, A_3\}$ ayant un angle obtus (resp. n'ayant que des angles aigus) dégénère en trois points alignés (resp. en deux point distincts) ils constituent une suite non bornée (resp. bornée) de systèmes unisolvants pour $n = 2,\ r = 1$.

Pour utiliser une suite de systèmes unisolvants dans une démonstration d'analyse, il convient de contrôler la croissance des normes de cette suite (cf. Appendice II).

<u>Expression intégrale du reste de la formule de Lagrange</u>

Considérons un système unisolvant $\mathcal{A}$ et deux points B et M de $\mathcal{E}$. Appliquons l'opérateur $\mathcal{L}_{\mathcal{A}}$ aux deux membres du développement de Taylor $f = T_B f + R_B f$ d'une fonction f de classe $\mathcal{E}^{r+1}(\mathcal{E};\mathbb{B})$ (où plus généralement de classe $\mathcal{E}^{r+\mathrm{Lip}}(\mathcal{E};\mathbb{B})$). En tenant compte de $\mathcal{L}_A T_B f = T_B f$, on trouve $\mathcal{L}_{\mathcal{A}} f(M) = T_B f(M) + \mathcal{L}_{\mathcal{A}} R_B f(M)$.

Remplaçant M par B, il vient

I.4 $$\mathcal{L}_{\mathcal{A}} f(B) - f(B) = \sum p_i(B)\cdot R_B f(A_i) .$$

Utilisons maintenant l'expression intégrale du reste de la formule de Taylor, ou $\widehat{BA}$ représente un <u>arc quelconque</u> joignant B à A_i (cf. [7]). On obtient :

I.5 $$\mathcal{L}_{\mathcal{A}} f(B) - f(B) = \sum_{i\in N} p_i(B)\int_{\widehat{BA_i}} D_X^{r+1} f\left[\frac{\overset{r}{\circledcirc}\,\overline{XA_i}}{r!}\circledcirc dX\right] .$$

Cette formule est connue, sous diverses formes, dans le cas des fonctions d'une seule variable: dans ce cas, on peut regrouper les termes de la somme sous un signe d'intégration unique avec, pour intégrande, le produit de $\frac{d^{r+1}}{dx^{r+1}} f$ par une certaine fonction-spline.

Dans [7] nous étudions systématiquement des restes intégraux analogues, pour les fonctions de plusieurs variables. Nous montrons notamment pourquoi la formule précédente ne se laisse pas réduire, en général, à une intégrale unique; le long d'une seule 1-chaine d'intégration.

L'inconvénient majeur de l'interpolation de Lagrange tient à ce qu'elle ne s'applique qu'à des systèmes de noeuds particuliers.

Si l'on substitue à $\mathcal{P}_N$ un autre espace vectoriel de dimension $\binom{N+n}{n}$ constitué par d'autres fonctions, le même phénomène se retrouve. Un célèbre théorème de Haar (cf. [3]) affirme qu'il est impossible de concilier, dans le problème de l'interpolation des fonctions de plus d'une variable, l'existence et l'unicité pour tout système de noeuds.

Pour remédier à cette situation, deux stratégies sont concevables.

Ou bien, on renonce à l'existence. On ne cherche à interpoler les fonctions que sur des systèmes unisolvants. C'est <u>l'interpolation de Lagrange</u> et c'est presque uniquement dans cette voie que quelques tentatives ont été faites pour résoudre des problèmes particuliers (cf. par exemple [4] p.11 et 12 ainsi que les Appendices I et II).

Ou bien on renonce provisoirement à l'unicité. Nous allons développer maintenant une théorie où l'on utilise N points, tout à fait arbitrairement choisis pour projeter les fonctions de classe $\mathcal{C}^r$ sur un sous-espace à N dimensions de l'espace $\mathcal{P}_N$ (à $\binom{N+n}{n}$ dimensions). Cette interpolation ne conserve pas les polynômes du $N^{\text{ième}}$ degré. Le choix du noyau et de l'image de ce projecteur n'est pas déterminé d'une façon unique. Mais grâce à des conventions spéciales on parvient à restituer l'unicité. C'est la théorie des <u>schémas d'interpolation</u>.

II. <u>Idéaux de codimension finie</u>

Une théorie de l'interpolation n'est satisfaisante que si elle s'étend aux cas où certains noeuds viennent à se confondre selon certains processus convenables. On obtient ainsi des interpolations <u>confluentes</u> (ou <u>osculatrices</u>). Il convient donc d'accepter des systèmes de noeuds d'interpolation généralisés.

On remarque que $f - \mathcal{L}_A f$ est une fonction qui s'annule sur l'ensemble A. Autrement dit, si $f \in \mathcal{E}^r(\mathbb{K};\mathbb{R})$, $f - \mathcal{L}_A f$ appartient à <u>un idéal de codimension finie N.</u> De même, une formule d'interpolation confluente conduira à un reste qui appartiendra à un idéal de codimension N, plus compliqué, (avec annulation non seulement de la fonction, mai aussi de certaines dérivées).

C'est là la motivation de la fin de ce chapitre.

Soit $\tilde{\mathbb{K}}(N)$ l'ensemble des idéaux fermés de $\mathcal{E}^r(\mathbb{K};\mathbb{R})$ de codimension finie N. En particulier $\tilde{\mathbb{K}}(1)$, ensemble des idéaux maximaux (nécessairement fermés) s'identifie à $\mathbb{K}$. Il est commode de parler, par abus de langage, indifféremment du point $m \in \mathbb{K}$, ou de l'idéal $m \in \tilde{\mathbb{K}}(1)$ des fonctions qui s'annulent au point m.

Nous supposerons par la suite que

II.1 $\qquad N \leq r+1$

Voici les phénomènes pathologiques qu'il s'agit de maîtriser, en faisant intervenir cette hypothèse :

Pour tout entier $h > 0$, l'idéal m^{k+1} n'est pas fermé dans $\mathcal{E}^r(\mathbb{K};\mathbb{R})$ pour r fini, et sa codimension est infinie: il existe en effet des fonctions "rugueuses" (sin 1/x par exemple) qui multipliées par tout élément de m^{k+1} donnent un produit de classe $\mathcal{E}^r$, appartenant a $\overline{m^{k+1}}$, <u>mais</u> non a m^{k+1} [5].

Lorsque $k < r$ (resp. $k \geq r$) l'adherence $\overline{m^{k+1}}$ est l'ideal des fonctions k-plates (resp. r-plates) au point m.

Si H est un hyperplan non fermé de $\overline{m^{r+1}}$ contenant l'idéal non fermé $(\overline{m^{r+1}}.m)$, H est un idéal non fermé de codimension finie.

<u>Proposition 1</u> <u>Si $\mathcal{J}$ est un idéal primaire ((i.e. contenu dans un seul idéal maximal</u> m <u>de</u> $\mathcal{E}^r(\mathbb{K};\mathbb{R})$), <u>de codimension</u> N, <u>on a</u> $m^N \subset \mathcal{J} \subset m$. <u>De plus, si</u> $\mathcal{J}$ <u>est fermé</u> $\overline{m^N} \subset \mathcal{J} \subset m$.

En effet, l'anneau $\mathcal{E}^r(\mathbb{K};\mathbb{R})/\mathcal{J}$ est un anneau local, d'idéal maximal $m/\mathcal{J}$, et c'est aussi un espace vectoriel de dimension N. Le lemme de Nakayama, appliqué à la chaîne d'idéaux $m^k/\mathcal{J}$ montre qu'il existe un plus petit entier k tel que $m^k \subset \mathcal{J}$. Mais toute chaîne strictement monotone de sous-espaces vectoriels est, au plus de longueur N. Ainsi $k \leq N$. Par conséquent $m^N \subset \mathcal{J}$ et $\overline{m^N} \subset \mathcal{J}$.

Dans le cas général, $\mathscr{I}$ admet une décomposition primaire $\mathscr{I} = \mathscr{I}_1 \cap \mathscr{I}_2 \dots \cap \mathscr{I}_k = \mathscr{I}_1 . \mathscr{I}_2 . \ \dots \ . \mathscr{I}_k$ où les $\mathscr{I}_\alpha$ sont contenus dans des idéaux maximaux m_α distincts. Soit N_α la codimension de $\mathscr{I}_\alpha$. L'ensemble des points m_α s'appelle le spectre Spec $(\mathscr{I})$.

Proposition 2 Si $\mathscr{I}$ est un idéal fermé de codimension N de $\mathscr{E}^r(\mathbb{K}; \mathbb{R})$, le quotient $\mathscr{E}^r(\mathbb{K}; \mathbb{R})/\mathscr{I}$ est isomorphe à la somme directe des quotients $\mathscr{E}^r(\mathbb{K}; \mathbb{R})/\mathscr{I}_\alpha$. Par conséquent $N = \sum_{\alpha \leq k} N_\alpha$. Si l'on pose $m(\mathscr{I}) = \bigcap_{\alpha \leq k} m_\alpha$ et $n(\mathscr{I}) = \bigcap m_\alpha^{N_\alpha}$, on a

$$\text{II.2} \qquad \overline{n(\mathscr{I})} \subset \mathscr{I} \subset m(\mathscr{I}) .$$

Posons $\mathscr{K}_\alpha = \bigcap_{\beta \neq \alpha} \mathscr{I}_\beta$. On peut construire une partition de l'unité $u + u' = 1$ où u et u' sont des polynômes appartenant respectivement à $n(\mathscr{I}_\alpha)$ et $n(\mathscr{K}_\alpha)$; ainsi $\mathscr{I}_\alpha + \mathscr{K}_\alpha = \mathscr{E}^r(\mathbb{K}; \mathbb{R})$ et $\mathscr{I}_\alpha \cap \mathscr{K}_\alpha = \mathscr{I}$. Alors la suite exacte classique

$$0 \to \frac{\mathscr{E}^r(\mathbb{K}; \mathbb{R})}{\mathscr{I}_\alpha \cap \mathscr{K}_\alpha} \to \frac{\mathscr{E}^r(\mathbb{K}; \mathbb{R})}{\mathscr{I}_\alpha} \oplus \frac{\mathscr{E}^r(\mathbb{K}; \mathbb{R})}{\mathscr{K}_\alpha} \to \frac{\mathscr{E}^r(\mathbb{K}; \mathbb{R})}{\mathscr{I}_\alpha + \mathscr{K}_\alpha} \to 0$$

permet de conclure (par récurrence sur k), que $\mathscr{E}^r(\mathbb{K}; \mathbb{R})/\mathscr{I}$ est somme directe de ses sous-algèbres ponctuelles.

Si $\mathscr{I}$ est fermé, $n(\mathscr{I}) \subset \mathscr{I}$ implique $\overline{n(\mathscr{I})} \subset \mathscr{I}$.

Définition Deux sous-espaces $\mathbb{E}$ et $\mathscr{F}$ de $\mathscr{E}^r(\mathbb{K}; \mathbb{R})$ sont transverses si $\mathbb{E} + \mathscr{F} = \mathscr{E}^r(\mathbb{K}; \mathbb{R})$.

Théorème 1. L'espace $\mathscr{P}_{N-1}$ des polynômes de degré $\leq N - 1$ est transverse à tout idéal $\mathscr{I} \in \widetilde{\mathbb{K}}(N)$, et même à l'idéal $\overline{n(\mathscr{I})}$ correspondant.

Cela signifie, que pour tout $f \in \mathscr{E}^r(\mathbb{K}; \mathbb{R})$ et $\mathscr{I} \in \widetilde{\mathbb{K}}(N)$, il existe un polynôme P de degré $\leq N - 1$ tel que $f - P \in \overline{n(\mathscr{I})}$, et à plus forte raison $f - P \in \mathscr{I}$.

La démonstration procède par récurrence sur le nombre k des points distincts du spectre de $\mathcal{I}$.

Si $k = 1$, $\mathcal{I}$ est primaire, $\overline{n(\mathcal{I})}$ est l'idéal des fonctions $N - 1$ plates au point m. Pour $f \in \mathcal{E}^r(\mathbb{K}; \mathbb{R})$ on peut choisir P égal au développement de Taylor de f, d'ordre $N - 1$, au point m.

Supposons maintenant le théorème démontré jusqu'à l'entier $k - 1$. Soient $\mathcal{I}' = \mathcal{I}_1 \cap \mathcal{I}_2 \cap \ldots \cap \mathcal{I}_{k-1}$ et $N' = N - N_k = N_1 + N_2 + \ldots + N_{k-1}$. Alors $\mathcal{P}_{N'-1}$ est transverse à l'idéal $\overline{n(\mathcal{I}')}$. Pour tout $f \in \mathcal{E}^r(\mathbb{K}; \mathbb{R})$, il existe donc un polynôme $P' \in \mathcal{P}_{N'-1}$ tel que $f - P'$ soit $N_\alpha - 1$ plat en chacun des points m_α avec $\alpha \neq k$.

<u>Lemme 1</u> <u>Il existe un polynôme</u> $Q \in \mathcal{P}_{N'}$, <u>égal à</u> 1 <u>au point</u> m_k, <u>et</u> $(N_\alpha - 1)$ <u>plat en chacun des points</u> m_α (pour $\alpha \neq k$).

Il suffit de former un produit de N' polynômes du 1er degré, qui prennent tous la valeur 1 au point m_k ; parmi ceux-ci, on en choisira au moins N_α qui s'annulent au point m_α, pour $\alpha \leq k - 1$.

Ce polynôme est "inversible, modulo $\mathcal{I}_k$". Il en résulte que la multiplication par Q est une application linéaire injective de $\mathcal{P}_{N_k-1}$ dans $\mathcal{P}_{N-1}$, qui transforme tout sous-espace de P_{N_k-1} transverse à $\overline{m_k^{N_k}}$ (et en particulier $\mathcal{P}_{N_k-1}$) en un sous-espace de $\mathcal{P}_{N-1}$ transverse à $\overline{m_k^{N_k}}$. Il existe donc un polynôme P'', de degré $< N_k$, tel que $(f - P') - QP'' \in m_k^{N_k}$. Or $f - P'$, Q et par conséquent QP'', sont $N_\alpha - 1$ plats en chacun des points m_α où $\alpha \leq k - 1$. Donc $f - (P' + QP'') \in \overline{n(\mathcal{I})}$. Comme $P' + QP'' \in \mathcal{P}_{N-1}$ le théorème est démontré.

<u>Remarque</u> Si $p < N - 1$, l'espace $\mathcal{P}_p$ n'est plus transverse en général à $\mathcal{I} \in \hat{\mathbb{K}}(N)$. Il suffit de choisir $\mathcal{I} = m_1 \cap m_2 \cap \ldots \cap m_N$ où les points m_α sont distincts et <u>alignés</u> pour s'en convaincre. Tout polynôme de degré $p < N - 1$ qui s'annule en $p + 1$ points distincts alignés, s'annule identiquement sur cette droite.

<u>Proposition 3</u> <u>Tout idéal</u> $\mathscr{I} \in \tilde{\mathbb{K}}(N)$ <u>est l'adhérence d'un ideal engendré par des polynômes de degré</u> $\leq N$. (<u>On dit aussi qu'il est topologiquement engendré par des polynômes</u> $\in \mathscr{P}_N$.)

Nous allons d'abord mettre en évidence le lemme suivant:

<u>Lemme 2</u> <u>Etant donnés des points</u> m_α $(\alpha \leq k)$ <u>affectés d'entiers</u> $N_\alpha > 0$ <u>tels que</u> $\sum N_\alpha = N$, <u>alors l'idéal fermé des fonctions qui sont</u> $N_\alpha - 1$ <u>plates en</u> m_α <u>(pour</u> $\alpha \leq k$) <u>est topologiquement engendré par des polynômes appartenant à</u> $\mathscr{P}_N$.

En effet, cet idéal est topologiquement engendré dans $\mathscr{E}^r(\mathbb{K}\,;\,\mathbb{R})$ par tous les polynômes qui sont produits de N facteurs du premier degré, parmi lesquels N_α au moins s'annulent au point m_α.

Mais le théorème précédent affirme que toute fonction $f \in \mathscr{E}^r(\mathbb{K}\,;\,\mathbb{R})$ est somme d'un polynôme de degré $\leq N - 1$ et d'un élément de $\overline{n(\mathscr{I})}$. En appliquant ceci à $f \in \mathscr{I}$ la proposition est établie.

<u>Proposition 4</u> <u>Une condition nécessaire et suffisante pour qu'un idéal</u> $\mathscr{I}$ <u>de codimension finie soit fermé est que</u> $\overline{n(\mathscr{I})} \subset \mathscr{I}$.

La nécessité a été établie (prop.2).

Réciproquement, supposons $\overline{n(\mathscr{I})} \subset \mathscr{I}$. Comme $\overline{n(\mathscr{I})}$ est de codimension finie dans $\mathscr{E}^r(\mathbb{K}\,;\,\mathbb{R})$, il l'est aussi dans $\mathscr{I}$. Donc $\mathscr{I}$ est somme directe d'un sous-espace de dimension finie et d'un sous-espace fermé: par conséquent $\mathscr{I}$ est fermé (cf. Bourbaki E.V.T. Chap. 1 §2 Coroll.4).

III <u>Topologie grassmannienne sur</u> $\tilde{\mathbb{K}}(N)$ Soit $\mathscr{F}$ un sous-espace de dimension finie tel que $\mathscr{P}_N \subset \mathscr{F} \subset \mathscr{E}^r(\mathbb{K}\,;\,\mathbb{R})$. Pour tout $\mathscr{I} \in \mathbb{K}_N$, $\mathscr{I} \cap \mathscr{F}$ est un sous-espace vectoriel de codimension N, dans $\mathscr{F}$. Il appartient donc à une certaine grassmannienne $\mathscr{G}(\mathscr{F})$, construite sur l'espace $\mathscr{F}$. Nous écrirons $\mathscr{G}$ au lieu de $\mathscr{G}(\mathscr{P}_N)$.

<u>Proposition 5</u> <u>L'application</u> $\mathscr{I} \longmapsto \mathscr{I} \cap \mathscr{F}$ <u>de</u> $\tilde{\mathbb{K}}_N$ <u>dans</u> $\mathscr{G}(\mathscr{F})$ <u>est injective.</u>

En effet, $\mathcal{I} \cap \mathcal{J} = \mathcal{I}' \cap \mathcal{J}$, implique $\mathcal{I} \cap \mathcal{S}_N = \mathcal{I}' \cap \mathcal{S}_N$ et les idéaux $\mathcal{I}$ et $\mathcal{I}'$ qui sont topologiquement engendrés par les mêmes éléments (cf. Prop.3), sont égaux.

<u>Remarque</u> La même conclusion serait en défaut pour l'application $\mathcal{I} \mapsto \mathcal{I} \cap \mathcal{S}_{N-1}$, dans le cas $n = 1$. Pour tout idéal $\mathcal{I}$ de codimension N formé des fonctions d'une seule variable, $\mathcal{I} \cap \mathcal{S}_{N-1}$ se réduit à $\{0\}$.

L'injection $\mathcal{I} \mapsto \mathcal{I} \cap \mathcal{S}_N$ permet d'identifier $\tilde{\mathbb{K}}_N$ à une partie de $\mathcal{G}$: on définit ainsi la <u>topologie grassmannienne</u> de $\tilde{\mathbb{K}}_N$, induit par celle de $\mathcal{G}$.

<u>Théorème 2</u> <u>$\tilde{\mathbb{K}}_N$ est compact pour la topologie grassmannienne</u>.

Utilisant la compacité de $\mathcal{G}$, nous pouvons extraire de toute suite d'idéaux appartenant à $\tilde{\mathbb{K}}_N$, une sous-suite $\mathcal{I}_i$ telle que $\mathcal{I}_i \cap \mathcal{S}_N$ converge, au sens de $\mathcal{G}$, vers un sous-espace $\mathcal{H}$ de codimension N, dans $\mathcal{S}_N$. Il s'agit de prouver qu'il existe un idéal $\mathcal{I} \subset \tilde{\mathbb{K}}(N)$ tel que $\mathcal{H} = \mathcal{I} \cap \mathcal{S}_N$.

Nous allons d'abord construire le "candidat-spectre" de $\mathcal{I}$. En extrayant éventuellement une sous-suite, on peut faire converger Spec $\mathcal{I}_i$, au sens de la métrique de Hausdorff, (cf. Bourbaki Top. Géné. Chap. IX, §2 Exerc. 7) vers un ensemble fini $\mathbb{F} = \{m_1, m_2, \ldots, m_k\}$ où $k \leq N$. En effet :

<u>Lemme 3</u> <u>L'ensemble des parties finies de $\mathbb{K}$ dont le cardinal est inférieur ou égal à N est compact, pour la topologie de Hausdorff</u>.

Il suffit de prouver que si des $\mathbb{F}_i$, avec Card $\mathbb{F}_i \leq N$, convergent vers $\mathbb{F}$ au sens de la métrique de Hausdorff, le cardinal de $\mathbb{F}$ est inferieur ou égal à N. Dans le cas contraire, il existerait $N + 1$ boules disjointes centrées sur $\mathbb{F}$ qui devraient contenir individuellement au moins un point d'un ensemble $\mathbb{F}_i$. Or il est impossible de répartir N objets dans $N + 1$ tiroirs sans que l'un au moins soit vide.

$\mathbb{F} = \{m_1, m_2, \ldots, m_k\}$ où $k \leq N$ est le candidat-spectre cherché.

Nous allons affecter les points m_α de coefficients N_α tel que $\sum N_\alpha = N$: ce qui permettra de construire le candidat $\overline{n(\mathcal{I})}$.

Considérons pour cela un voisinage de $\mathbb{F}$, réunion de k boules disjointes $\mathcal{B}(m_\alpha ; \rho)$ de centre m_α et de rayon ρ. Pour i assez grand, le spectre de $\mathcal{I}_i$ sera inclus dans ce voisinage. On affectera le point m_α du coefficient N_α, somme des coefficients analogues pour $\mathcal{I}_i$ relatifs aux points de $\mathrm{Spec}(\mathcal{I}_i) \cap \mathcal{B}(m_\alpha; \rho)$.

Comme l'ensemble des décompositions d'un entier N en somme d'entiers strictement positifs est fini, on peut, en extrayant éventuellement une sous-suite, faire en sorte que le coefficient N_α affecté à m_α ne dépende pas de $\mathcal{I}_i$.

Nous sommes donc en mesure de définir le candidat $\overline{n(\mathcal{I})}$, avant d'avoir défini $\mathcal{I}$ lui-même. Ce sera l'ensemble des fonctions de $\mathcal{E}^r(\mathbb{K} ; \mathbb{R})$ qui sont $N_\alpha - 1$ plates au point m_α, pour $\alpha \leq k$.

Prouvons maintenant que $\mathcal{H}$ contient $\overline{n(\mathcal{I})} \cap \mathcal{S}_N$.

Remarquons, dans ce but, que sauf pour une infinité dénombrable de directions d'hyperplans H de $\mathcal{E}$, on peut trouver une droite transverse à H, telle que la réunion des spectres des idéaux $\mathcal{I}_i$ se projette injectivement sur cette droite. Cela permet de construire une suite de polynômes de $n(\mathcal{I}_i) \cap \mathcal{S}_N$ dépendant d'une seule variable (i.e. constant sur tout hyperplan parallèle à H). qui converge vers un generateur du sous-espace des polynômes de cette variable appartenant à $n(\mathcal{I}) \cap \mathcal{S}_N$ (d'après la continuité des racines d'une équation algebrique à une inconnue).

Cette limite appartient aussi à $\mathcal{H}$, puisque $n(\mathcal{I}_i) \cap \mathcal{S}_N \subset \mathcal{I}_i \cap \mathcal{S}_N$. En faisant varier H, on obtient suffisamment de polynômes pour engendrer le sous-espace vectoriel $\mathcal{S}_N \cap \overline{n(\mathcal{I})}$ de $\mathcal{S}_N$ (cf. Lemme 2). Ainsi $\overline{n(\mathcal{I})} \cap \mathcal{S}_N \subset \mathcal{H}$.

Posons maintenant $\mathcal{I} = \mathcal{H} + \overline{n(\mathcal{I})}$: cette somme d'un sous-espace fermé et d'un sous-espace de dimension finie est <u>fermée</u> dans $\mathcal{E}^r(\mathbb{K}; \mathbb{R})$.

Vérifions que $\mathcal{H} = \mathcal{I} \cap \mathcal{S}_N$ et que $\mathcal{I}$ est de codimension finie :

$$(\mathcal{I} \cap \mathcal{S}_N) = \left(\mathcal{H} + \overline{n(\mathcal{I})}\right) \cap \mathcal{S}_N = \mathcal{H} + \left(\overline{n(\mathcal{I})} \cap \mathcal{S}_N\right) = \mathcal{H} .$$

Si $\mathcal{Y}$ est un supplémentaire de $\mathcal{H}$ dans $\mathcal{S}_N$ (c'est un sous-espace à N dimensions), $\mathcal{Y}$ est aussi un supplémentaire de $\mathcal{I}$ dans $\mathcal{E}^2(\mathbb{K} ; \mathbb{R})$. Car

$$\mathcal{E}^r(\mathbb{K} ; \mathbb{R}) = \mathcal{S}_N + \overline{n(\mathcal{I})} = (\mathcal{Y} + \mathcal{H}) + \overline{n(\mathcal{I})} = \mathcal{Y} + \mathcal{I}$$

et

$$\mathfrak{J} \cap \mathscr{I} = (\mathfrak{J} \cap \mathscr{P}_N) \cap \mathscr{I} = \mathfrak{J} \cap \mathscr{H} = \{0\} .$$

Le théorème sera démontré, lorsqu'on aura établi le lemme :

Lemme 4 $\mathscr{I}$ est un idéal de $\mathscr{E}^r(\mathbb{K}; \mathbb{R})$.

Pour tout $f \in \mathscr{E}^r(\mathbb{K}; \mathbb{R})$ et $\phi \in \mathscr{I}$, il s'agit de prouver $f\phi \in \mathscr{I}$. Mais en utilisant les décompositions

$$\mathscr{E}^r(\mathbb{K}; \mathbb{R}) = \mathscr{P}_N + \overline{n(\mathscr{I})} \quad \text{et} \quad \mathscr{I} = \mathscr{H} + \overline{n(\mathscr{I})},$$

il suffit de supposer $f \in \mathscr{P}_N$ et $\phi \in \mathscr{H}$, puisque $\overline{n(\mathscr{I})}$ est un idéal.

Par définition $\phi \in \mathscr{H}$ est limite d'une suite $\phi_i \in \mathscr{I}_i \cap \mathscr{P}_N$. Et par conséquent $f\phi = \lim(f\phi_i)$. Mais les $f\phi_i$ sont des polynômes de degré $\leq 2N$.

Pour les étudier, considérons la suite des $\mathscr{I}_i \cap \mathscr{P}_{2N} \in \mathscr{G}(\mathscr{P}_{2N})$. En extrayant éventuellement une sous-suite, on obtient un sous-espace $\mathscr{H}' \subset \mathscr{P}_{2N}$, limite pour la topologie de $\mathscr{G}(\mathscr{P}_{2N})$ des $\mathscr{I}_i \cap \mathscr{P}_{2N}$.

On démontre, comme pour $\mathscr{I}$, que $\mathscr{H}' + \overline{n(\mathscr{I})}$ est un sous espace fermé de $\mathscr{E}^r(\mathbb{K}; \mathbb{R})$ de codimension N. Mais comme, évidemment $\mathscr{H} \subset \mathscr{H}'$, il en résulte que $\mathscr{H} + \overline{n(\mathscr{I})} \subset \mathscr{H}' + \overline{n(\mathscr{I})}$, et par conséquent

$$\mathscr{H}' + \overline{n(\mathscr{I})} = \mathscr{H} + \overline{n(\mathscr{I})}$$

puisqu'ils ont la même codimension N.

Ceci posé, comme $f\phi_i \in \mathscr{P}_{2N} \cap \mathscr{I}_i$, on a, à la limite,

$$f\phi \in \mathscr{H}' \subset \mathscr{H}' + \overline{n(\mathscr{I})} = \mathscr{H} + \overline{n(\mathscr{I})} = \mathscr{I} .$$

Ce qui achève la démonstration du lemme, et donc du théorème.

Remarques

1. Un tel théorème serait faux, si $r + 1 < N$. Par exemple, dans le cas des fonctions d'une variable, considérons un idéal $\mathscr{I}_i$ formé de fonctions qui s'annulent en N points distincts. C'est bien un idéal fermé de codimension N. Considérons maintenant une suite de tels idéaux, tels que le spectre de $\mathscr{I}_i$ tende, au sens de Hausdorff, vers un ensemble réduit à un seul point 0. Alors $\mathscr{I}_i \cap \mathscr{P}_N$ converge vers l'ensemble $\mathscr{H}$ des polynômes de $\mathscr{P}_N$ qui sont $N - 1$ plats en 0. Mais l'adhérence de l'idéal engendré par $\mathscr{H}$ est l'ensemble des fonctions r-plates à l'origine, dont la codimension est $r \neq N$.

2. $\tilde{\mathbb{K}}(N)$ muni de la topologie définie grâce à l'injection $\mathscr{I} \mapsto \mathscr{I} \cap \mathscr{F}$ de $\tilde{\mathbb{K}}(N)$ dans $\mathscr{G}(\mathscr{F})$ est également compact, comme le prouve la même argumentation. Comme ces deux topologies sont comparables, elles sont identiques. La topologie grassmannienne de $\tilde{\mathbb{K}}(N)$ ne dépend pas du choix de l'espace $\mathscr{F}$, de dimension finie, contenant $\mathscr{P}_N$

3. L'espace $\tilde{\mathbb{K}}(N)$ s'identifie à une partie fermée de $\mathscr{G}$ dont il serait intéressant de faire une étude plus détaillée. Contentons nous de définir l'ouvert $\operatorname{Reg}\left(\tilde{\mathbb{K}}(N)\right)$ des idéaux "réguliers"; ce sont ceux qui sont l'intersection de N idéaux maximaux distincts, correspondant à des points m_α intérieurs a $\mathbb{K}$.

$\operatorname{Reg}\left(\tilde{\mathbb{K}}(N)\right)$ est manifestement un sous-ensemble ouvert de $\tilde{\mathbb{K}}(N)$. (Il paraît vraisemblable que $\operatorname{Reg}\left(\tilde{\mathbb{K}}(N)\right)$ est dense dans $\tilde{\mathbb{K}}(N)$.)

$\operatorname{Reg}\left(\tilde{\mathbb{K}}(N)\right)$ est une variété différentielle de dimension $N \times n$ (localement homéomorphe a $\mathscr{E}^N \simeq \mathbb{R}^{N\times n}$).

On remarquera que $\mathscr{G}$ est une variété compacte à $N \times \left\{\binom{N+n}{n} - N\right\}$ dimensions; la dimension de $\operatorname{Reg}\left(\tilde{\mathbb{K}}(N)\right)$ est inférieure, puisque $\binom{N+n}{n} \geqslant N + n$; l'égalité n'a lieu que pour $n = 1$ ou $N = 1$.

4. La "strate" la plus simple dans $\tilde{\mathbb{K}}(N)$, après $\operatorname{Reg}\left(\tilde{\mathbb{K}}(N)\right)$ est constituée par les idéaux, intersections de $(N - 2)$ idéaux maximaux et d'un idéal primaire de codimension 2, contenue dans un idéal maximal distinct des précédents. On vérifie que la dimension de cette strate est $(N - 2)n + 2n - 1 = Nn - 1$, et que cette strate appartient bien à l'adhérence de $\operatorname{Reg}\left(\tilde{\mathbb{K}}(N)\right)$.

5. On définit $\tilde{\mathscr{E}}(N)$ l'ensemble des idéaux fermés de codimension N de $\mathscr{C}^r(\mathscr{E}; \mathbb{R})$ (munie de la topologie de la convergence uniforme d'ordre r sur tout compact de $\mathscr{E}$).

On a $\tilde{\mathbb{K}}(N) \subset \tilde{\mathscr{E}}(N) \subset \mathscr{G}$, ce qui permet de munir $\tilde{\mathscr{E}}(N)$ de la topologie grassmannienne.

Nous utiliserons plus loin la facile proposition suivante :

<u>Proposition 6</u> <u>Le groupe des translations de $\mathscr{E}$ opère continuement sur</u> $\tilde{\mathscr{E}}(N)$.

Soit $T_{\vec{V}}$ la translation de $\mathscr{E}$ définie par le vecteur $\vec{V}$. Si f est une fonction définie sur $\mathscr{E}$, on pose $\tau_{\vec{V}} f = f \circ T_{(-\vec{V})}$.

La proposition résulte d'une comparaison de $\mathcal{S} \cap \mathcal{P}_N$ et $\tau_{\vec{V}} \mathcal{S} \cap \mathcal{P}_N$ basée sur l'inégalité (III 1) ci dessous valable pour $\|\vec{V}\| \leq 1$.

Si $\mathbb{K}$ est un cube d'arête a, et $\mathbb{K}'$ le cube concentrique d'arête $a + 1$, on utilisera le théorème des accroissements finis et le fait que sur $\mathcal{P}_N$ les deux normes $\|\cdot\|_{\mathbb{K}}^r$ et $\|\cdot\|_{\mathbb{K}'}^{r+1}$ sont équivalentes (avec une constante C). Si $P \in \mathcal{P}_N$:

(III 1) $\|P - \tau_{\vec{V}} P\|_{\mathbb{K}} \leq \|\vec{V}\| \cdot \|P\|_{\mathbb{K}'}^{r+1} \leq C \cdot \|\vec{V}\| \cdot \|P\|_{\mathbb{K}}^r$.

Utilisant dans $\mathcal{G}$ la distance définie par $d(\mathcal{S}, \mathcal{S}') = \rho(\mathcal{S} \cap \mathcal{B}, \mathcal{S}' \cap \mathcal{B})$ où $\mathcal{B}$ est la boule unité de $\mathcal{P}_N$, et ρ la distance de Hausdorff, on prouve ainsi que

$$d(\mathcal{S}, \tau_{\vec{V}} \mathcal{S}) \leq C\|\vec{V}\| .$$

L'application $(\mathcal{S}, V) \to \tau_{\vec{V}} \mathcal{S}$ de $\tilde{\mathbb{K}}(N) \times \mathbb{E}$ dans $\tilde{\mathbb{K}}(N)$ est donc uniformément lipschitzienne par rapport à $\vec{V}$, avec un coefficient de Lipschitz indépendant de $\mathcal{S}$.

IV Dualité

Désignons par θ l'injection canonique de $\mathcal{P}_N$ dans $\mathcal{E}^r(\mathbb{K}; \mathbb{R})$. Sa transposée ${}^t\theta$ est une surjection de l'espace $\mathcal{D}'^r(\mathbb{K})$ des distributions d'ordre $\leq r$ sur le dual $\mathcal{P}_N^*$.

On sait que les éléments de $\mathcal{P}_N^*$ ne sont pas <u>localisables</u>: deux distributions à supports disjoints peuvent avoir la même image par ${}^t\theta$. Cependant :

<u>Proposition 7</u> <u>Pour tout $\mathcal{S} \in \tilde{\mathbb{K}}_{(N)}$ ${}^t\theta$ induit une bijection de $\mathcal{S}^{\perp}$, sous-espace des distributions orthogonales à $\mathcal{S}$, sur le sous-espace de $\mathcal{P}_N^*$ orthogonal à $\mathcal{P}_N \cap \mathcal{S}$.</u>

En effet, soit Δ une distribution orthogonale à $\mathcal{S}$. Si ${}^t\theta(\Delta) = 0$, c'est que Δ est aussi orthogonale à $\mathcal{P}_N$, donc a $\mathcal{S} + \mathcal{P}_N = \mathcal{E}^r(\mathbb{K}; \mathbb{R})$ (Théorème 1). Ainsi $\Delta = 0$.

Inversement, on prolonge d'une façon unique toute forme linéaire sur $\mathcal{P}_N$ orthogonale à $\mathcal{S} \cap \mathcal{P}_N$, en une distribution qui s'annule sur $\overline{n(\mathcal{S})}$. Cette distribution est orthogonale à $(\mathcal{S} \cap \mathcal{P}_N) + \overline{n(\mathcal{S})} = \mathcal{S}$.

<u>Remarque</u> Si $\mathscr{s} \in \widetilde{\mathbb{K}}(N)$, $\mathscr{s}^{\perp}$ est formé de distributions moléculaires d'ordre $\leq N - 1 \leq r$. Cela résulte de l'inclusion

$$\overline{m(\mathscr{s})^N} \subset \overline{n(\mathscr{s})} \subset \mathscr{s} .$$

V <u>Multijets</u>

<u>Définition</u> L'espace $\check{\mathbb{K}}(N)$ des <u>co-multijets</u> d'ordre N, sur $\mathbb{K}$, est le sous-ensemble de $\widetilde{\mathbb{K}}(N) \times \mathscr{D}'(\mathbb{K})$ des couples $(\mathscr{s}, \Delta)$ tels que $\Delta \in \mathscr{s}^{\perp}$.

D'après la remarque précédente l'ensemble $\check{\mathbb{K}}(N)$ est inclus dans $\widetilde{\mathbb{K}}(N) \times \mathscr{D}'^{r}(\mathbb{K}) \subset \widetilde{\mathbb{K}}(N) \times \mathscr{D}'(\mathbb{K})$.

On peut munir $\check{\mathbb{K}}(N)$ de diverses topologies, induites par ces produits, où $\widetilde{\mathbb{K}}(N)$ est muni de la topologie grassmannienne, et $\mathscr{D}'(\mathbb{K})\big($ou $\mathscr{D}'^{r}(\mathbb{K})\big)$ de l'une des topologies usuelles qu'envisage la théorie des distributions.

En fait, il est commode de définir encore une autre topologie, utilisant l'application :

$(\mathscr{s}, \Delta) \longmapsto (\mathscr{s}, {}^{t}\theta\Delta)$ de $\check{\mathbb{K}}(N)$ dans le produit $\widetilde{\mathbb{K}}(N) \times \mathscr{P}^{*}_{N}$.

Cette application est injective (prop. 7) et $\check{\mathbb{K}}(N)$ s'identifie alors à une partie fermée de $\widetilde{\mathbb{K}}(N) \times \mathscr{P}^{*}_{N}$, qui est lui-même un sous-ensemble fermé de la variété différentielle $\mathscr{G} \times \mathscr{P}^{*}_{N}$, à

$$N \times \left\{\binom{N+n}{n} - N\right\} \times \binom{N+n}{n} \text{ dimensions.}$$

<u>Théorème 3</u> <u>Les topologies induites par</u> $\widetilde{\mathbb{K}}(N) \times \mathscr{D}'^{r}(\mathbb{K})$ <u>faible par</u> $\widetilde{\mathbb{K}}(N) \times \mathscr{D}'(\mathbb{K})$ <u>faible ou fort, et par</u> $\widetilde{\mathbb{K}}(N) \times \mathscr{P}^{*}_{N}$ <u>sont identiques</u>.

Il n'en est plus de même pour $\widetilde{\mathbb{K}}(N) \times \mathscr{D}'^{r}(\mathbb{K})_{fort}$. Ainsi dans le cas $N = 2$, $r = 1$, et $n = 1$ $\dfrac{f(a+h) - f(a)}{h} - f'(a)$ tend vers 0 (lorsque h tend vers 0) pour toute $f \in \mathscr{C}^{1}(\mathbb{K}; \mathbb{K})$. Mais cette convergence n'est pas uniforme lorsque f décrit la boule unité $\|f\|^{1}_{\mathbb{K}} \leq 1$.

On sait que les topologies faibles et fortes de $\mathcal{D}'(\mathbb{K})$ induisent la topologie faible de $\mathcal{D}'^{r}(\mathbb{K})$ sur sa boule unité.

Désignons par $\mathcal{C}$ et $\mathcal{C}'$ les topologies induites sur $\check{\mathbb{K}}(N)$ par $\tilde{\mathbb{K}}(N) \times \mathcal{P}_N^*$ et par $\tilde{\mathbb{K}}(N) \times \mathcal{D}'^{r}_{\text{faible}}$. Il suffit donc de prouver que $\mathcal{C}$ et $\mathcal{C}'$ sont identiques, et comme $\mathcal{C}$ est moins fine que $\mathcal{C}'$, il reste à démontrer que toute suite $(\mathcal{s}_i, \Delta_i)$ qui est $\mathcal{C}$-convergente est aussi $\mathcal{C}'$-convergente.

<u>Lemme 5</u> <u>Si</u> $(\mathcal{s}_i, \Delta_i)$ <u>est</u> $\mathcal{C}$ <u>-convergent, la suite des distributions</u> Δ_i <u>est bornée.</u>

A priori, dans un contexte plus général, une suite de distributions Δ_i peut être bornée par rapport aux polynômes de degré $\leq N$ $\left(\text{i.e. } \langle P, \Delta_i\rangle\right.$ est bornée pour tout $\left.P \in \mathcal{P}_N,\right)$ sans être bornée par rapport à tout $f \in \mathcal{C}^r(\mathbb{K}; \mathbb{R})$. La suite des multiples entiers d'une distribution non-nulle orthogonale à $\mathcal{P}_N$ fournit le contre-exemple.

Raisonnons par l'absurde. Supposons que $(\mathcal{s}_i, \Delta_i)$ soit convergente pour $\mathcal{C}$, vers $(\mathcal{s}, \Delta) \in \check{\mathbb{K}}(N)$, et qu'il existe une fonction $f \in \mathcal{C}^r(\mathbb{K}; \mathbb{R})$ telle que $\langle f, \Delta_i\rangle$ ne reste pas bornée. Par extraction d'une sous-suite, on peut supposer que $\langle f, \Delta_i\rangle$ tende vers l'infini et que $\|f\|^r_{\mathbb{K}} = 1$.

Considérons maintenant l'espace $\mathcal{F}$, engendré par $\mathcal{P}_N$ et $\{f\}$, de dimension finie. Soit ψ l'injection canonique de $\mathcal{F}$ dans $\mathcal{C}^r(\mathbb{K}; \mathbb{R})$.

L'injection $(\mathcal{s}, \Delta) \mapsto (\mathcal{s}, {}^t\psi\Delta)$ de $\check{\mathbb{K}}(N)$ dans $\tilde{\mathbb{K}}(N) \times \mathcal{F}^*$ permet de définir sur $\check{\mathbb{K}}(N)$ une topologie $\mathcal{C}''$, intermédiaire entre $\mathcal{C}$ et $\mathcal{C}'$.

Soit Σ la sphère unité de $\mathcal{F}^*$ $\Big($munie de la norme induite par la norme de $\mathcal{D}'^{r}(\mathbb{K})\Big)$. Σ est compacte puisque $\mathcal{F}^*$ est de dimension finie.

Considérons maintenant les distributions $\delta_i = (1/\langle f, \Delta_i\rangle) . \Delta_i$. La suite des ${}^t\theta(\delta_i)$ tend manifestement vers 0, dans $\mathcal{P}_N^*$ mais ${}^t\psi(\delta_i)$ reste sur la sphère unité de $\mathcal{F}^*$ puisque $|\langle f, \delta_i\rangle| = 1$.

Une sous-suite de $(\mathcal{s}_i, \delta_i)$ devrait converger vers $(\mathcal{s}, \delta)$ $\left(\text{ou } \|\delta\|_{\mathcal{F}^*} = 1\right)$, au sens de la topologie $\mathcal{C}''$ et vers 0 pour la topologie $\mathcal{C}$. D'où la contradiction, d'après le théorème de comparaison de deux topologies compactes comparables, sur un même ensemble.

Utilisant ce lemme, on en déduit que si $(\mathscr{s}_i, \Delta_i)$ converge, au sens de la topologie $\mathscr{C}$, les Δ_i restent dans une boule faiblement compacte $\mathscr{B}$ de $\mathscr{D}'^r(\mathbb{K})$.

L'ensemble $\check{\mathbb{K}}(N) \cap \left(\tilde{\mathbb{K}}(N) \times \mathscr{B}\right)$ est compact pour la topologie $\mathscr{C}'$. Sur cet ensemble la topologie $\mathscr{C}'$ est identique à la topologie séparée moins fine $\mathscr{C}$.

Donc la suite $(\mathscr{s}_i, \Delta_i)$ converge vers $(\mathscr{s}, \Delta)$, pour la topologie $\mathscr{C}$.

<u>Corollaire 1</u> <u>Soit</u> $\mathscr{s}_i$ <u>une suite d'éléments de</u> $\tilde{\mathbb{K}}(N)$ <u>convergant vers</u> $\mathscr{s}$ <u>et</u> f_i <u>une suite de fonctions, appartenant à</u> $\mathscr{E}^r(\mathbb{K}, \mathbb{R})$, <u>telle que</u> $\|f_i - f\|^r_{\mathbb{K}}$ <u>tende vers</u> 0 <u>et</u> $f_i \in \mathscr{s}_i$ <u>pour tout</u> i. <u>Alors</u> $f \in \mathscr{s}$.

En effet, soit $\Delta \in \mathscr{s}^\perp$. Par définition de $\check{\mathbb{K}}(N)$ et en vertu du théorème précédent, il existe une suite Δ_i de distributions convergent faiblement vers Δ, telle que $(\mathscr{s}_i, \Delta_i)$ converge vers $(\mathscr{s}, \Delta)$ dans l'espace topologique $\check{\mathbb{K}}(N)$. Mais $\langle f_i, \Delta_i \rangle = 0$. D'après un théorème classique [12] on a $\langle f, \Delta \rangle = \lim \langle f_i, \Delta_i \rangle = 0$. Ainsi f appartient au biorthogonal de $\mathscr{s}$, donc à $\mathscr{s}$ qui est supposé fermé.

<u>Définitions</u> Si $\mathscr{s} \in \tilde{\mathbb{K}}(N)$, le <u>multijet</u> de $f \in \mathscr{E}^r(\mathbb{K}; \mathbb{R})$ en $\mathscr{s}$ est la classe de f modulo $\mathscr{s}$: ce multijet est complètement défini par ses valeurs sur $\mathscr{s}^\perp$.

A toute fonction $f \in \mathscr{E}^r(\mathbb{K}; \mathbb{R})$, on associe son <u>extension</u> aux multijets $mj^{N-1}(f)$; c'est la fonction $mj^{N-1}(f) : \check{\mathbb{K}}(N) \to \mathbb{R}$ définie par

$$mj^{N-1}(f)(\mathscr{s}, \Delta) = \langle f, \Delta \rangle \qquad (\text{où } \Delta \in \mathscr{s}^\perp).$$

Pour pouvoir calculer $mj^{N-1}f$, il est nécessaire que f admette des dérivées (non necessairement continues) jusqu'à l'ordre N - 1 au moins.

Considérons maintenant la restriction de $mj^{N-1}(f)$ à la partie régulière $\mathrm{Reg}\left(\check{\mathbb{K}}(N)\right)$ de $\tilde{\mathbb{K}}(N)$ située au-dessus de $\mathrm{Reg}\left(\tilde{\mathbb{K}}(N)\right)$.

On peut définir $mj^{N-1}f$ restreint à $\mathrm{Reg}\,\check{\mathbb{K}}(N)$, pour toute fonction numérique f définie sur $\mathbb{K}$.

<u>Théorème 4</u> <u>Une condition nécessaire et suffisante pour qu'une fonction</u> f <u>soit de classe</u> $\mathscr{E}^{N-1}(\mathbb{K}; \mathbb{R})$ <u>est qu'elle admette une extension aux multijets réguliers, qui soit uniformément continue sur toute partie précompacte de</u> $\mathrm{Reg}\left(\check{\mathbb{K}}(N)\right)$.

Ce résultat se trouve déja, dans le cas des fonctions d'une variable, dans [9].

Si $f \in \mathcal{C}^{N-1}(\mathbb{K} ; \mathbb{R})$, la continuité de $mj^{N-1}(f)$ n'est qu'une traduction du théorème précédent: la topologie induite sur $\check{\mathbb{K}}(N)$ par le produit de $\widetilde{\mathbb{K}}(N)$ et de $\mathcal{D}'^{N-1}(\mathbb{K})$ faible est précisément celle qui assure cette continuité.

Inversement, soit Δ une droite de $\mathcal{E}$ qui rencontre $\mathbb{K}$ suivant un segment. Si f est une fonction numérique admettant une extension aux multijets réguliers qui se prolonge par continuité à $\check{\mathbb{K}}(N)$, il en est de même pour la restriction de f à $\Delta \cap \mathbb{K}$. Appliquant le résultat cité [9] relatif aux fonctions d'une variable on en déduit que f admet des dérivées partielles jusqu'à l'ordre $N - 1$.

Ces dérivées partielles sont continues sur $\mathbb{K}$ en vertu de la proposition 6. Donc $f \in \mathcal{C}^{N-1}(\mathbb{K} ; \mathbb{R})$.

Remarque Ainsi pour démontrer qu'une fonction f est de classe $\mathcal{C}^{N-1}$ il suffit de démontrer la continuité uniforme d'une fonction $mj^{N-1}f$ dans $\mathrm{Reg}(\check{\mathbb{K}})$.

On notera le contraste avec la théorie des jets. Un champ continu de polynômes de degré $\leq N - 1$, défini sur $\mathbb{K}$, n'est pas nécessairement un champ taylorien.

VI Schémas d'interpolation Soit $\mathcal{G}_N$ la grassmannienne des sous-espaces de dimension N dans $\mathcal{P}_N$. On définit un schéma d'interpolation par une application continue $\mathfrak{s} \longmapsto \mathcal{J}(\mathfrak{s})$ de $\widetilde{\mathbb{K}}(N)$ dans $\mathcal{G}_N$, telle que $\mathcal{J}(\mathfrak{s})$ soit un supplémentaire de $\mathfrak{s}$ dans $\mathcal{C}^r(\mathbb{K} ; \mathbb{R})$.

Par exemple, on peut munir $\mathcal{P}_N$ d'une structure euclidienne (dont le choix est évidement soumis à un certain arbitraire) et choisir $\mathcal{J}(\mathfrak{s})$ perpendiculaire a $\mathfrak{s} \cap \mathcal{P}_N$.

On construit ainsi un espace fibré de base $\widetilde{\mathbb{K}}(N)$, dont la fibre au dessus du point $\mathfrak{s} \in \widetilde{\mathbb{K}}(N)$ est l'espace vectoriel $\mathcal{J}(\mathfrak{s})$. Cet espace fibré est isomorphe à $\check{\mathbb{K}}(N)$.

Pour tout $f \in \mathcal{C}^r(\mathbb{K}; \mathbb{R})$ et $\mathfrak{s} \in \widetilde{\mathbb{K}}(N)$, il existe une décomposition directe

$$f = S_{\mathfrak{s}}f + \mathcal{R}_{\mathfrak{s}}f$$

où $S_{\mathfrak{s}}f \in \mathcal{J}(\mathfrak{s})$ s'appelle le polynôme d'interpolation de f sur $\mathfrak{s}$ et $\mathcal{R}_{\mathfrak{s}}f \in \mathfrak{s}$ est le reste. L'opération $S_{\mathfrak{s}}$ est un projecteur de $\mathcal{C}^r(\mathbb{K} ; \mathbb{R})$ sur $\mathcal{J}(\mathfrak{s}) \subset \mathcal{P}_N$.

Théorème 5 L'application $(f, \mathcal{A}) \to S_{\mathcal{A}} f$ de $\mathcal{C}^r(\mathbb{K}; \mathbb{R}) \times \hat{\mathbb{K}}(N)$ dans $\mathcal{P}_N$ est continue.

C'est une conséquence facile du théorème 3.

D'ailleurs, une démonstration explicite se trouve en quelque sorte dans ([4] p.112 Prop. VII). Certes, le contexte n'est pas tout à fait le même, mais formellement les mêmes arguments s'appliquent ici. En particulier, on y obtient, grâce au théorème de Banach-Steinhaus le résultat suivant:

Proposition Il existe une constante K, ne dépendant que de $\mathbb{K}$ et de r telle que

$$\|S_{\mathcal{A}} f\|^r_{\mathbb{K}} \leq K . \|f\|^r_{\mathbb{K}} .$$

Remarque Ces résultats s'étendent facilement à l'espace $\mathcal{C}^r(\mathbb{K}; \mathbb{A})$ où $\mathbb{A}$ est un espace de Banach, à condition d'utiliser la théorie du produit tensoriel topologique complété $\hat{\hat{\otimes}}_\epsilon$ (cf. [13] exposés 7 et 10). En effet l'espace $\mathcal{C}^r(\mathbb{K}; \mathbb{A})$ est isomorphe à $\mathcal{C}^r(\mathbb{K}; \mathbb{R}) \hat{\hat{\otimes}}_\epsilon \mathbb{A}$.

Nous renonçons ici à recopier ces démonstrations détaillées: elles seraient inutiles aux lecteurs familiers avec la théorie du produit $\hat{\hat{\otimes}}_\epsilon$, et incompréhensibles aux autres.

VII Comparaison des deux procédés d'interpolation

Supposons, pour fixer les idées, que l'on se propose d'utiliser des systèmes de $N = \binom{n+r}{r}$ noeuds d'interpolation pour étudier les propriétés differentielles d'une fonction f. La méthode de Lagrange (notée $(\mathcal{L})$) ne sera utilisable que si $\mathcal{A}$ est unisolvant. Elle ne fournira des valeurs approchées que des derivées partielles d'ordre $\leq$ r, et cette approximation ne sera bonne que si $\|\mathcal{A}\|$ est faible.

Par exemple, dans le cas $n = 2$, $r = 1$, seuls les triangles ayant trois angles aigus, et un diamètre faible fourniront de bous renseignements sur les dérivées partielles premières de f.

La méthode des schémas d'interpolation (notée (S)) sera utilisable sans restriction et fournira des renseignements sur les dérivées jusqu'à l'ordre $N - 1$, strictement plus grand que r (si $n > 1$). Les systèmes de noeuds A qui étaient efficaces dans la méthode $\mathcal{L}$, pour l'étude des dérivées d'ordre $\leq r$ de f, resteront à peu près aussi efficacesdans la méthode S, pour l'étude des mêmes dérivées. Mais, ce sont précisement les systèmes dégénérés qui fourniront (par la méthode S) les renseignements précis concernant les dérivées d'ordre compris entre r et $N - 1$.

Soit, par exemple A une systeme de N points alignés sur une droite Δ.

$S_A f$ sera un polynôme de degré $N - 1$, dont la définition précise pourra varier avec le choix du schéma d'interpolation. Mais de toute façon, la restriction de $S_A f$ à la droite Δ, sera un polynôme d'interpolation, de degré $N - 1$, relatif à N points distincts. Cette restriction est l'unique polynôme de Lagrange classique, dans le cadre des fonctions d'une variable.

La dérivée $N - 1$ - ième d'un tel polynôme s'appelle une <u>différence divisée</u>, et l'on sait qu'elle fournit une bonne approximation de la dérivée $N - 1$ - ième de f.

Ainsi la comparaison des deux procédés tourne entièrement à l'avantage de la méthode S.

VIII <u>Différences divisées</u> Dans la théorie des fonctions d'une variable la différence divisée de f, relative à $A = \{A_1, A_2, \ldots, A_N\}$, est définie par

$$\text{VIII 1} \qquad \Delta_A f = D^{N-1}(\mathcal{L}_A f) .$$

Cette formule garde un sens dans le cas des fonctions de plusieurs variables, lorsque A est unisolvant.

Mais malheureusement, ses précieuses propriétés algorithmiques (auxquelles elle doit son nom) ne se généralisent pas.

Dans le cas n= 1, les deux polynômes du premier degré

$$p_1(x) = \frac{A_{N-1} - x}{A_{N-1} - A_N} \qquad p_2(x) = \frac{x - A_N}{A_{N-1} - A_N}$$

constituent une partition de l'unité. Si l'on pose

$$\mathcal{A}' = \{A_1, A_2, \ldots, A_{N-1}\} \quad \text{et} \quad \mathcal{A}'' = \{A_1, A_2, \ldots, A_{N-2}, A_N\}$$

on en déduit aussitôt

$$\text{VIII 2} \qquad \mathcal{L}_{\mathcal{A}} f = p_1 \times \mathcal{L}_{\mathcal{A}'} f + p_2 \times \mathcal{L}_{\mathcal{A}''} f \quad ,$$

et par la formule de Leibnitz

$$\text{VIII 3} \qquad \Delta_{\mathcal{A}} f = \frac{\Delta_{\mathcal{A}''} f - \Delta_{\mathcal{A}'} f}{A_{N-1} - A_N} \quad .$$

La différence d'ordre N s'obtient bien par division d'une différence de deux différences d'ordre N - 1 .

La formule VIII.2 ne subsiste pas dans la méthode de Lagrange pour plusieurs variables où des systemes de N - 1 et de N points ne sont pas simultanément utilisables.

Il serait utile de rechercher s'il existe des schémas d'interpolation pour lesquels le calcul de tous polynômes d'interpolation relatif à N points peut s'effectuer par récurrence sur un nombre inférieur de noeuds.

Appendice I

Etude différentielle de l'interpolation de Lagrange

Soit $\mathcal{A} = \{A_1, A_2, \ldots, A_N\}$ un système unisolvant. Si on range les points dans un ordre spécifié, on obtient un élément $(A_1, A_2, \ldots, A_N) \in \mathcal{E}^N$. Tout élément $(B_1, B_2, \ldots, B_N) \in \mathcal{E}^N$ suffisamment voisin du précédent provient d'un $\mathcal{B}$, qui est aussi unisolvant.

Nous allons considérer un ouvert $\Omega \subset \mathcal{E}^N$ tel que chacun de ses éléments provienne d'un système unisolvant.

Un problème important d'analyse numérique est la majoration de l'erreur commise dans le calcul de $\mathcal{L}_{\mathcal{A}} f$, lorsqu'on remplace le système de noeuds $\mathcal{A}$ par un système voisin.

Une conjecture plausible est qu'il existe une inégalité du type $\left(\text{pour } f \in \mathcal{C}^2(\mathbb{K}\,;\,\mathbb{A})\right)$.

App.1 $$\|\mathcal{L}_{\mathcal{A}} f - \mathcal{L}_{\mathcal{B}} f\|^r_{\mathbb{K}} \leq K \,.\, d(\mathcal{A}, \mathcal{B}) \,.\, \|f\|^r_{\mathbb{K}}$$

où $d(\mathcal{A}, \mathcal{B})$ désigne la distance dans Ω, et K une constante.

Cette distance peut s'exprimer par

$$d(\mathcal{A}, \mathcal{B}) = \operatorname*{Max}_{i \leq N} \|A_i B_i\| .$$

Mais une définition plus raisonnable fait intervenir le groupe $\mathcal{J}_N$ des permutations des N indices

$$d(\mathcal{A}, \mathcal{B}) = \operatorname*{Inf}_{\sigma \in \mathcal{J}_N} \left(\operatorname*{Max}_{i \leq N} \|A_i B_{\sigma(i)}\| \right) .$$

Nous montrons ailleurs [7] que ce point de vue est erroné : ce n'est pas ainsi qu'il convient d'évaluer la distance de deux systèmes unisolvants. Cependant les distances indiquées ci-dessus donnent un résultat correct (quitte à adopter un facteur K très grand) si l'on reste dans un compact de Ω, tel que $\mathcal{A}$ et $\mathcal{B}$ soient suffisamment voisins.

Pour le voir, nous allons considérer, pour $f \in \mathcal{C}^r(\mathbb{K}\,;\,A)$, $\mathcal{A}$ et $\mathcal{B}$ comme des points appartenant à la variété différentielle $\Omega \subset \mathcal{E}^N$, et montrer que l'application $(\mathcal{A}, M) \longmapsto \mathcal{L}_{\mathcal{A}} f(M)$ est dérivable. Nous réserverons le symbole D (resp. ∇) pour désigner la dérivation partielle par rapport à $M \in \mathcal{E}$, (resp. par rapport à $\mathcal{A} \in \mathcal{E}^N$).

Le plan tangent à Ω au point $\mathcal{A}$ s'identifie à $\mathbb{E}^N$. Un "vecteur tangent" s'écrira donc $(\vec{V}_1, \vec{V}_2, \ldots, \vec{V}_N)$. Il sera commode de considérer plutôt un <u>champ de vecteurs</u> $\tilde{V}$ sur $\mathcal{E}$ et de lui faire correspondre, au point $\mathcal{A} \in \mathcal{E}^N$, le "vecteur tangent" $\tilde{V}(\mathcal{A}) = \left(\tilde{V}(A_1), \tilde{V}(A_2), \ldots, \tilde{V}(A_N)\right)$.

<u>Théorème 6</u> <u>L'application</u> $\mathcal{L}f : \mathcal{A} \longmapsto \mathcal{L}_{\mathcal{A}} f$ <u>de</u> Ω <u>dans</u> $\mathcal{P}_N(\mathbb{A})$ <u>est de classe</u> $\mathcal{C}^r$ (<u>si</u> f <u>l'est) et sa dérivée première s'exprime par</u>

<u>App</u>.2 $$(\nabla_{\mathcal{A}} \mathcal{L}f)[\tilde{V}] = \mathcal{L}_{\mathcal{A}}\left\{\left(D^1(f - \mathcal{L}_{\mathcal{A}} f)\right)[\tilde{V}]\right\} .$$

Explicitons la signification du second membre: si $f \in \mathcal{C}^r(\mathbb{K}\,;\mathbb{A})$, il en est de même pour $f - \mathcal{L}_{\mathcal{A}} f$: sa dérivée $D^1(f - \mathcal{L}_{\mathcal{A}} f)$ est de classe $\mathcal{C}^{r-1}\left(\mathbb{K}\,; \mathcal{L}(\mathbb{E}\,;\mathbb{A})\right)$: on peut contracter cette dérivée avec le champ de vecteurs $\tilde{V}$, et obtenir ainsi une application

$$M \longmapsto \left(D^1_M(f - \mathcal{L}_{\mathcal{A}} f)\right)[\tilde{V}_M]$$

de classe $\mathcal{C}^{r-1}(\mathbb{K}\,;\mathbb{A})$ (si l'on suppose, comme nous le faisons, que $\tilde{V}$ est de classe $\mathcal{C}^\infty$).

On obtient le second membre de la formule en calculant le polynôme de Lagrange de cette fonction sur le système unisolvant $\mathcal{A}$.

<u>Démonstration</u> Au lieu de calculer directement la dérivée $\nabla_{\mathcal{A}}$ nous allons fixer tous les points A_i de $\mathcal{A}$, sauf A_j et commencer par calculer la dérivée partielle $\nabla_{j\mathcal{A}}$.

Pour cela nous supposons que $A_i = B_i$ pour $i \neq j$. Considérons alors les N fonctions $p_i : \mathcal{B} \to p_i\{\mathcal{B}\}$ (base de Lagrange) définies sur Ω, et aux valeurs dans $\mathcal{P}_r$.

<u>Lemme</u> <u>La</u> $j^{\text{ième}}$ <u>dérivée partielle de</u> p_i <u>est donnée par</u>

$$\nabla_{j\mathcal{A}} p_i = - p_j\{\mathcal{A}\} \times D_{A_j}\left(p_i\{\mathcal{A}\}\right) .$$

Examinons d'abord le cas, plus facile, où $i = j$. On remarque aussitôt que si B_j est suffisamment proche de A_j, on a

$$p_j\{\mathcal{B}\} = p_j\{\mathcal{A}\}/p_j\{\mathcal{A}\}(B_j) .$$

En effet le second membre est bien un polynôme qui s'annule aux divers points $B_i = A_i$ $i \neq j$ et qui prend la valeur 1 en B_j.

Dérivons cette formule par rapport à B_j. Il vient

$$\nabla_{j\mathcal{B}} p_j = -p_j\{\mathcal{A}\} \times \frac{D_{B_j} p_j\{\mathcal{A}\}}{[p_j\{\mathcal{A}\}(B_j)]^2} .$$

En remplaçant B_j par A_j on obtient le résultat. Dans le cas où $j \neq i$ on remarque que

$$p_i\{\mathcal{B}\} = p_i\{\mathcal{A}\} - p_j\{\mathcal{A}\} \times \frac{p_i\{\mathcal{A}\}(B_j)}{p_j\{\mathcal{A}\}(B_j)}$$

représente bien un polynôme nul aux points B_k, avec $k \neq i$ (et en particulier pour $k = j$) et qu'il prend la valeur 1 au point $B_i = A_i$.

En effectuant la dérivée par rapport à B_j, puis en faisant $B_j = A_j$ le lemme est établi.

Nous sommes maintenant en mesure de calculer

$$\nabla_{j\mathcal{B}} \mathcal{L} f = \nabla_{j\mathcal{B}} \left\{ \sum_{i=0}^{N} f(B_i) p_i\{\mathcal{B}\} \right\} .$$

Tous les B_i sont fixés, sauf B_j. Ainsi $f(B_j)$ fournira la contribution $p_j\{\mathcal{A}\} . D_{A_j} f$ à l'égalité suivante:

$$\nabla_{j\mathcal{A}} \mathcal{L} f = p_j\{\mathcal{A}\} D_{A_j} f - p_j\{\mathcal{A}\} \sum_{i=0}^{N} f(A_i) D_{A_j} p_i\{\mathcal{A}\} ,$$

qui s'écrit encore

$$\nabla_{j\mathcal{A}} \mathcal{L} f = p_j\{\mathcal{A}\} . (D_{A_j} (f - \mathcal{L}_{\mathcal{A}} f)) .$$

Passons maintenant des dérivées partielles à la dérivée totale ∇ et calculons

$$(\nabla_A \mathcal{L} f)[\vec{V}_1, \vec{V}_2, \ldots, \vec{V}_N] \quad = \quad \nabla_A \mathcal{L} f[\tilde{V}] \quad .$$

On trouve

$$\nabla_A \mathcal{L} f[\tilde{V}] \quad = \quad \sum_{j=0}^{N} p_j\{A\} \cdot \left(D_{A_j}(f - \mathcal{L}_A f)[\tilde{V}(A_j)]\right) \; .$$

On reconnait dans le second membre l'expression du polynôme de Lagrange de la fonction (à valeurs dans $\mathbb{A}$) obtenue en contractant $D^1(f - \mathcal{L}_A f)$ avec $\tilde{V}$. La formule (App.2) est démontrée. On peut calculer successivement les dérivées d'ordre supérieur de $A \to \mathcal{L}_A f$ en dérivant le second membre de (App.2) par la formule des fonctions composées. Si f est de classe $\mathcal{C}^r$, cette opération peut être répetée r fois.

<u>Corollaire</u> <u>Si</u> A <u>et</u> B <u>peuvent être joints par un segment</u> $[A, B]$ <u>entièrement contenu dans</u> $\Omega \subset \mathcal{E}^N$ <u>et si</u> M <u>est une constante telle que</u> $\|\mathcal{C}\| \leq M$ <u>pour tout</u> $\mathcal{C} \in [A, B]$ <u>alors</u>

$$\|\mathcal{L}_A f - \mathcal{L}_B f\|^r_{\mathbb{K}} \leq M(1 + M) d(A, B) \cdot \|f\|^r_{\mathbb{K}}$$

<u>où</u>

$$d(A, B) \leq \sum_{i \in N} \|A_i B_i\| \, .$$

C'est une conséquence immédiate de la formule des accroissements finis.

Appendice II

Sur l'inégalité de Łojasiewicz

Une fonction f définie sur une boule compacte de rayon R $\mathbb{K} \subset \mathcal{E}$, y satisfait à une inégalité de Łojasiewicz d'exposant $\alpha > 0$ s'il existe une constante C telle que, pour tout $M \in \mathbb{K}$,

App. II.1
$$|f(M)| \geqslant C\left(d(M)\right)^{\alpha}$$

où $d(M)$ est la distance du point M à l'ensemble $\mathbb{F} = \{ f^{-1}(0) \}$ des zéros de f.

La démonstration du théorème suivant a été mise au point en collaboration avec J.C. Tougeron.

Théorème   Lorsque $r \geqslant \alpha > 0$, une fonction $f \in \mathcal{C}^{r+1}(\mathbb{K}; \mathbb{R})$ satisfaisant à une inégalité de Łojasiewicz d'exposant α n'a aucun zéro d'ordre strictement supérieur à r.

(a) **Méthode de démonstration**   Raisonnons par l'absurde: supposons que P est un point frontière de l'ensemble des zéros d'ordre $\geqslant r+1$ de f (i.e. on suppose que f est r-plate en P).

Cette r-platitude s'exprime au moyen de la distance de M à P, alors que la minoration (App.II.1) fait intervenir la distance de M à $\mathbb{F}$. Il s'agit d'analyser les relations entre ces deux distances et d'en tirer une contradiction.

Pour traduire le fait que $\mathbb{F}$ est l'ensemble des zéros de f on effectuera une interpolation relative à un système unisolvant dont les noeuds appartiennent à $\mathbb{F}$: le polynôme d'interpolation est alors nul, et toute majoration du reste de la formule de Lagrange fournit, en réalité, une majoration de f.

Nous construirons une suite $\{N_j\}$ de points convergent vers P, et nous lui associerons une suite $A(j)$ de tels systèmes unisolvants: pour majorer la suite

$$f(N_j) = f(N_j) - \mathcal{L}_{A(j)} f(N_j) \; ,$$

il sera nécessaire de contrôler les normes de $A(j)$.

La construction de $A(j)$ s'obtiendra à partir d'un système unisolvant A, fixé une fois pour toute, auquel on fera subir une suite d'homothéties, suivies de légères perturbations : ces dernières auront pour objet d'amener le support de $A(j)$ dans $\mathbb{F}$.

(b) Préliminaires sur l'interpolation de Lagrange

Soit $A = \{A_i\}$ un système unisolvant de $\mathcal{E}$ de $m = \binom{n+r}{r}$ points. Un nombre $\epsilon > 0$ est A-admissible si tout système de m points $\tilde{A} = \{\tilde{A}_i\}$ satisfaisant à $\|A_i\tilde{A}_i\| < \epsilon$ (pour tout $i \leq m$) est encore unisolvant: on dira que $\tilde{A}$ s'obtient à partir de A par une ϵ-perturbation.

Choisissons une fois pour toute un tel A et un ϵ A-admissible tel que la boule compacte $\mathbb{K}$, de rayon R, contienne non seulement les noeuds de A, mais aussi les noeuds de tous les $\tilde{A}$ obtenus par ϵ-perturbation à partir de A.

Posons

$$\|A\|_{\mathbb{K}} = \sup_{i \leq m \text{ et } M \in \mathbb{K}} |p_i(M)|$$

où $\{p_i\}$ est la base de Lagrange de A, et

$$\|A\|_{\mathbb{K},\epsilon} = \sup \|\tilde{A}\|_{\mathbb{K}}$$

où $\tilde{A}$ décrit l'ensemble de tous les ϵ-perturbés de A.

Cette norme est invariante par homothétie: d'une façon précise, si λA (resp. $\lambda\mathbb{K}$) désigne la transformée de A (resp. $\mathbb{K}$) par une homothétie de rapport λ, on vérifie que λA est unisolvant, que $\lambda\epsilon$ est λA-admissible, et que

App. II.2 $$\|\lambda A\|_{\lambda\mathbb{K};\lambda\epsilon} = \|A\|_{\mathbb{K},\epsilon} .$$

(c) Démonstration du théorème Partons d'une suite de points $M_j \notin \mathbb{F}$ convergent vers P, et désignons par $\mathcal{B}(j)$ la boule fermée $\mathcal{B}(M_j; 1/2\,\|PM_j\|)$ de centre M_j et de rayon $1/2\,\|PM_j\|$. La fonction $M \mapsto d(M)$ restreinte a $\mathcal{B}(j)$ y atteint son maximum ρ_j en un point $N_j \in \mathcal{B}(j)$ au moins :

App. II.3 $$\rho_j = d(N_j) \geq d(M_j) > 0 .$$

Par hypothèse, il existe deux constantes C et C' telles que pour tout $M \in \mathbb{K}$

$$C\big(d(M)\big)^{\alpha} \leq |f(M)| \leq C'\|PM\|^{r+1} .$$

On en déduit que

$$d(M)/\|PM\| \leq (C'/C)^{1/\alpha}\|PM\|^{(r+1-\alpha)/\alpha} .$$

Utilisant le fait que $1/2\|PM_j\| \leq \|PN_j\| \leq 3/2\|PM_j\|$, on en conclut que $d(N_j)/\|PM_j\|$ tend vers 0 lorsque j augmente indéfiniment. En particulier il existe un indice j_0 à partir duquel

$$\rho_j \times (R/\epsilon) \leq 1/2\|PM_j\| .$$

Dans ces conditions, il existe un boule $\mathbb{K}(j)$ de rayon $\rho_j \times (R/\epsilon)$ telle que $N_j \in \mathbb{K}(j) \subset \mathcal{B}(j)$.

Par une homothétie de rapport ρ_j/ϵ on peut transformer $\mathbb{K}$ en $\mathbb{K}(j)$; soit $\mathcal{A}'(j) \subset \mathbb{K}(j)$ l'image de $\mathcal{A}$ par une telle homothétie. On note que ρ_j est $\mathcal{A}'(j)$-admissible, puisque $\rho_j = \epsilon \times (\rho_j/\epsilon)$, (d'après App. II.2). Mais, par définition de ρ_j, les points de $\mathcal{B}(j)$, et en particulier les noeuds de $\mathcal{A}'(j)$ sont situées a une distance inférieure à ρ_j de $\mathbb{F}$. Il est donc possible de ρ_j-perturber $\mathcal{A}'(j)$ pour obtenir un système unisolvant $\mathcal{A}(j) \subset \mathbb{K}(j) \cap \mathbb{F}$, tel que $\|\mathcal{A}(j)\|_{\mathbb{K}(j)} \leq \|\mathcal{A}\|_{\mathbb{K},\epsilon}$: le second membre est indépendant de j.

Ceci posé, d'après l'expression (I.4) du reste de la formule d'interpolation de Lagrange,

$$f(N_j) = f(N_j) - \mathcal{L}_{\mathcal{A}(j)}f(N_j) = \sum p_i^j(N_j) \left[T_{N_j} f(A_i^j) - f(A_i^j) \right] .$$

Comme $|p_i^j(N_j)| \leq \|\mathcal{A}(j)\|_{\mathbb{K}(j)} \leq \|\mathcal{A}\|_{\mathbb{K},\epsilon}$, $\|N_j A_i^j\| \leq (2R\rho_j)/\epsilon$,

et puisqu'il existe une constante K telle que

$$|T_{N_j} f(A_i^j) - f(A_i^j)| \leq K\|A_i^j N_j\|^{r+1} ,$$

on obtient une majoration $|f(N_j)| \leq C''(\rho_j)^{r+1}$ où C'' est une constante.

Rapprochant de (App. II.1), appliqué aux points N_j, on aboutit à

$$C\rho_j^{\alpha} \leq C''\rho_j^{r+1}$$

ce qui est absurde, puisque les $\rho_j \neq 0$ tendent vers 0 et $\alpha < r + 1$.

REFERENCES

[1] N. Bourbaki, "Topologie générale". Hermann, Paris.

[2] N. Bourbaki, "Espaces vectoriels topologiques". Hermann, Paris.

[3] Ph. J. Davis, "Interpolation and approximation". Blaisdell, New York, 1963

[4] G. Glaeser, Etude de quelques algèbres Tayloriennes. Journal d'Analyse Mathématique 6 (1958), 1-124.

[5] G. Glaeser, Multiplicateurs rugueux des fonctions différentiables. Annales de l'Ecole Normale Supérieure 79 (1962) 61-67.

[6] G. Glaeser, "Calcul différentiel" (Fonctions de plusieurs variables.) Encyclopaedia Universalis - Paris, 1969.

[7] G. Glaeser, Géométrie des distributions à support fini. Seminaire L. Schwartz-Joulaomi Ecole Polytechnique Paris (1971).

[8] B. Malgrange, "Ideals of differentiable functions". (Tata Institute, Bombay) Oxford University Press, 1966.

[9] J. Merrien, Prolongateurs de fonctions différentiables d'une variable réelle. Journal Maths Pures et Appliquées 45 (1966), 291-309.

[10] H. Salzer, Some new divided difference algorithms for two variables. In "On numerical approximation" Proceedings of a Symposium, University of Wisconsin (Madison) 1959.

[11] A. Sard, "Linear Approximation". Amer. Math. Soc. 1963.

[12] L. Schwartz, "Théorie des distributions". Hermann, Paris, 1950.

[13] L. Schwartz, Produits tensoriels topologiques. Séminaire de la Faculté des Sciences de Paris, 1953.

[14] J.F. Steffensen, "Interpolation". Chelsea Publ.Co., 1927.

NORMAL FORMS FOR ANALYTIC MATRIX VALUED FUNCTIONS

Samir Khabbaz and Gilbert Stengle

1. Introduction

The purpose of this paper is to study the existence of normal forms for analytic matrix valued functions under the similarity $A \to S^{-1}AS$ and, to a lesser extent, under the transformation $A \to S^{-1}AS - S^{-1}\frac{dS}{dX}$. We are mainly concerned with functions of a single real or complex variable. Here, as elsewhere, there is a profound difference between one variable and several, and it is only in the case of a single variable that we have identified suitable normal forms. However we will attempt to put the several variable case in perspective with some remarks and results. Normal forms under similarity have been studied in a number of recent works (see [1] through [5] of the bibliography). Most of these investigations proceed under strong hypotheses which permit correspondingly strong, simple normal forms to be achieved. In contrast we will obtain weak forms, for example: block triangular, under barely more than necessary conditions. Our results unify and extend previous results and are obtained by methods which are comparatively more algebraic and geometric and less analytic in nature. We remark that the subtleties of this problem are sometimes overlooked in the literature, particularly if an oversimplified normal form is found to be convenient in the inner workings of a larger problem.

Our exposition is organized in the following way. In Section 2 we discuss conventions and prove a basic Lemma. In Section 3 we obtain three normal forms, a block triangular form and two forms related to the companion or rational normal form. Section 4 contains illustrative remarks and examples. In Section 5 we discuss the role of local and global considerations in the several variable case, illustrating the role of topological considerations in obtaining global results with a theorem on real projective spaces. We conclude in Section 6 with allied results in the theory of differential equations, obtaining and applying some normal forms for analytic linear systems.

We express our gratitude to Gerhard Rayna for several illuminating discussions.

2. Conventions. A Basic Lemma

We consider analytic functions $A : X \to L(n,C)$ or $L(n,R)$ where X is either a noncompact connected Riemann surface or the real line, and $L(n,C)$, alternatively $L(n,R)$, is the ring of complex, alternatively real, $n \times n$ matrices. We will favor the case of the Riemann surface for the sake of simplicity in exposition, but our main results also apply to analytic functions on the real line as well as other situations - see the discussion of Section 5, and the appendix to this section.

We find it natural to regard the matrix valued function A in a number of different ways. To begin with, A can be viewed as a bundle map acting on the trivial complex vector bundle, $X \times C^n$, with the natural projection map $p : X \times C^n \to X$. Then $A(x,v) = (x,A(x)v)$ where v is an element of C^n written as a column vector. Or instead we can emphasize the local situation by considering A to be a sheaf endomorphism of the analytic sheaf of $\mathcal{O}(X)$ modules, $\mathcal{O}^n(X)$, where $\mathcal{O}(X)$ is the sheaf of rings of germs of analytic functions on X. Here A maps a germ v_x of $\mathcal{O}^n(X)$ at x onto $(A)_x v_x$ where $(A)_x$ is the germ of A at x. Next, let $M(X)$ be the field of global meromorphic functions on X. Then, to the extent to which methods of ordinary linear algebra apply to the study of matrix valued functions, it is useful to regard A as an ordinary linear transformation on the vector space $M^n(X)$ over $M(X)$. Finally the point of view which is most germane to the problems at hand is to consider A as an endomorphism of the module of global analytic vector valued functions $H^n(X)$ over the ring $H(X)$ of global analytic functions on X, that is, as an endomorphism of the module of global sections of $\mathcal{O}^n(X)$ over the ring of global sections of $\mathcal{O}(X)$.

A problem with which we must come to grips is the following. Suppose that in the course of doing linear algebra over $M(X)$ we obtain some subspace M_0 of $M^n(x)$ of dimension m. Then M_0 has as a basis over $M(X)$, m column-vector valued functions $v_1, \dots v_m$, which, without loss of generality, we can suppose to be analytic. This basis determines a singular or _weak subbundle_ S of dimension m of $X \times C^n$ in which at each point x we take the fiber to be the subspace $V(x)$ of C^n spanned by $v_1(x), \dots v_m(x)$. If the dimension of $V(x)$ is constant as x varies over X, then S is actually an analytic subbundle of $X \times C^n$. However it can happen that at an isolated set of points in X the dimension of $V(x)$ is less than m. We call such points _singularities_ of the weak subbundle. The question then arises as to whether a singularity x_0 is _removable_ in the sense that a beter choice of $v_1, \dots v_m$ exists which leads to a weak bundle with no singularity at x_0. The following Lemma shows that all singularities are removable in this sense. This property is, in fact, the signal distinguishing feature of the single variable analytic case and fails badly in more general situations.

Lemma. _Any weak sub-bundle_ S _of_ $X \times C^n$ _is contained in a unique sub-bundle_ $\overline{S}$ _of_ $X \times C^n$ _having the same dimension. All singularities of_ S _are simultaneously removable. If_ S_1, S_2 _are weak bundles and_ A _is an endomorphism of_ $X \times C^n$, _then_ $AS_1 \subset S_2$ _implies_ $A\overline{S}_1 \subset \overline{S}_2$. _We shall say that_ $\overline{S}$ _is spanned by_ S.

Proof. Let S be a weak sub-bundle of $X \times C^n$ determined by $\{v_1, \dots, v_m\}$. The proof will proceed by induction on m. Suppose first that $m = 1$, then we may assume that $v = v_1$ is not identically zero. Hence given any point w of X, it has a neighbourhood N_w in X such that $v(x)$ may be represented on N_w in the form $v(x) = x^s f_w(x)$, with f depending on w, and $f_w(x) \neq 0$ for $x \in N_w$. Now take for $\overline{S}$ the subset $\{(x, cf_w(x)) | x \in X$ and $c \in C\}$ of $X \times C^n$, provided with the fibering $p : \overline{S} \to X$ defined by projecting on the first coordinate.

The local product structure on $\overline{S}$ may be given by

$$g_w : N_w \times C \to p^{-1}(N_w) = \{(x, cf_w(x)) | c \in C \text{ and } x \in N_w\}$$

where $g(x, c) = (x, cf_w(x))$.

This proves the lemma for $m = 1$. Now assume that $k \geq 1$ and that the lemma is valid for $m \leq k$. Let S_1 be the vector bundle spanned by the weak vector bundle determined by $\{v_1, \ldots, v_k\}$. Then it follows from the theory of analytic vector bundles over a holomorphically complete manifold that with respect to the usual Hermitian metric on C^n, the bundle $X \times C^n$ is analytically equivalent to the Whitney sum $S_1 \oplus S_2$, where S_2 is again trivial. (For the case of a real analytic manifold or the case in which C is replaced by R see the appendix at the end of this section.)

To complete the induction step, write v_{k+1} analytically in the form $v_{k+1} = v^1 + v^2$ with $v^i \in S_i$. By the case $m = 1$, write $S_2 = \overline{\{v^2\}} \oplus S_3$, where $\overline{\{v^2\}}$ is the subbundle of S_2 spanned by the (perhaps weak) bundle determined by $\{v^2\}$. Then clearly we may set $\overline{S} = S_1 \oplus \overline{\{v^2\}}$. This concludes the existence part of the lemma. The uniqueness follows. With respect to a fixed Hermitian metric, the analytic subbundles of $X \times C^n$ are in one-one correspondence with the set of analytic projections. Since questions concerning uniqueness arise only at isolated points, the uniqueness part of the lemma follows from the continuity of the projections. The last statement $A\overline{S}_1 \subset \overline{S}_2$ follows similarly.

The preceding lemma is a special case of the following:

<u>Lemma</u>. <u>Let</u> $\alpha : V^m \to V^n$ <u>be an analytic homomorphism of vector bundles over a (not necessarily non-compact) Riemann surface</u> W. <u>Then the kernel</u> K <u>and image</u> I <u>of</u> α <u>define unique vector subbundles</u> $\overline{K}$ <u>and</u> $\overline{I}$ <u>of</u> V^m <u>and</u> V^n <u>having the same dimensions as</u> K <u>and</u> I <u>respectively, and such that</u> $\overline{K} \subset K$ <u>and</u> $\overline{I} \supset I$.

<u>Proof</u>. The uniqueness part follows from the statement about the dimensions. Hence it suffices to define $\overline{K}$ and $\overline{I}$ locally.

To do this consider the associated analytic sheaf homomorphism

$$S(\alpha) : S(V^m) \to S(V^n)$$

where $S(V^m)$ for instance is the sheaf of germs of analytic sections of V^m. Now $S(V^m)$ and $S(V^n)$ are coherent analytic sheaves which are locally free. Since W is one-dimensional the subsheaves $S(K) = \text{kernel } S(\alpha)$ and $S(I) = \text{Image } S(\alpha)$ are also locally free, cf [6]. Hence locally each of $S(K)$ and $S(I)$ has a basis of sections. These basis elements may be interpreted as local sections of K and I respectively. If for instance the sections of K arising this way are denoted according to our previous notation by $v_1, \ldots, v_k$, then $\{v_1, \ldots, v_k\}$ may be desingularized as in the preceding lemma to define $\bar{K}$ locally. The remaining details are obvious.

It may be pointed out that the preceding lemma is not a consequence of the one-one correspondence between isomorphism classes of analytic vector bundles and isomorphism classes of analytic locally free sheaves. To see this consider the trivial complex line bundle $X \times C$. This has no one-dimensional locally free subbundles. Yet the sheaf $S(X \times C)$ of germs of analytic sections of $X \times C$ has proper subsheaves isomorphic to it. Assume X is the complex plane, and consider the map $\alpha : X \times C \to X \times C$ defined by $\alpha(z, c) = (z, zc)$. Then it is easy to see, of [6] and the proof of the lemma, that $S(\alpha)\,(S(X \times C))$ is a locally free subsheaf of $S(X \times C)$ which is free since $H^1(X, \mathcal{G}_p) = 0$, $\mathcal{G}_p$ being the sheaf of germs of holomorphic invertible $p \times p$ matrix-valued functions.

Appendix on Analytic Vector Bundles

Our results hold equally well for matrix valued functions $A : Y \to L(n,R)$ or $A : Y \to L(n,C)$, Y being the real line or the unit circle.

The validity of the theory in these cases depends on the equivalence between the theory of real (alternatively complex) analytic vector bundles on a real analytic manifold W and the theory of real (alternatively complex) continuous vector bundles over W. This means that if an analytic vector bundle is topologically trivial then it is analytically trivial, and that any continuous vector bundle is topologically isomorphic to an analytic one. For a holomorphically complete manifold X the corresponding equivalence is a famous theorem of H. Grauert. For Y real analytic, the complex case is derived from this in [11]. Here we will outline a proof due to John Mather for the remaining case.

We prove:

<u>Theorem:</u> <u>The theory of real analytic vector bundles over a real analytic manifold W is equivalent to the theory of continuous real vector bundles on</u> W.

Proof. The assertion is equivalent to showing that any continuous mapping $f : W \to G$, G being the Grassmann manifold of k-planes in R^t, can be approximated arbitrarily closely by an analytic mapping. To see how this is possible, note first that by a result of N. Steenrod we may assume that f is differentiable. Now a theorem of H. Whitney states that: Given an open subset U of the real m-dimensional euclidean space R^m, a differentiable map $g : U \to R^n$, a covering $\{C_i\}_{i=1}^{\alpha}$ of U by coordinate neighbourhoods, a sequence $\{k_i\}_{i=1}^{\alpha}$ of integers, and a sequence $\{\epsilon_i\}_{i=1}^{\alpha}$ of positive numbers, there exists an analytic function $\bar{g} : U \to R^n$ such that on C_i $g-\bar{g}$ as well as all its mixed partial derivatives of orders up to k_i are bounded in absolute value by ϵ_i, $i=1,2,\ldots$. By another theorem of Whitney we embed W and G analytically in some R^m and R^n by means of maps $i : W \to R^m$ and $j : G \to R^n$. Then W and G have open neighbourhoods $N(W)$ and $N(G)$ in R^m and R^n which fiber analytically over W and G respectively. Denote the projections of these fibrations by $p : N(W) \to W$ and $q : N(G) \to G$. Now we may use Whitney's theorem to approximate the map $j \circ f \circ p : N(W) \to R^n$ sufficiently closely by an analytic map h, for the composition $q \circ h \circ i$ to be well defined and as close to our original map f as we please.

3. An invariant sub-bundle theorem. Normal forms

We now show how linear algebra over $M^n(X)$ combines with the Lemma of Section 2 to yield concrete results. The following theorem shows that given a matrix function A, corresponding to any global analytic factor of its characteristic equation $p(\lambda) = \det(\lambda I-A) = 0$, there is an invariant sub-bundle of $X \times C^n$.

Theorem 3.1 (Invariant subbundle theorem). Suppose that $A : X \to L(n,C)$ is analytic and that its characteristic polynomial $p(\lambda)$ has the form $q(\lambda)r(\lambda)$ where q and r are monic polynomials in λ of degrees k and $n-k$ respectively with analytic coefficients. Then $X \times C^n$ has an analytic sub-bundle S of dimension k which is invariant under A and such that when the restriction of A to S is *considered as a function into* $L(k,C)$, *it has characteristic polynomial* $q(\lambda)$. *Further, if the resultant of* q *and* r *is not identically zero, then* S *is unique*.

Proof. Considered as a linear transformation on the vector space $M^n(X)$ over the field $M(X)$, A has an invariant subspace of dimension k on which its characteristic polynomial is q. Let S_0 be a weak bundle determined by any basis for this subspace and let $S = \overline{S}_0$ according to the basic Lemma. If the resultant of q and r is not identically zero, then q and r are relatively prime over $M(X)$ and it follows from ordinary linear algebra that the subspace corresponding to q is unique. Since this subspace determines S on a dense set of points in X,S is also unique.

We now use this theorem to prove the following

Theorem 3.2. (Block Triangular Form). *Let* $A : X \to L(n,C)$ *be analytic and suppose that its characteristic polynomial has the form* $p(\lambda) = p_1(\lambda) \ldots p_n(\lambda)$, *each* p_i *being monic with analytic coefficients. Then there exists a nowhere singular matrix function* $T : X \to GL(n,C)$ *such that* $T^{-1}AT$ *has upper block triangular form, the k-th diagonal block being a matrix function with characteristic polynomial* p_i.

Proof. It clearly suffices to consider the case of only two factors $p = qr$. Let S be an invariant subbundle corresponding to q as in the preceding theorem and let $T_1 \ldots T_k$ be a set of trivializing sections. Choose any bundle complementary to S and let $T_{k+1} \ldots T_n$ be trivializing sections for it. Then the matrix function T having T_k as its k-th column is a matrix function with the desired properties.

This theorem can be specialized to the triangulization theorem proved in [5].

Theorem 3.3. (Triangular Form). *If the characteristic polynomial of* A *can be resolved into linear analytic factors, then* A *is analytically similar to a triangular matrix function*.

We now use similar arguments to come as close to obtaining rational or companion normal forms as the problem will permit. As in ordinary matrix theory, the advantage of these forms is that they can be obtained without any restrictions on the characteristic equation. We recall from the ordinary theory of vector spaces that the simplest instance of such a rational form arises when the characteristic equation of A is irreducible. In this case there is a cyclic vector v such that $v, Av, \ldots, A^{n-1}v$ span the space. Choosing these vectors as a basis, A (regarded as a linear transformation) becomes a matrix with 1's on the main subdiagonal, entries in the last column and 0's elsewhere. We call this form subtriangular companion form. It is the upper subtriangular aspect of this form, that is, vanishing entries below the main subdiagonal, that survives in the following result.

Theorem 3.4 (Subtriangular Form). Let $A : X \to L(n, C)$ be analytic. Let $p = p_1 \ldots p_n$ be a factorization of its characteristic polynomial into irreducible factors. Then there is an analytic $T : X \to GL(n,C)$ such that $T^{-1}AT$ has upper block triangular form in which the k-th diagonal block is an upper subtriangular matrix with characteristic polynomial p_k.

Proof By the previous theorem we can assume that A has upper block triangular form corresponding to the global irreducible factors of its characteristic equation $p = 0$. It is easy to see that to prove the theorem it suffices to reduce the diagonal blocks to subtriangular form. In other words we need only consider the case where p is globally irreducible. This means that p is irreducible over $M(X)$. Hence, reverting again to linear algebra over $M^n(X)$, A has a "cyclic" vector v such that $v, Av, \ldots, A^{n-1}v$ span $M^n(X)$. Let S_k be weak bundles determined by $v, Av, \ldots, A^{k-1}v$, $k = 1, 2, \ldots n$. Then the bundles $\bar{S}_1, \bar{S}_2, \ldots, \bar{S}_n = X \times C^n$ satisfy $\bar{S}_k \subset \bar{S}_{k+1}$, $A\bar{S}_k \subset \bar{S}_{k+1}$ since the weak bundles S_k satisfy similar inclusions. Choose v_k to be a analytic nonvanishing section of any line bundle in $\bar{S}_k$ complementary to $\bar{S}_{k-1}$. Then Av_k is in the $H(X)$ module of analytic sections of $\bar{S}_{k+1}$. Since $v_1, \ldots, v_{k+1}$ generate this module, Av_k is a linear combination of these with analytic coefficients. This is equivalent to the assertion that the matrix T with v_k as its k-th column reduces A to the required form.

Further reductions depend on the nature of the entries on the main subdiagonal. We illustrate this by deriving the following local result from the previous theorem.

Remark : The special case when $n = 1$ in Theorem 3.4 was first obtained by G. Rayna by means of a constructive proof.

Theorem 3.5. (Local Subtriangular Form). Let A be an analytic matrix germ at $x = 0$. Let $p = p_1 \dots p_n$ be a factorization of its characteristic polynomial into irreducible factors over the ring of analytic germs at $x = 0$. Then there is an invertible analytic matrix germ T such that $T^{-1}AT$ has an upper block triangular form in which the characteristic polynomial of the k-th diagonal block is p_k. Moreover the diagonal blocks are each subtriangular, the subdiagonal elements have the form x^k for some integer k (depending on the entry). Also all elements above the main subdiagonal in the column of this entry are polynomials of degree less than k.

Proof. We suppose that A has the form of the previous theorem. If the block triangular form is actually triangular, that is, all the blocks are one dimensional, then there are no subdiagonal entries and there is nothing to prove. We therefore suppose that the first q blocks have dimension higher than one. Let $n_1, \dots, n_q$ be the dimensions of these blocks and let e_i be the column vector with zero entries except for a 1 in the i-th place. Then the subtriangular form of A means that

$$Ae_i = \sum_{j=1}^{i} a_{ij} e_j + a_i e_{i+1} \quad \begin{cases} i \leq n_1 + n_2 + \dots + n_q \\ i \neq n_1, n_1 + n_2, \dots \end{cases}$$

Moreover each a_i is not identically zero. For if some $a_i \equiv 0$ then the diagonal block containing it would be properly block triangular. This would in turn imply that its characteristic polynomial is reducible, contradicting the preliminary normal form of A. Thus we can suppose that $a_i = \alpha_i x^{k_i}$, $\alpha_i(0) \neq 0$. The above relation can then be put in the form

$$Ae_i = \sum_{j \leq i} \beta_{ij} e_j + x^{k_i} \left(\sum_{j \leq i} \alpha_{ij} e_j + \alpha_i e_{i+1} \right)$$

where the β_{ij} are polynomials of degree less than k_i and (we reiterate) α_i is invertible. We now observe that a similar set of relations with the same k_i's must persist among the elements of any set of independent columns $e_1', \dots, e_n'$ for which e_i' is contained in the span $e_1, \dots, e_i$. Suppose, proceeding inductively, that $e_1' \dots e_m'$ have been determined so that $\alpha_{ij} = 0$, $\alpha_i = 1$ for $i < m$ (this describes the desired normal form entry wise). Then if $m \neq n_1, n_1 + n_2, \ldots,$ we have

$$Ae'_m = \sum_{j \le m} \beta'_{mj} e'_j + x^{k_m} \Big(\sum_{j \le m} \alpha_{mj} e_j + \alpha_m e_{m+1} \Big)$$

Since α'_m is a unit the choice $e'_{m+1} = \sum'_{j \le \ell} \alpha'_{mj} e_j + \alpha'_m e_{m+1}$ is acceptable. Then in terms of $e'_1 \ldots e'_{m+1}$, Ae_m has the desired form. If $m = n_1, n_1 + n_2, \ldots,$ simply let $e'_m = e_m$. Determining $e'_1, \ldots, e'_{n_1 + \ldots + n_q}$ according to this process and letting $e'_k = e_k$ for $k > n_1 + \ldots + n_q$ we find that the germ T having e'_k as its k-th column satisfies the requirements of the theorem.

We note that in the special case in which each subdiagonal of x is 1, the entries above are zero and we nearly obtain one of the usual block companion forms of ordinary matrix theory. We will return to this result in Section 6 when we consider similar normal forms for systems of linear differential equations.

4. Remarks and Examples. Block Diagonal Forms

We now assume that the characteristic polynomial $p(\lambda)$ of A has the form $p(\lambda) = q(\lambda)r(\lambda)$ where $q(\lambda,x)$ and $r(\lambda,x)$ are relatively prime except at an isolated set of points, or equivalently, where $q(\lambda)$ and $r(\lambda)$ are relatively prime as polynomials over the field $M(X)$. In this case A, regarded as an element of $L(n,M(x))$, has a pair of complementary invariant subspaces uniquely determined by q and r. As we have seen in the previous section, we can associate invariant bundles S_q and S_r of A (regarding it now as a bundle map) with each invariant subspace. However we cannot expect that these derived bundles will be complementary, and in general, if $q(\lambda,x)$ and $r(\lambda,x)$ fail to be relatively prime, the fibers of S_q and S_r at x will fail to span C^n. However in this circumstance it is obviously not necessary that the fibers fail to be complementary. The following standard construction from ordinary matrix theory is relevant here.

Since $q(\lambda)$ and $r(\lambda)$ are relatively prime over the ring $H(x)$, there exist polynomials $\alpha(\lambda)$ and $\beta(\lambda)$ in $H(X)(\lambda)$ such that the resultant $R(q,r)$ of q and r is given by

$$\alpha(\lambda)q(\lambda) + \beta(\lambda)r(\lambda) = R(q,r).$$

Moreover $q(\lambda,x)$ and $r(\lambda,x)$ are relatively prime if and only if $R(x) \neq 0$. It follows from this that the meromorphic matrices $\pi_q = R^{-1}\alpha(A)q(A)$ and $\pi_r = R^{-1}\beta(A)r(A)$ commute with A, are analytic at x if $R(x) \neq 0$, and satisfy the relations $\pi_q^2 = \pi_q$, $\pi_r^2 = \pi_r$, $\pi_q\pi_r = \pi_r\pi_q = 0$, $\pi_q + \pi_r = I$, where I is the identity matrix. If it is the case that S_q and S_r are complementary, then the projections on S_q and S_r are everywhere analytic and agree with π_q and π_r on a dense set in x, hence everywhere. This means that π_q and π_r must be analytic on X. Conversely if π_q and π_r are analytic, then the previously mentioned properties imply that they are projections on complementary invariant subbundles corresponding to q and r.

Here it is also appropriate to note again the significance of the fact that X supports only trivial analytic complex vector bundles. For if S_q and S_r are complementary, then choosing the columns of a matrix function T to be independent global sections in S_q, then in S_r we obtain a T for which $T^{-1}AT$ has a block diagonal form corresponding to the factorization $p = qr$ in the sense that the diagonal blocks have characteristic polynomials q and r. Conversely if such a T exists its columns span two complementary invariant bundles which must coincide with S_q and S_r.

Summarizing these observations we have the following

Theorem 4.1. (Decomposition Criterion). Let $A : X \to L(n,C)$ be analytic and assume that the characteristic polynomial $p(\lambda)$ has the form $p(\lambda) = q(\lambda)r(\lambda)$ where $R(q,r) \not\equiv 0$. Let S_q and S_r be the unique invariant subbundles associated with q and r. Let π_q and π_r be the projections on the invariant subspaces of $M^n(X)$ corresponding to q and r. Then the following three conditions are equivalent.

1. $X \times C^n$ is the direct sum of S_q and S_r
2. The meromorphic matrices π_q and π_r have removable singularities at each zero of $R(q,r)$.
3. There is an analytic similarity reducing A to a block diagonal form corresponding to the factorization $p = qr$.

A special case of this is the following result of [3].

Theorem 4.2. (Block Diagonal Form). Let $A : X \to L(n,C)$ be analytic and assume that the characteristic polynomial $p(\lambda)$ has the form $p(\lambda) = q(\lambda)r(\lambda)$ where $R(q,r)$ never vanishes. Then there is an analytic similarity reducing A to a block diagonal form corresponding to the given factorization.

The following three examples illustrate both the scope and the limitations of the previous analysis.

Example 1. $A = \begin{bmatrix} 0 & 1 \\ 0 & x \end{bmatrix}$. In this case $p = \lambda(\lambda-x)$. Let $q = \lambda$, $r = \lambda-x$. Then $R(q,r) = x$, $\alpha = 1$, $\beta = -1$, $\pi_q = x^{-1}\begin{bmatrix} 0 & 1 \\ 0 & x \end{bmatrix}$, $\pi_r = x^{-1}\begin{bmatrix} x & -1 \\ 0 & 0 \end{bmatrix}$. Also nonzero sections in S_q and S_r are $\begin{bmatrix} 1 \\ x \end{bmatrix}$ and $\begin{bmatrix} 1 \\ 0 \end{bmatrix}$ which fail to span C^2 at $x = 0$. Here it is obviously the coalescing of the unique invariant subspaces of $A(x)$ at $x = 0$ that produces the singular behaviour.

Example 2.

$$A = \begin{bmatrix} 0 & 1 & 0 \\ 0 & 0 & 1 \\ 0 & -x^3 & x^2+x \end{bmatrix}, \quad p(\lambda) = \lambda(\lambda-x)(\lambda-x^2). \text{ Let}$$

$q = \lambda$, $r = (\lambda-x)(\lambda-x^2)$. Then $R(q,r) = x^3$. The invariant bundles corresponding to the linear factors of p are spanned by

$$v_1 = \begin{bmatrix} 1 \\ 0 \\ 0 \end{bmatrix}, \quad v_2 = \begin{bmatrix} 1 \\ x \\ x^2 \end{bmatrix} \text{ and } v_3 = \begin{bmatrix} 1 \\ x^2 \\ x^4 \end{bmatrix}.$$

The sections v_2 and v_3 determine a weak invariant bundle corresponding to $r(\lambda)$. However this weak bundle is singular at $x = 0$ and $x = 1$ where the span of v_2 and v_3 is only one dimensional. The triangular change of sections $v_2' = v_2$,

$$v_3' = x^{-1}(x-1)^{-1}(v_3-v_2) = \begin{bmatrix} 0 \\ 1 \\ x(x+1) \end{bmatrix}$$

leads to better sections, v_2', v_3', which realize the invariant bundle S_r as a non-singular weak bundle.

In order that the reader does not grant too willingly the existence of $\bar{S}$ for any weak bundle S, and the consequences which flow from it, such as the existence of S_q for any factor $q(\lambda)$ of $p(\lambda)$, we cite the following example (also see Example 4 in the next Section).

Example 3. Let $A : R \to L(2, C)$ be given by

$$A(t) = \begin{cases} e^{-1/t^2} \begin{bmatrix} \cos^2 1/t & \cos 1/t \sin 1/t \\ \cos 1/t \sin 1/t & \sin^2 1/t \end{bmatrix} & t \neq 0 \\ 0 & t = 0 \end{cases}$$

Then A is infinitely differentiable. For $t \neq 0$, the invariant lines of $A(t)$ are spanned by $v_1 = \begin{bmatrix} \cos 1/t \\ \sin 1/t \end{bmatrix}$ and $v_2 = \begin{bmatrix} -\sin 1/t \\ \cos 1/t \end{bmatrix}$ corresponding to characteristic roots $e^{-1/t}$ and 0. If $q = \lambda - e^{-1/t^2}$, $r = \lambda$, then $R(q,r)$ vanishes only at $t = 0$. Here v_1 defines a weak bundle corresponding to q in the sense that it spans an ordinary line bundle corresponding to q on the dense set $R - \{0\}$. But it is evident from the behaviour of v_1 near zero that the singularity at zero cannot be removed, and that there is no differentiable (or continuous for that matter) invariant line bundle of A corresponding to q.

5. Remarks on the Case of Several Variables. The Role of Topological Considerations.

We continue the discussion of the previous section with the following example which illustrates the local complications in the case of matrix functions of several variables and which continue to show the extent and limits of our previous analysis.

Example 4.

$$A(x,y) = \begin{bmatrix} x^2 & xy \\ xy & y^2 \end{bmatrix}, \quad p = \lambda(\lambda - x^2 - y^2),\ q = \lambda,\ r = \lambda - x^2 - y^2 ,$$

$R(q, r) = x^2 + y^2$. If we restrict A to the set where $x^2 + y^2 \neq 0$, A has unique invariant line bundles S_q and S_r with nonvanishing sections

$$u = \begin{bmatrix} x \\ y \end{bmatrix} \quad \text{and} \quad v = \begin{bmatrix} y \\ -x \end{bmatrix} .$$

But neither of these can be the restriction to the dense set $C^2 - \{x^2 + y^2 = 0\}$ of a bundle over all of C^2 (bear in mind that the fact that this would be so in the one variable case is the simplest instances of the basic Lemma). This can be seen in many ways, of which we select two. Firstly, if we approach $(0, 0)$ along the real spiral $x = e^{-1/2t^2} \cos 1/t,\ y = e^{-1/2t^2} \sin 1/t$ as $t \to 0$, we obtain exactly Example 3 of the previous section. Secondly the sheaf $\mathcal{S}_q$ of germs of analytic sections of S_q would have to be the image of A considered as an endomorphism of $\mathcal{O}^2(C^2)$. This image is generated by the columns of A,

$$a_1 = \begin{bmatrix} x^2 \\ xy \end{bmatrix} \quad \text{and} \quad a_2 = \begin{bmatrix} xy \\ y^2 \end{bmatrix}.$$

On any neighbourhood of $(0, 0)$ both generators are

needed since neither is an analytic multiple of the other. But $ya_1 - xa_2 = 0$, so that $\mathcal{S}_q$ is not locally free and could not be the sheaf of germs of analytic sections of any vector bundle. The same considerations show that if we restrict A to $C^2 - \{(0,0)\}$, it then has invariant line subbundles even though there is still a large set on which $x^2 + y^2 = R(q,r)$ vanishes. For on $C^2 - \{(0,0)\}$ either a_1 or a_2 is a local generator for $\mathcal{S}_q$ (similarly for $\mathcal{S}_r$). In this situation however the invariant line bundles fail to be complementary on $\{x^2 + y^2 = 0\} - \{(0,0)\}$. This can read in the projections π_q and π_p

$$\pi_q = (x^2+y^2) \begin{bmatrix} x^2 & xy \\ xy & y^2 \end{bmatrix}, \quad \pi_r = (x^2+y^2)^{-1} \begin{bmatrix} -y^2 & xy \\ xy & -x^2 \end{bmatrix}$$

which have poles on this set. We note that the points $\{x^2 + y^2 = 0, x \neq 0$ or $y \neq 0\}$ are distinguished from $\{(0,0)\}$ by the following properties of π_q. Any point in $\{x^2 + y^2 = 0, x \neq 0$ or $y \neq 0\}$ has a neighbourhood on which there is a scalar function α such that $\alpha\pi_q$ is analytic and everywhere of rank 1 (pick $\alpha = x^2 + y^2$ or $\alpha = x \pm iy$) while no such α can be found on a neighbourhood of (0,0). It is easy to see that this property characterises the removability of singularities of weak bundles in the sense of Section 2 in terms of their associated meromorphic projections.

We now turn from this example, in which we have emphasized local considerations, to an augmented version of the decomposition criterion, Theorem 4.1, in which we will be obliged to reconsider global questions. The discussion which precedes Theorem 4.1 must be qualified in the following three ways. To begin with, we can no longer take for granted the existence of invariant subbundles corresponding to any factor of the characteristic equation (as the preceding example shows). Secondly, the existence of a normalizing analytic similarity transformation requires certain bundles to be trivial. In the case of a noncompact Riemann surface all complex vector bundles are trivial. In contrast, higher dimensional domains which support nontrivial vector bundles are very common. Thirdly, we must now distinguish between topological and analytic triviality, notions which coincide on a noncompact Riemann surface, but not in general. We therefore restrict our attention to a holomorphically complete complex manifold N or to a real analytic manifold N on which the two notions still coincide (in order to retain a usable criterion for analytic triviality). For a proof of the latter assertion see [11].

We wish to give a criterion that embodies everything useful that we can say here about the decomposition problem without making further progress in the local situation (what, for example, is a suitable class of local normal forms?). To this purpose we reconsider the discussion of Section 4. Suppose we have A with a characteristic polynomial $p(\lambda) = p_1(\lambda)\ldots p_k(\lambda)$ where the $p_r(\lambda)$ are relatively prime over $M(N)$. Then polynomials $\gamma_r(\lambda)$ over $M(N)$ can be found so that $\sum \gamma_j(\lambda)p_j(\lambda) = 1$. The meromorphic matrices $\pi_{p_j} = \gamma_j(A)p_j(A)$ are projections on invariant subspaces of $M^n(X)$ corresponding to $p_j(\lambda)$. We can again associate a weak bundle S_{p_j} with each π_{p_j} in the following way. On a holomorphically complete manifold a meromorphic function is always the quotient of two global analytic functions. Writing $\pi_{p_j} = f_j^{-1}G_j$ where $f_j : X \to C$ and $G_j : X \to L(n,C)$ are analytic, we can take the fiber of S_{p_j} at x to be $G_j(x)C^n$ in complete analogy with the single variable case. We observed in the discussion of Example 4 that the singularities of S_{p_j} will be removable if and only if for each $x_0 \in N$ there is a neighbourhood U of x and an analytic $f : U \to C$ for which $f\pi_{p_j}$ has constant rank on U. Let us express this by saying that any meromorphic matrix function with this property is suitable at x_0. We then have the following extension of Theorem 4.1. which again summarizes the preceding remarks.

Theorem 5.1. Let N be a holomorphically complete manifold or a real analytic manifold of arbitrary dimension. Let $A : N \to L(n,C)$ be analytic and assume that the characteristic polynomial has the form $p(\lambda) = p_1(\lambda)\ldots p_k(\lambda)$, where p_i has degree n_i and $R(p_i,p_j) \not\equiv 0$ for $i \neq j$. Let π_{p_j} be the projection on the invariant subspace of $M^n(N)$ corresponding to $p_j(\lambda)$ and let S_{p_j} be an invariant weak bundle associated with $p_j(\lambda)$. Then S_{p_j} determines an invariant subbundle $\bar{S}_{p_j}$ if and only if π_{p_j} is suitable at each point of N. Moreover suppose that N has the following property: if $N \times C^n$ is the Whitney sum of k vector bundles of dimension $n_1, \ldots, n_k$, then each summand is trivial. Then the following three conditions are equivalent:

1. $\bar{S}_{p_j}$ exists and $N \times C^n = \sum_{j=1}^{k} \oplus \bar{S}_{p_j}$.

2. The meromorphic matrix functions π_{p_j} have only removable singularities.

3. There is an analytic similarity reducing A to a block diagonal form corresponding to the factorization $p(\lambda) = p_1(\lambda) \ldots p_k(\lambda)$.

We observe that this criterion implies the following result of Hsieh and Sibuya [3].

Corollary 5.2. Let N be a polycylinder. Let $A : N \to L(n,C)$ be analytic with characteristic equation $p = qr$ where $R(q,r)$ is never zero. Then there is an analytic similarity transforming A to a block diagonal form corresponding to the given factorization.

In the next two Corollaries the topological ingredients are more pronounced. The first is Theorem C of [4] generalized to allow removable singularities.

Corollary 5.3. Let N be 2-connected. Let $A : N \to L(n,C)$ be analytic. Suppose that the characteristic polynomial $p(\lambda)$ factors into linear factors locally, that no factors coincide identically, and that the corresponding local projections have only removable singularities. Then A is analytically similar to a diagonal matrix function.

Proof. The proof is that of the above reference [4]. Briefly, the simple connectedness of N assures that the local factorizations fit together to give a global factorization, then 2-connectedness implies that N supports only trivial complex line bundles.

Corollary 5.4. Suppose that N has the homotopy type of $P_m(C)$, the m-dimensional complex projective space, and that $A : N \to L(n,C)$ is analytic, where $m \geq n$. Suppose that the characteristic polynomial $p(\lambda)$ again satisfies the local conditions of Corollary 5.3. Then A is analytically similar to a diagonal matrix function.

Proof. Again the simple connectedness of $P_m(C)$ ensures that p factors globally. Further, we need only observe that the condition in the hypotheses of Theorem 5.1 concerning the triviality of the summands in a direct sum decomposition of $N \times C^n$ can be established as in the proof of Theorem 5.5 below.

We conclude our discussion of the higher dimensional situation with a more elaborate illustration of the role of topological considerations in obtaining global results. Since the methods involved are indifferent to any refined aspects of the local structure such as analyticity, we consider the case of a continuous matrix function (although the same arguments apply for instance also to the differentiable case).

Theorem 5.5. Let $A : P^m \to L(n,C)$ be continuous (differentiable) where P^m is m-dimensional real projective space and $m \geq 2n$. Assume that the discriminant of p, the characteristic polynomial of A, never vanishes. Then A is continuously (differentiably) similar to a diagonal matrix.

Proof. Consider the space

$$B = \{(x,\alpha) \mid x \in P^m \text{ and } p(\alpha,x) = 0\}$$

as a subset of $P^m \times C$. Since the roots of $p(\lambda)$ are locally unique this is a covering space of P^m with projection $\pi_1(x,\alpha) = x$. We show that it is trivial. If not, a connected component will have the form $S^m \xrightarrow{\varphi} P^m$ with S^m being the standard m-sphere and the map φ being the one which identifies antipodal points. Denote the restriction of the inclusion $i : B \to P^m \times C$ to S^m by $i|S^m$, and let f be this map followed by the projection π_2 from $P^n \times C \to C$. Then we have the commutative diagram

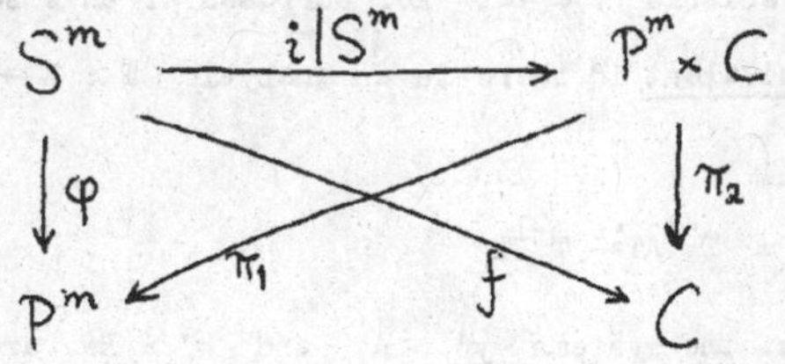

Since the discriminant of p never vanishes, this allows us to define an antipode preserving map $g : S^m \to S^1$ by $g(x) = \frac{f(x)-f(-x)}{|f(x)-f(-x)|}$. However such a map from S^m to S^k is known to exist only if $k \geq m$. Thus $B \to P^m$ is a trivial covering surface, which is equivalent to $p(\lambda)$ factoring globally into linear factors. The pointwise distinctness of these factors permits us to construct invariant line bundles as before. Finally the triviality of these line bundles follows from the condition $m \geq 2n$ upon examining the Chern classes of the set of line bundles, see [7], or directly from the structure of the ring $K_C(P^n)$ of equivalence classes of complex vector bundles on P^m, see [8].

6. Normal Forms for Systems of Linear Differential Equations

The theory of ordinary differential equations, particularly the asymptotic theory, requires analysis of matrix valued functions. For numerous applications of this kind see [9], also [5], where the transformation of matrix functions by similarity is the main tool. Here instead, we consider the direct application of the arguments of Sections 1 and 2 to the transformation $A \to T^{-1}AT - T^{-1}T'$ which results from transforming the linear vector system of differential equations

$$y' = A(x)y$$

by the change of dependent variable $y = Tw$. For purposes of this section we will say that A and B are <u>equivalent</u> if there is an analytic $T : X \to GL(n,C)$ for which

$$B = T^{-1}AT - T^{-1}T'$$

We also say in this case that the systems $y' = Ay$ and $w' = Bw$ are <u>equivalent</u>.

The most basic result of this section is the following

<u>Theorem 6.1.</u> (Subtriangular Form for Linear Systems). <u>Let</u> X <u>be as in Section 2. Let</u> $A : X \to L(n,C)$ <u>be analytic. Then the system</u> $y' = Ay$ <u>is equivalent to a system</u> $w' = Bw$ <u>in which</u> B <u>is upper subtriangular.</u>

<u>Proof</u>. We imitate the proof of Theorem 3.4. We begin by performing differential algebra over the field $M(X)$. Let $\tilde{A}$ be the differential operator on $M^n(X)$ (or $H^n(X)$) defined by $\tilde{A}v = Av - v'$. Following a standard argument we can choose any v_1 in $H^n(X)$ and continue the sequence $v_1, \tilde{A}v_1, \ldots \tilde{A}^{m_1 -1} v_1$, until $\tilde{A}^{m_1} v_1$ is dependent as an element of $M^n(X)$ on the previous elements. Continuing in this way we find a basis, call it $u_1, \ldots u_n$, in $H^n(X)$ for $M^n(X)$ of the form

$$v_1, \tilde{A}v_1 \ldots \tilde{A}^{m_1 -1} v_1, v_2, \tilde{A}v_2 \ldots \tilde{A}^{m_2 -1} v_2, v_3 \ldots \tilde{A}^{m_q -1} v_q$$

generated by cyclic vectors $v_1, v_2, \ldots v_q$ of orders $m_1, m_2, \ldots, m_q$ where $m_1 + m_2 + \ldots + m_q = n$. In essence this is a well known argument for transforming A into a strict block companion form with meromorphic entries since the matrix U with columns $u_1, u_2 \ldots u_n$ induces such a transformation. However these columns need not be independent at each point and in general U^{-1} is only meromorphic. Instead we recall the proof of Theorem 3.4. Let S_k be a weak bundle determined by $u_1 \ldots u_k$, $k = 1, \ldots, n$, and let t_k be a nonvanishing section of a line bundle in $\bar{S}_k$ complementary to $\bar{S}_{k-1}$. Again we have $\bar{S}_k \subset \bar{S}_{k+1}$. The remaining vital step is to show that $\tilde{A}\,\bar{S}_k \subset \bar{S}_{k+1}$. In this case however the prior inclusion $\tilde{A}\,S_k \subset S_{k+1}$ is not self evident since $\tilde{A}$ is not an endomorphism of $M^n(X)$. Specifically $\tilde{A}$ fails to commute with scalar multiplication in $M^n(X)$. Instead of relying on linearity we must use the following property of first order differential operators. Namely for $\alpha \in H(X)$ and $u \in H^n(X)$ we have

$$\tilde{A}\,\alpha u = \alpha \tilde{A}u - \alpha' U.$$

This implies (using the fact also that $\tilde{A}$ is additive) that

$$\tilde{A} \text{ span } (u_1, \ldots, u_m) \subset \text{span } (u_1 \ldots u_m, \tilde{A}u_1, \ldots \tilde{A}u_m)$$

If we restrict $\bar{S}_1, \ldots, \bar{S}_n$ to the dense set where none of the S_k are singular we have

$$\begin{aligned} \tilde{A}\,\bar{S}_k = \tilde{A} \text{ span } (u_1 \ldots u_k) &\subset \text{span } (u_1 \ldots u_k, \tilde{A}u_1 \ldots \tilde{A}u_k) \\ &\subset \text{span } (u_1 \ldots u_{k+1}) = \bar{S}_{k+1} \end{aligned}$$

It follows by continuity that $\lambda \bar{S}_k \subset S_{k+1}$ (λ is a continuous operator on the module of global analytic sections of $\bar{S}_k$ if these are given any natural kind of topology. See [10] for example.) In particular λt_k is an analytic section of $\bar{S}_{k+1}$. Since $t_1, \ldots t_{k+1}$ generate the module of global analytic sections of $\bar{S}_{k+1}$ we necessarily have

$$\lambda t_k = \sum_{j \leq k+1} b_{jk} t_j$$

where $b_{jk} \in H(X)$. Letting t_k be the k-th column of T, this expresses (columnwise) the relation $AT-T' = TB$, that is, A is equivalent to B where B has the desired subtriangular form.

If we use the reasoning which leads from Theorem 3.4 to Theorem 3.5 we obtain the more specific normal form of the Theorem.

Theorem 6.2. (Local Subtriangular Form for Linear Systems).

Let $A : X \to L(n,C)$ be analytic at $x = 0$. Then on some neighbourhood of $x = 0$ the system $y' = Ay$ is equivalent to $w' = Bw$ where B is upper block triangular, each diagonal block of B being subtriangular with entries on its main subdiagonal of the form x^k for some integer k (depending on the entry) and each element of B above such an entry being a polynomial of degree less than k.

Proof. The proof is that of Theorem 3.5 with one slight modification. It is easy to verify that if a basis $v_1, \ldots, v_n$ satisfies "subtriangular" relations of the form

$$A_{v_i} - v_i' = \sum_{j \leq i} \beta_{ji} v_j + \beta_i x^{k_i} v_{i+1} \qquad i = 1, \ldots, n$$

then any basis of sections $v_1^*, \ldots, v_n^*$ related to $v_1, \ldots, v_n$ by a triangular change of basis (that is $v_k^* \subset \text{span}(v_1, \ldots, v_k)$ for each k) satisfies similar relations with the same exponents k_i. Since it is this property that underlies the inductive argument in Theorem 3.5, we obtain the same normal form here. We now consider systems with a finite number of singular points $x_1, \ldots, x_q$ having the form

$$(x-x_1)^{h_1} \cdots (x-x_q)^{h_q} y' = A(X)y,$$

where the h_i are nonnegative integers. It is possible to achieve additional simplifications for such systems because it is natural to extend the class of transformations, allowing T in $T^{-1}AT-T^{-1}T'$ to be any meromorphic transformation which does not introduce new singular points or increase any of the h_i's. If we understand equivalence of singular systems in this sense we have the following immediate consequence of Theorem 6.1.

Theorem 6.3. (Global Subtriangular Form for Singular Systems).
Let $A : X \to L(n,C)$ be analytic. Then the system $(x-x_1)^{h_1}\dots(x-x_q)^{h_q}y' = Ay$ is equivalent to a system $(x-x_1)^{h_1}\dots(x-x_q)^{h_q}w' = Bw$ where B is upper block triangular, each diagonal block of B being subtriangular with subdiagonal entries which do not vanish at $x_1,\dots,x_q$.

Proof. We assume that A has been transformed to subtriangular form. Then the similarity induced by the diagonal matrix

$$T = \operatorname{diag}\left\{1, (x-x_i)^{-k_2}, (x-x_i)^{-k_2-k_3}\dots(x-x_i)^{-k_2-k_3\dots-k_n}\right\}$$

has the following properties. The subdiagonal entry in the j-th row of A is multiplied by $(x-x_i)^{-k_j}$. If the k_j's are non-negative integers, then each entry of A above the main subdiagonal is multiplied by a nonnegative power of $(x-x_i)$. The matrix $(x-x_1)^{h_1}\dots(x-x_q)^{h_q}T^{-1}T'$ is an analytic diagonal matrix (here it is essential that x_i really is a singular point, i.e. $(h_i > 0)$). This means that we can put negative integral powers of $(x-x_i)$ into the subdiagonal entries at will without disturbing the subtriangular form. In particular we can remove all zeros at the singular points of the nonidentically vanishing subdiagonal entries, thereby obtaining the conclusion of the Theorem.

If we now take for granted that the singular system has the normal form of this theorem, and if further we restrict the system to any subset of X excluding the zeros of the nonidentically vanishing subdiagonal entries we can use the argument of Theorem 6.2 to obtain the following specific normal form.

Theorem 6.4. (Semilocal Subtriangular Form for Singular Systems). *Let the singular system* $(x-x_1)^{h_1}\cdots(x-x_q)^{h_q}y' = A(x)y$ *be specified by* $A : X \to L(n,C)$ *where* A *is analytic and the* h_k*'s are positive integers. Then there is an open connected neighbourhood of* $\{x_1,\ldots,x_q\}$ *in* X *on which this system is equivalent to* $(x-x_1)^{h_1}\cdots(x-x_q)^{h_q}y' = B(x)y$ *where* B *is block upper triangular, the diagonal blocks being subtriangular companion matrices, and each superdiagonal block having zero entries except possibly in its last column.*

Proof. We restrict the problem to a connected neighbourhood X' of the singular points on which the nonidentically vanishing subdiagonal elements do not vanish and then use the argument of Theorem 6.2 in the special case in which these subdiagonal entries are invertible in $H(X')$.

In conclusion we remark that some of the main difficulties in the asymptotic analysis of such systems occur in problems in which $A(x_i)$ is nilpotent and of low rank. But here this difficulty has virtually disappeared in the normal form of Theorem 6.4. In fact we have the following pleasant dichotomy. Either there is only one diagonal block, in which case $A(x_i)$ has rank at least $n-1$ at each x_i, or else there is more than one diagonal block in which case the problem reduces to the study of lower dimensional systems. We will apply these forms to asymptotics in a subsequent investigation.

Bibliography

[1] Wasow, W. On holomorphically similar matrices, J. Math. Anal. Appl. 4, 202-206.

[2] Sibuya, Yosutaka. Some global properties of matrices of functions of one variable, Math. Ann. 67-77. (1965).

[3] Hsieh, P. F. and Sibuya, Y. A global analysis of matrices of functions of several variables, J. Math. Anal and Appl. 14, 332-340 (1966).

[4] Khabbaz, S. and Stengle, G. An application of K-theory to the global analysis of matrix valued functions, Math. Ann. 179, 115-122 (1969).

[5] Khabbaz, S. and Stengle, G. Global triangulization of analytic matrix valued functions with applications to the asymptotic theory of differential equations. To appear.

[6] Gunning, R. C. Lectures on vector bundles over Riemann surfaces, Mathematical Notes, Princeton University Press (1967).

[7] Hirzebruch, F. Topological methods in algebraic geometry, Springer-Verlag, New York, (1966).

[8] Adams, J. F. Vector fields on spheres, Ann. Math. 75, 603-632 (1962).

[9] Wascow, W. Asymptotic Expansions for Ordinary Differential Equations, Interscience, New York, (1965).

[10] Gunning, R. and Rossi, H. Analytic functions of several complex variables. Prentice-Hall Inc., (1965).

[11] Khabbaz, S. The equivalence of the rings of continuous and analytic complex vector bundles on a real-analytic manifold, Bol. Soc. Math. Mexicana. 49-54 (1968).

GEOMETRIC ASPECTS OF THE SINGULAR SOLUTIONS OF CERTAIN DIFFERENTIAL EQUATIONS

S.A. Khabbaz

0. Introduction

Owing to the presence of difficulties stemming from insufficient information on the geometry of analytic sets, most treatments of differential equations confine themselves to the study of differential equations of the form $F(x, y, p) = 0$, $p = \frac{dy}{dx}$, under the assumption that one may solve $F = 0$ for p in terms of x and y in a differentiable manner, and of course the quickest way of assuring this is to assume that $\partial F/\partial p \neq 0$. As a result the study of singular solutions of differential equations has suffered in recent years, and as often as they are introduced they are seldom brought to a satisfying conclusion. To be sure this is not intentional, for their occurrence in the study of partial differential equations, see [6], and in differential geometry (in the theory of envelopes) is well known. Add to this that some differential equations such as $F = p^2 + y^2 = 0$, of §2, possess only the singular solution. Rather, the lack of any coherent theory makes a satisfactory exposition impossible. For a resumé of some of the older results we refer to [4] and the references in it on the subject. The more recent attacks on the problem such as in the works of E.R. Kolchin and J.F; Ritt have come from the direction of Ritt algebras, a subject which was developed with this problem in mind. However, the old results comprise those along analytic lines consisting largely of particular cases, and those along the lines of differential geometry which are primarily concerned with singular solutions as equations of envelopes of families of curves. And the new ones, cf [8 - 11], comprise attacks on the problem from an ideal theoretic point of view, an approach which like the corresponding one in algebraic geometry, leaves many geometric questions unanswered.

For a brief sample of the type of result obtainable in this way the reader may conveniently consult [7], particularly pages 51 - 52.

Here, inspired by recent progress made by Lojasiewicz, Milnor and Thom in the study of real varieties such as in [1, 2, 3, 12, 13], we study a class of real analytic first order differential equations from a point of view similar to theirs. We show as a very special case that when the differential equation is of even degree $m + 1$, its leading coefficient having only isolated real zeros, and $F = 0$ represents a non-singular surface, then the maximal highest order singular solutions (i.e. those of order m, see §1) of $F = 0$, $p = y'(x)$, fall into disjoint curves; and that an upper bound on their number in the case in which F is a polynomial of total degree n is given by $n(n - m)$. A statement of the principal results is given by §3. Further results are given in the last section.

1. Conventions and Definitions to be used throughout.

We shall consider a first order differential equation of the form $F(x, y, p) = 0$, $p = \frac{dy}{dx}$. We shall assume that F is a real analytic function in x, y, and p.

Set $F_x = \partial F/\partial x$, $F_y = \partial F/\partial y$ and $F_p = \partial F/\partial p$, and let $F_i = F_{p^{(i)}}$ denote the i-th partial derivative of F with respect to p, F_0 being just F. If $F_p(x_0, y_0, p_0) = 0$, then in general the fundamental theorems of differential equations governing uniqueness, existence and parametrizability of solutions near (x_0, y_0) do not apply, and the solutions near (x_0, y_0) fail to form a one-parameter family which is proper, in the sense that the dependence on the parameter is not one-one. We recall the usual definition, c.f. [4] or [7] :

Definitions : (1) <u>A singular solution</u> of $F = 0$, $p = y'(x) = \frac{dy}{dx}$ is a real valued differentiable (C^1) function $y = f(x)$ defined on a nondegenerate interval D which may be open, half open or closed, bounded or unbounded, at every point x of which $F(x, y(x), y'(x)) = 0$ and $F_p(x, y(x), y'(x))$ is zero. The first condition says that $y(x)$ is a solution, and the second that it is singular.

(2) A singular solution $y(x)$ is said to be <u>maximal</u>, provided that it is maximal

with respect to its domain of definition. We remark that by Zorn's Lemma every singular solution is contained in a maximal one.

(3) We shall say that a solution $y(x)$ passes (or goes) through a point x_0 or $(x_0, y(x_0))$ provided that x_0 is an <u>interior</u> point of the domain of definition of $y(x)$. When we refer to the curve $y(x)$ we will of course mean the set of pairs $(x, y(x))$ in the (x, y) plane.

(4) A singular solution $y(x)$ is said to be of <u>order m</u> provided that each of $F_1, F_2, \ldots, F_m$ vanishes at all points of the form $(x, y(x), y'(x)), x \in D$ and that F_{m+1} does not have this property. Thus m is the largest integer for which

$$\sum_{i=0}^{m} F_i^2 (x, y(x), y'(x)) = 0 \text{ for each } x \text{ in } D.$$

Again every singular solution of order m is contained in a maximal one such. We remark also that for the differential equations of class C that we shall consider (cf. §3), the domain of definition of a maximal singular solution of order m will turn out to be open.

When F is a polynomial in p of degree k in p and $m \geq k$, there are no singular solutions of order m. Further, any plane curve on which all the coefficients $a_i(x, y)$ of

$$F = a_k(x, y)\, p^k + a_{k-1}(x, y)\, p^{k-1} + \ldots + a_0(x, y)$$

vanish is a singular solution, but one which is not of finite order. Such solutions form the set of all singular solutions of infinite order. They are not interesting since they do not involve the derivative, and one may exclude their occurrence by requiring that F be irreducible (over the real field) and that it be of degree at least one in p.

Thus if we assume F irreducible over the real field, then if F has degree one in p, there are no singular solutions, and if F has degree two in p, every singular solution must have order one.

2. Examples

(a) The equation $y^2 + (y')^2 = 0$ possesses only one maximal (real) solution, and this is singular. The solution is $y = 0$.

(b) Singular solutions most commonly arise as envelopes. Clairaut's equation $y - xy' - (y')^2 = 0$, cf. [4], arises upon replacing the parameter p in the one-parameter family $F(x, y, p) = y - px - p^2$ of straight lines by another parameter, namely p_0 gets replaced by the slope y'_0 of the line in the family corresponding to p_0. Besides being satisfied by the members of the original family, Clairaut's equation has another solution, the "singular one", given by

$x = -2p$, which is just $F_p = 0$, and

$y = -2p^2 + p^2 = -p^2$, obtained by replacing x by $-2p$ in $F = 0$.

This is just the envelope $y = \frac{-x^2}{4}$ of the family of lines.

(c) In contrast, Serret's equation $F = y - 2px - p^2$, $p = y'(x)$, has no singular solution, although it has the same form as does Clairaut's equation. For the locus $y = -x^2$ obtained as above is not a solution.

(d) The equation

$$p^2 \prod_{i=1}^{n-2} (y - x + i) + (p - 1)^2 \prod_{i=1}^{n-2} (y - i) = 0$$

has degree two in p and total degree n. It has $2(n - 2)$ maximal singular solutions, namely the lines $y - x + i = 0$ and $y - i = 0$.

3. The Main Theorems

Henceforth let m denote a fixed positive <u>odd</u> integer, let R denote the real field, and let R^n denote the n-dimensional euclidean space. I shall presently define a class C of functions of the form $F(x, y, p)$, where F is a real analytic function of the three real variable x, y and p. This class will contain all functions of the form

$$F = a_k(x, y)\, p^k + a_{k-1}(x, y)\, p^{k-1} + \dots + a_0(x, y)$$

where :

(1) Each a_i is a real analytic function of x and y,

(2) the real surface defined in R^3 by $F = 0$ has no singular points,

(3) the leading coefficient a_k has only isolated zeros in R^2,

and

(4) the set $\{(x, y) \in R^2 | .F(x, y, p) = 0$ has more than $m + 1$ real solutions for $p\}$ consists of isolated points. Here a solution of multiplicity t is counted t times.

Note that because of (3) condition (4) is satisfied whenever $k \leq m + 1$. An example of an F which is of degree four in p and which is in the class C for $m = 1$ is provided by $F = p^4 + y^2p^2 + x$. The only point (x, y) for which $F(x, y, p) = 0$ has more than two real roots is $(x, y) = (0, 0)$, where $F(0, 0, p) = 0$ has a zero of order four.

To describe the class C it is convenient to make the following :

<u>Definition</u> : We shall say that a subset A of R^3 is <u>locally bounded at</u> $(x_0, y_0) \in R^2$, provided there exists an $\epsilon > 0$ such that $A \cap \{(x, y, p) \in R^3 | (x - x_0)^2 + (y - y_0)^2 < \epsilon\}$ is bounded in R^3.

Definition of the class C : The functions F in the class C are subject to C_1, C_2 and C_3 :

(C_1) : F is real analytic.

(C_2) : The function $\sum_{i=0}^{m} F_i^2 + (F_x^2 + F_y^2)$ has no real zeros.

(C_3) : If (x_0, y_0) is a point such that $F(x_0, y_0, p) = 0$ has a real root p_0 of multiplicity at least $m + 1$, i.e. $\sum_{i=0}^{m} F_i^2(x_0, y_0, p_0) = 0$, then (x_0, y_0) has a neighbourhood N in R^2 such that for each $(x, y) \in N$ with $(x, y) \neq (x_0, y_0)$ conditions C_{3a} and C_{3b} below hold :

C_{3a} : If $F(x, y, p)$ has a real solution for p, then the set $F = 0$ is locally bounded at (x, y) ,

C_{3b} : If $F(x, y, p) = 0$ has a real solution for p of multiplicity <u>exactly</u> $m + 1$, then it has no other (real) solutions for p .

(Observe that $F = 0$ need not be locally bounded at (x_0, y_0) itself, or at a point (x, y) for which $F(x, y, p) = 0$ has no real solutions for p). Also note that C_1, C_2, C_{3a} and C_{3b} follow respectively from (1), (2), (3) and (4) of the preceding paragraph.

In what follows we shall be concerned with a differential equation of the form $F(x, y, p) = 0$, $p = \frac{dy}{dx}$, where F is in the class C. Note in particular that the following theorems apply to any second degree differential equation with a non-singular F whose leading coefficient $a_2(x, y)$ has only isolated real zeros, and that for such equations a singular solution is the same as a first order singular solution.

Our first main result is the following :

Theorem A : Let F be in the class C. Then through any point there can pass at most one maximal singular solution of order m of the differential equation $F(x, y, p) = 0$, $p = \frac{dy}{dx}$.

Theorem A raises the question of how many maximal singular solutions of order m are there for polynomial functions of the form

$F = a_k(x, y)p^k + a_{k-1}(x, y)\, p^k + \dots + a_0(x, y)$? Estimates on this number are provided by the next theorem. When the total degree n of F is two the differential equation $p^2 + y(y + 1) = 0$, $p = y'(x)$, shows that the estimate in it is the best possible.

Theorem B : Let F be a real polynomial which is of total degree n in x, y and p, and which is in class C. Then the maximal singular solutions of order m of the differential equation $F = 0$, $p = \frac{dy}{dx}$ are all open and disjoint, and their number does not exceed $n(n - m)$.

Further results and estimates are given in the last section.

4. The Proofs of Theorems A and B .

Conventions : In what follows a set such as $F = 0$ will often be denoted simply by just F ; the word solution will always mean singular solution of order m ; and the lemmas will pertain to the differential equation

$$F(x, y, p) = 0, \quad p = y'(x) , \quad F \in C .$$

Lemma 1. The set

$$B = \left(\sum_{i=0}^{m} F_i^2 \right) - F_{m+1} ,$$

where the "-" denotes the usual difference of sets, is a one-dimensional real analytic manifold without singularities, or is empty. Further, $F_x^2 + F_y^2$ does not vanish on $\sum_{i=0}^{m} F_i^2 = 0$.

Proof : This follows from condition C_2. The matrix whose rows are the gradients of F and F_m has rank two on B.

Lemma 2. Let S denote the set of isolated points of $B \cap (F_x + pF_y)$. Then $V = [B \cap (F_x + pF_y)] - S$ is a one-dimensional analytic manifold (perhaps empty). Further F_y does not vanish on $\left(\sum_{i=0}^{m} F_i^2 \right) \cap (F_x + pF_y)$.

Proof : This follows from Lemma 1 and condition C_2.

The following lemma is essentially known, cf. [4], and is the key to subsequent lemmas. It allows us to pass from differential equations to algebraic geometry. Since it is not recorded in the literature in a form suitable for our purposes we shall reproduce it, giving the proof for the reader's convenience. Note that by Lemma 2 we may assume that curves in V are nonsingular. See also Definition 3 of §1.

Lemma 3. The set V is such that every curve in it projects (under $(x, y, p) \to (x, y)$) onto a solution. Conversely, given any solution $y(x)$, $x \in D$, its representative

$$\bar{y}(x) = \{(x, y(x), y'(x)) \in R^3 \mid x \in D\}$$

is contained in $\bar{V}$, the closure of V.

Proof : We prove the converse first. $\bar{y}$ is contained in $\bar{B}$, the closure of B, since y is a solution. Since B is, by lemma 1, one dimensional analytic, it suffices by continuity to show that $\bar{y}$ is contained in $F_x + pF_y$ near points of it which are in B. Let $(x_0, y_0, p_0) = w_0$ be such a point. Then near w_0, $\bar{y}$ lies in $F_m = 0$. Since $F_{m+1}(w_0) \neq 0$, we may use the implicit function theorem to solve for p differentiably in terms of x and y near (x_0, y_0), and therefore in terms of x along $\bar{y}$ near w_0. Hence we may differentiate $F = 0$ implicitly with respect to x near w_0 to obtain :

$$F_x + \frac{dp}{dx} F_p + \frac{dy}{dx} F_y = 0 .$$

Since F_p is zero along $\bar{y}$, we obtain $F_x + \frac{dy}{dx} F_y = 0$. Since $(x, y(x), p)$ is in $\bar{y}$ if and only if $p = y'(x)$, this shows that $\bar{y}$ near w_0 is in $F_x + p F_y = 0$.

Next, let L be a smooth curve in V through $w_0 = (x_0, y_0, p_0)$. Since L lies in $F_m = 0$ and $F_{m+1}(x_0, y_0, p_0) \neq 0$, L projects under $(x, y, p) \to (x, y)$ in a one-one manner onto a smooth curve $y(x)$ which we denote by K. Now if the tangent to K is not vertical at (x_0, y_0), we may solve for y as a function of x near this point, and since $F_{m+1}(w_0) \neq 0$, we may solve $F_m = 0$ for p as a function of (x, y) on L near w_0, and therefore as a function of x. Thus L can be parametrized by means of x as a parameter in the form $(x, y(x), p(x))$. We will check that this $y(x)$ is a singular solution. The parametrization allows us to differentiate $F = 0$ with respect to x to obtain $F_x + \frac{dp}{dx} F_p + \frac{dy}{dx} F_y = 0$. Since $F_p = 0$ along L, we obtain $F_x + y' F_y = 0$. Recalling that $F_y \neq 0$ along L, cf. Lemma 2, this together with $F_x + p F_y = 0$ imply that $p = y'$ along the curve, so that it is a solution. That it is singular of order m is immediate. For the proof to be complete we must show that the tangent to K at (x_0, y_0) cannot be vertical. To see this, note that by Lemma 2, $F = 0$ is nonsingular at w_0, and since $F_p(w_0) = 0$, the hyperplane H in R^3 which is tangent to $F = 0$ at w_0 must be parallel to the p-axis. Hence its intersection with the (x, y)-plane must be the line tangent to K at (x_0, y_0). If this line were vertical, it would follow (since H is already parallel to the p-axis) that H must also be parallel to the y-axis. This would contradict $F_y(w_0) \neq 0$.

Remark : In the proof of Lemma 3 we use only differentiability and condition C_2.

Definition : In the context of Lemma 3, the curve $\bar{y}(x)$ is called the representative in $\bar{V}$ of $y(x)$, and $y'(x)$ is called the height of $(x, y(x), y'(x))$ and also of $(x, y(x))$.

Lemma 4. Let (x_0, y_0) be a point in R^2, and let N be a circular neighbourhood for it as in condition C_3. Let S denote either a circle in R^2 with centre (x_0, y_0) lying within N, or let it be a straight line in R^2 not containing (x_0, y_0). Let Q denote the surface in R^3 determined by S, i.e. $Q = \{(x, y, p) | (x, y) \in S\}$. Suppose that Q intersects $F = 0$ transversally at two points $a = (x_1, y_1, p_1)$ and $b = (x_2, y_2, p_2)$ of B with $(x_i, y_i) \in N$ for $i = 1, 2$, cf. Lemma 1. Assume that there is a curve in $Q \cap F$ joining a to b. Then there is another curve in $Q \cap F$ joining a to b, and intersecting the first in finitely many points. Further, the projection of the component of $Q \cap F$ containing a and b into S all lies in just one of the pieces of S determined by (x_1, y_1) and (x_2, y_2).

Proof : To fix matters consider the case in which S is a straight line, and look first at what happens near (x_1, y_1, p_1) say. By transversality $F \cap Q$ contains a non-singular curve C through this point. Since $F_p(x_1, y_1, p_1) = 0$, the line $x = x_1, y = y_1$ is tangent to C at (x_1, y_1, p_1). Further, condition C_{3b} implies that this line intersects F only at (x_1, y_1, p_1). Now using the fact that $m + 1$ is even, it is easy to see that the projection of a sufficiently small neighbourhood of (x_1, y_1, p_1) in C into the line S lies all on one side of (x_1, y_1). This also shows that the component of $Q \cap F$ containing a and b projects (under $(x, y, p) \to (x, y)$) onto the segment of S determined by (x_1, y_1) and (x_2, y_2). The remaining assertion follows from this, the fact that $F \cap Q$ is closed, the local boundedness in assumption C_{3a}, the fact that analytic sets possess locally finite triangulations [10], and that in these triangulations every point has a neighbourhood which is homeomorphic to the cone over a complex of even Euler characteristic. The last fact is a recently proved theorem of P. Deligne and D. Sullivan. In the one-dimensional case it is a classical theorem on the local parametrizability of one-dimensional analytic sets, proved in the algebraic case in [1], see also [12].

Lemma 5. There are no two solutions y_1 and y_2 defined on a finite open interval $D = (x_0, x_1)$ which are different arbitrarily near x_0 and having the properties :

$$\lim_{x\to x_0} y_1(x) = \lim_{x\to x_0} y_2(x) = y_0 \neq \pm\infty$$

<u>and</u> $$\lim_{x\to x_0} y_1'(x) = \lim_{x\to x_0} y_2'(x) = p_0 \neq \pm\infty .$$

A similar assertion holds of course with x_0 being replaced by x_1.

<u>Proof</u> : Deny this. By [12, 13] the projection of $\bar{V}$ into the (x,y)-plane may be triangulated by means of a locally finite triangulation, the set $\bar{V}$ being one-dimensional. Hence we may assume that x_1 is sufficiently close to x_0 so that $y_1(x) \neq y_2(x)$ holds for each $x \in D$. Then Lemmas 2 and 3 imply that $(x_0, y_0, p_0) \in \bar{V}$, that both representatives $\bar{y}_i$ terminate at (x_0, y_0, p_0), and that $F_y(x_0, y_0, p_0) \neq 0$. Then by the implicit function theorem, a neighbourhood N of (x_0, y_0, p_0) in the surface $F = 0$ is mapped diffeomorphically onto a small disc T in the (x,p)-plane about (x_0, p_0) under the map $f : (x, y, p) \to (x, p)$. The portions of the representative curves $\bar{y}_i$ which lie in N correspond in T under f to two curves y_i^*, i = 1, 2. Since $y_1(x) \neq y_2(x)$ for any $x \in D$, $\bar{y}_1$ does not intersect $\bar{y}_2$, and since f is one-one we have that y_1^* does not intersect y_2^*. Hence there exists a number a, $x_0 < a < x_1$, such that the line $x = a, y = 0$ intersects y_1^* and y_2^* in points z_1^* and z_2^* with $z_1^* \neq z_2^*$, and such that the segment of this line between the z_i^*'s corresponds in N to a curve $\bar{S}$ which is strictly increasing (with respect to p) and joins the points $\bar{z}_i \in \bar{y}_i$ corresponding to the z_i^* under the above diffeomorphism. Now by Lemma 4 applied with Q being the plane $x = a$, there is in $F = 0$ another curve $\bar{S}'$ which is different from $\bar{S}$ in the plane $x = a$ which also joins $\bar{z}_1$ to $\bar{z}_2$. We will show that this contradicts f being one-one.

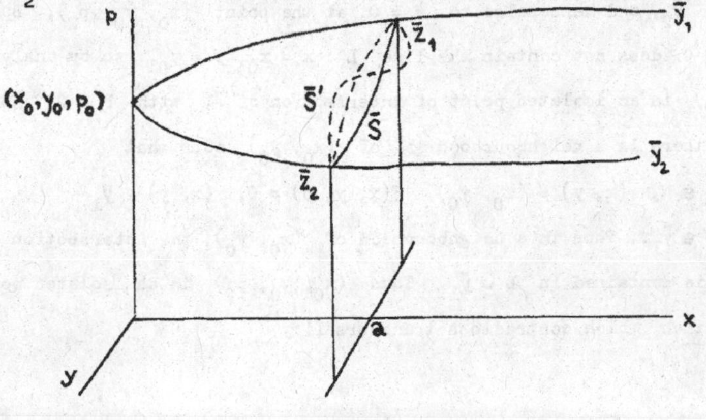

In the figure the curves $\bar{S}$ and $\bar{S}'$ lie in the plane $x = a$. The solid curve joining $\bar{z}_1$ to $\bar{z}_2$ is $\bar{S}$; and the dotted ones represent possibilities for $\bar{S}'$, arising by Lemma 4. By choosing a close enough to x_0, we may assume that the plane $x = a$ intersects $F = 0$ as well as $\bar{y}_i$ transversally at $\bar{z}_i$, $i = 1, 2$, and $\bar{S}' \neq \bar{S}$.

We may assume that N is small enough so that $F_y \neq 0$ on it. Suppose that the height of $\bar{z}_1$ is greater than that of $\bar{z}_2$. If $\bar{S}$ intersects $\bar{S}'$ strictly between the end points, then near this intersection f is obviously not one-one. If not, then $\bar{S}'$ is either all above or all below $\bar{S}$. If it is above for instance, then it is easy to see that since $\bar{S}$ is increasing, f cannot be one-one near z_2, for in this case both curves are above z_2, start from z_2, and lie in the plane $x = a$.

<u>Lemma 6. Let (x_0, y_0) be a point for which $F = 0$ does not contain the line $x = x_0$, $y = y_0$. Then given any two solutions through (x_0, y_0), there is a neighbourhood of it in which they coincide.</u>

<u>Note</u> In reading this proof and the next, it is helpful to keep in mind some of the standard facts about analytic sets, such as those at the end of the proof of Lemma 4.

<u>Proof</u> : Let y_1 and y_2 be two singular solutions passing through (x_0, y_0) with slopes p_1 and p_2. By Lemma 5 we may assume that $p_1 \neq p_2$.

Now by C_2 or by Lemma 2, the surface $F = 0$ is locally euclidean near each (x_0, y_0, p_i), $i = 1, 2$. Since $p_1 \neq p_2$ the surface

$$M = \{(x, y, p) : y = y_1(x)\}$$

is transverse to $\bar{y}_2$, and hence also to $F = 0$, at the point (x_0, y_0, p_2). But by hypothesis, $F = 0$ does not contain the line $L : x = x_0, y = y_0$, so by analyticity (x_0, y_0, p_2) is an isolated point of intersection of L with $F = 0$. And by condition C_{3b} there is a neighbourhood N of (x_0, y_0) such that

$$(x, y) \in N, \quad (x, y) \neq (x_0, y_0), \quad F(x, y, p) = 0, \quad (x, y) \in y_1$$

imply $(x, y, p) \in \bar{y}_1$. Thus in a neighbourhood of (x_0, y_0), the intersection of M with $F = 0$ is contained in $L \cup \bar{y}_1$. Thus (x_0, y_0, p_2) is an isolated point of this intersection, which contradicts transversality

<u>The proof of Theorem A.</u> We must show that if $y_1(x)$ and $y_2(x)$ are two solutions through (x_0, y_0), there is a neighbourhood of this point on which they coincide. Assume this is not so. By Lemma 5 we may assume that $p_1 = y_1'(x_0) \neq y_2'(x_0) = p_2$, and by Lemma 6 we may assume that $F = 0$ contains the line L given by $x = x_0$, $y = y_0$. Further we may assume that y_1 and y_2 have only (x_0, y_0) in common. Let $\bar{y}_i$ be the representative of y_i, and let g be the function given by $g(x, y, p) = (x - x_0)^2 + (y - y_0)^2$. The proof will be carried out by investigating the topology of $E_r = F \cap (g - r)$ for sufficiently small $r > 0$. Let $\bar{y}_{ij}$ (respectively y_{ij}), $j = 1, 2$, be the two intersection points of E_r with $\bar{y}_i$ (respectively y_i), $i = 1, 2$. For small r it follows from the last part of Lemma 1 that E_r is one-dimensional. We will always work "above" the $N(x_0, y_0)$ of condition C_3, and assume r is small enough for the following to work.

Case 1. There are sufficiently small $r > 0$ for which $\bar{y}_{i1}$ and $\bar{y}_{i2}$ (say for $i = 1$) lie on the same component K of E_r:

For small r the cylinder $g = r$ will intersect each $\bar{y}_i$ transversally at points of B; and by Lemma 4 the projection of K into the (x, y)-plane must lie on the circle $(x - x_0)^2 + (y - y_0)^2 = r$ between y_{11} and y_{12}. Then since y_1 and y_2 intersect transversally at (x_0, y_0), condition C_{3b} implies that K also contains one of the points $\bar{y}_{2j}$, say $\bar{y}_{21}$. Now one obtains a contradiction by noting that the projection of K into the above circle must as before lie between every two of the three points y_{11}, y_{12}, and y_{21}. Or alternatively by a non-singularity argument using the fact that $F_x^2 + F_y^2$ does not vanish along L and noting that K must cross itself at $\bar{y}_{21}$ by C_3.

Case 2. For sufficiently small r the four points $\bar{y}_{ij}$ lie on precisely two components of E_r:

By case 1 the two components K_1 and K_2 must be such that K_1 contains $\bar{y}_{1i}$ and $\bar{y}_{2j}$, while K_2 contains the remaining points. By Lemma 4, K_1 contains a closed curve through $\bar{y}_{1i}$ and $\bar{y}_{2j}$ whose projection into the (x, y)-plane lies on the appropriate circle between y_{1i} and y_{2j}, and similarly for K_2. Recalling that $F_x^2 + F_y^2 \neq 0$ along L, it follows that if r is small enough, K_i has no singular points (x, y, p) with $p_1 < p < p_2$, if $p_1 < p_2$ say. Let p_3 be greater than the height (cf. the definition following Lemma 3) of every point in $\bar{y}_1 \cap (g \leq r) = \bar{y}_1 \cap \{(x, y, p) | g(x, y, p) \leq r\}$ and smaller than the height of every point in $\bar{y}_2 \cap \{(x, y, p) | g(x, y, p) \leq r\}$. There is such a p_3 for small r. Then the hyperplane $p = p_3$ intersects $K_1 \cup K_2$ in four distinct points. Letting $r \to 0$, it is not hard to see that this traces four different curves in the hyperplane $p = p_3$ all of which lie in $F = 0$ and tend to (x_0, y_0, p_3). Since surely $p = p_3$ and $F = 0$ have a nonsingular intersection at this point, this case cannot occur.

Case 3. The remaining case :

Consider first the case in which F is a polynomial, and $F = 0$ is nonsingular. Let Q_r denote the component of L in the set $F \cap (g \leq r)$. Then for sufficiently small r, the function g has no critical points on $Q_r - L$, cf. [1] section 2. In our case the $\bar{y}_{ij}$'s lie on more than two components of the boundary $\partial Q_r = Q_r \cap (g = r)$ of Q_r.

Choose a strictly decreasing sequence $r_i \to 0$, and consider the sequence of pairs $(Q_{r_0}, Q_{r_0} - Q_{r_i})$, for $i = 1, 2, 3, \ldots$. Using the singular homology theory with coefficients from the ring Z of integers and recalling that $F = 0$ has no singular points on the line $g = 0$, we have

$$H_1(Q_{r_0}, \bigcup_{i=1}^{\infty} (Q_{r_0} - Q_{r_i})) = Z .$$

On the other hand since Q_{r_0} is connected, $H_0(Q_{r_0}) = Z$, and by assumption we have that the rank of $H_0(\partial Q_{r_0})$ is at least three. Since g has no critical values in the interval $0 < r \leq r_0$, the inclusions $\partial Q_{r_0} \to (Q_{r_0} - Q_{r_i})$ are all homotopy equivalences. Thus the inclusion map $H_0(Q_{r_0} - Q_{r_i}) \to H_0(Q_{r_0} - Q_{r_j})$ is an isomorphism, and its rank is at least three. The naturality of the exact sequence of the pair $(Q_{r_0}, Q_{r_0} - Q_{r_i})$ then yields that the inclusion

$$H_1(Q_{r_0}, Q_{r_0} - Q_{r_i}) \to H_1(Q_{r_0}, Q_{r_0} - Q_{r_j})$$

is an isomorphism whose rank is at least two. Considering that

$$H_1(Q_{r_0}, \bigcup_{i=1}^{\infty} (Q_{r_0} - Q_{r_i})) = \lim_{\to} H_1(Q_{r_0}, Q_{r_0} - Q_{r_i})$$

and that the second member of this equation must have rank which is at least two, this leads to a contradiction, thus proving the theorem in the special case.

In the general case the above analysis still holds provided we restrict all our considerations to that portion of Q_r which is bounded by $\bar{y}_1$ and $\bar{y}_2$. That g will have no critical points on this portion of $F = 0$ or on each $\bar{y}_i$ within $0 < g \leq r$ may be deduced from the facts that F_p vanishes on L, that $F_x^2 + F_y^2$ does not vanish on L and that each $\bar{y}_i$ intersects L transversally. That the inclusions in the proof of the algebraic case are homotopy equivalences is established by sliding along trajectories in $F = 0$ which are integral curves of the vector field (properly normalized when necessary) constructed at a point w on Q by taking that component of the gradient of g at w which lies in the tangent plane to $F = 0$ at w. By standard technical modifications it can be arranged that vectors in this vector field which are based at a point of $\bar{y}_i$ are also tangent to it, so that our deformation takes place within our strip. This concludes the proof of Theorem A.

The next lemma supplements the description given in Theorem A of the maximal singular solutions, and will be used to prove Theorem B.

Lemma 7. Let $y(x)$, $x \in D$, D being determined by a and b with $-\infty \leq a < b \leq \infty$ be a maximal solution. Then if a or b, say a is finite, $\lim_{x \to a} y'(x)$ is infinite. Thus the representative $\bar{y}(x)$ of $y(x)$ is unbounded in either direction.

Proof : Suppose $\lim_{x \to a} y'(x) = p_0 \neq \pm \infty$. Then $\bar{y}$ terminates at a point (a, t, p_0) in $\bar{V}$. Then because locally an analytic set is a cone over a complex of even Euler characteristic, there is an analytic curve in V different from $\bar{y}$ which terminates at (a, t, p_0). This projects by Lemma 3 onto a singular solution y_1 easily seen to be different from y, and with the property $\lim_{x \to a} y_1'(x) = p_0$. See the remark preceding Lemma 3.

Lemma 5 then implies that y_1 is defined to the left of a. Hence these two solutions may be spliced together to yield one solution passing through (a, t) with slope p_0. The continuity of the derivative follows from the second part of Lemma 3.

<u>The proof of Theorem B</u> :

The statement about openness and disjointness follows from Theorem A and Lemma 7. In order to obtain the estimate, let $\{y_i\}$, $i \in I$ be the set of maximal singular solutions, and let $\bar{y}_i$ denote the representative in $\bar{V}$ of y_i for each i. By Theorem A and Lemma 7 the $\bar{y}_i$'s are disjoint and every $\bar{y}_i$ intersects every sufficiently large sphere S_t of radius t about the origin in R^3 in two points. Further, the definitions show that the rank of the matrix whose rows are gradient F and gradient F_m is two at all points of any one $\bar{y}_i$ excepting possibly a finite number from it. Recall that we are in the algebraic case. Note that we are not assuming that I is finite. Thus for a fixed i we may choose t_i so large that the singular points of $\bar{y}_i$ lie inside S_{t_i} and so that the function $r = x^2 + y^2 + z^2$ has no critical points on $\bar{y}_i$ outside of S_{t_i}. Now let $y_1, \ldots, y_k$ be members of $\{y_i\}$, $i \in I$. It suffices to show that $k \leq n(n - m)$. Pick a number $t > t_i$ for $1 \leq i \leq k$, and let $S_t \cap \bar{y}_i$ consist of p_{i1} and p_{i2}. (We need not worry about this intersection having more than two points as this would not affect our argument). Then at each point of curve $\bar{y}_i$, $1 \leq i \leq k$, outside S_t, the matrix whose rows are : gradient F, gradient F_m and gradient r has rank three. Now approximate F, F_m and r by real polynomials $\bar{F}$, $\bar{F}_m$ and $\bar{r}$ of the same degrees and which represent three complex surfaces which are in general position in the three dimensional complex cartesian space. Then Bezout's theorem, cf. [5], implies that the intersection of the latter surfaces consists of $2n(n - m)$ points. However if the approximation is close enough, the rank condition which holds at each p_{ij} implies that each p_{ij} lies closest to some point $\bar{p}_{ij}$ of this intersection and that the map $p_{ij} \to \bar{p}_{ij}$ is one-one. Since each y_i gives rise to two p_{ij}'s the estimate follows.

5. Further Estimates and Results.

We continue with the notations and conventions of previous sections. For simplicity, in this section we let

$$F = a_{m+1}(x, y)\, p^{m+1} + a_m(x, y)\, p^m + \ldots + a_0(x, y)$$

be a pseudo-polynomial which satisfies (1), (2), (3) and (4) as at the beginning of section (3). In this case one can give a more convenient description of the set of singular solutions. When $k = m + 1$ as I have (unnecessarily) assumed here, (4) is a consequence of (3).

To do this, let $R_{ij} = R(F_i, F_j)$ denote the resultant of F_i and F_j when these are regarded as polynomials in p, and set

$$W = [(\bigcap_{i=1}^{m} R(F, F_i)) \cap R(F, F_x + pF_y)] - R(F, F_{m+1}).$$

Then one has :

Theorem C : The closures of the sets W and $\{(x, y) |$ a singular solution of order m passes through $(x, y)\}$ are equal. If $F = 0$ is locally bounded at each point of R^2, then both sets are closed.

We omit the proof, since it follows easily from the previous analysis.

Next, set $\bar{R}_{ij} = R_{ij}/a_{m+1}(x,y)$ for $0 \leq i, j$. These expressions are analytic, and are polynomials in y or in y and x whenever this is true of F. The following theorem gives better estimates than those of Theorem A for the present class of functions.

Theorem D : (1) The number N of maximal singular solutions of order m of $F(x, y, p) = 0, p = y'(x)$, which are defined over the whole x-axis is bounded by the degree (possibly infinite) in y of $\bar{R}_{ij}$, $i \neq j$, $0 \leq i, j \leq m$. (If $m = 1$ for instance, and F is a polynomial (in all the variables) of total degree n, then this number does not exceed $2(n - 1)$.)

(2) If the surface $F = 0$ is locally bounded at each point of R^2 (such as when $a_{m+1}(x, y)$ has no real zeros) then each of its maximal singular solutions is unbounded in either direction, and N does not exceed twice the number of connected components of the surface. In this case, N is also bounded by the total degree (which is finite for polynomial F) of each $\bar{R}_{ij}$, $i \neq j$, $0 \leq i, j \leq m$.

Proof : The proof of (1) is best illustrated by proving the statement in parenthesis for the case when $m = 1$. In this case we may write $F = \alpha(x, y) p^2 + \beta(x, y) p + \gamma(x, y)$, so that $R(F, F_p) = \alpha(4\alpha\gamma - \beta^2)$ and $\bar{R}_{01} = 4\alpha\gamma - \beta^2$. Then α cannot vanish at all points of a singular solution curve having order $m = 1$. Also such curves belong to the closure of W and hence to $R_{01} = 0$. Hence they belong to $\bar{R}_{01} = 0$. Thus a bound on the number of maximal singular solutions defined on the whole x-axis may be obtained by counting the intersections of the line $x = 0$ in the (x, y)-plane with $4\gamma\alpha - \beta^2 = 0$. This intersection obviously has no more than $2(n - 1)$ points.

(2) Next, if $F = 0$ is locally bounded at each point of R^2, then Lemmas 3 and 7 show that each maximal solution y is unbounded in either direction, and determines by means of its representative $\bar{y}$ an unbounded component Y of $F = 0$. Applying the first part of the proof of Lemma 4 to hyperplane sections of $F = 0$ of the form $F \cap Q$ where Q is a plane of the form $x = a$, it is obvious that the projection of Y into the (x, y)-plane must all lie on one side of the curve $(x, y(x))$. Clearly this cannot happen for three such y's, since these partition the plane into four connected unbounded components.

This proves the first two statements.

To prove the last, we do as in the proof of Theorem B, excepting that in our case, since our solutions are unbounded in either direction, we may count them directly by intersecting them with a large enough circle in the (x, y)-plane. The conclusion of the theorem is obtained by intersecting $\bar{R}_{ij}$ which vanishes along our solutions with $x^2 + y^2 = r$. This completes the proof.

We conclude with some remarks.

Remark 1. It is clear from Lemma 3 that solutions are piece-wise analytic. They are already C^1. If one knows in addition that a solution is C^∞ it would follow that it is analytic, For it can be proved using the Malgrange preparation theorem that a C^∞ semi-analytic submanifold of R^n is analytic. This fact was pointed out to me by S. Łojasiewicz, who has been most generous with his time in explaining to me his basic results on semi-analytic sets. The proofs may be found in [14], and also in the paper by J. Bochnak in the same volume.

Remark 2. The conclusion of Theorem A is valid for a point (x_0, y_0) provided that conditions C_1 and C_2 hold only at all points of the form (x_0, y_0, p).

Remark 3. It is clear from Lemmas 3 and 7 that the projection of $\bar{V}$ into the (x, y)-plane is precisely the set of points through which solutions pass. This completes the description of the set of solutions given in Theorem C.

Remark 4. The fact that m is odd is essential in Theorem A. However although Theorem B uses Theorem A in its proof, this theorem and Lemmas 3 and 7 depend in an essential way only on conditions C_1 and C_2. Consider the case of an F as in this section. Theorem C shows that the singular solutions are contained in $\bar{W}$, the closure of W. Hence solutions through singular points of $\bar{W}$ may be continued (uniquely) along the natural analytic branches of this set, so that there is no trouble in sorting them out for counting them. Then to obtain estimates we can either use the process in the proof of Theorem B, or under assumptions such as local boundedness, use the method in the proof of the last part of Theorem D.

Bibliography

[1] J. Milnor, Singular points of complex hypersurfaces, Annals of Math. Studies Princeton University Press, Study 61, 1968.

[2] ______ On the Betti numbers of real varieties, Proc. A.M.S. 15 (1964)

[3] R. Thom, Sur l'homologie des varietes algebriques reelles, Differential and Combinatorial Topology, Princeton University Press 1965, 225-265.

[4] E.L. Ince, Ordinary Differential Equations, Dover, New York, 1956.

[5] B. van der Waerden, Einführung in die algebraische Geometrie, Springer, Berlin, 1939.

[6a] R. Courant & D. Hilbert, Methods of Mathematical Physics, Vol.II (Partial Differential Equations). Interscience (Wiley and Sons), 1965.

[6b] P.R. Garabedian, Partial Differential Equations, Wiley and Sons,Inc., third edition, 1967.

[7] I. Kaplansky, An introduction to differential algebra, Herman, Paris, 1957.

[8] E.R. Kolchin, Singular solutions of algebraic differential equations and a lemma of Arnold Shapiro, Topology 3 (supplement 2) 1965, 309-318.

[9a] ______ Algebraic matric groups and the Picard-Vessiot theory of homogeneous linear ordinary differential equations, Ann. of Math. 49, (1948), 1-42.

[9b] E.R. Kolchin, Existence theorems connected with the Picard-Vessiot theory of homogeneous linear ordinary differential equations, Bull. Amer. Math. Soc. 54, (1948), 927-932.

[10] J.F. Ritt, Differential equations from the algebraic standpoint, Amer. Math. Soc. Coll. Pub., Vol.14, New York, 1932.

[11] ________ Differential Algebra, Amer. Math. Soc. Coll. Pub. vol.33, New York, 1950.

[12] S. Lojasiewicz, Ensembles semi-analytiques, Institute des Hautes Etudes Scientifiques, Bures-Sur-Yvette, France, 1965.

[13] ____________ Triangulation of semi-analytic sets, Annali della Scuola Normale Superiore di Pisa, Serie III, Vol.XVIII, Fasc. IV (1964).

[14] J. Siciak, Studia Math. Vol.35, (1970).

[15] R. Thom, Sur les Equations Différentielles multiformes et leurs intégrales singulières. (Manuscript).

MORSE RELATIONS FOR CURVATURE AND TIGHTNESS

Nicolaas H. Kuiper

Abstract In chapter I we define the critical points of a continuous function $\varphi : X \to \mathbb{R}$ on a compact metric space X, and the Morse polynomial for an isolated critical point. We reproduce the proof of the Morse relations for a function with isolated critical points for this more general set-up. In chapter II we define polynomial, absolute and alternating curvature measures of certain maps or embeddings of X into E^N in terms of critical points and we obtain Morse relations for these curvature measures. In chapter III we discuss maps (and sets) in E^N of minimal total absolute curvature (tight maps) and we recall some results on tight embeddings of smooth, piecewise linear and topological manifolds.

I. Morse relations

A point $x \in X$ in a metrisable topological space X (we have in mind manifolds with or without boundary and simplicial complexes) is called <u>non critical</u> for a continuous function $\varphi : X \to \mathbb{R}$, if there is for some neighbourhood U of x in X a homeomorphism (call it a <u>chart</u>) $\pi : U \to V \times J$, J an open set in $\mathbb{R}$, and a commutative diagram with p_1 and p_2 the projections onto the factors V and J:

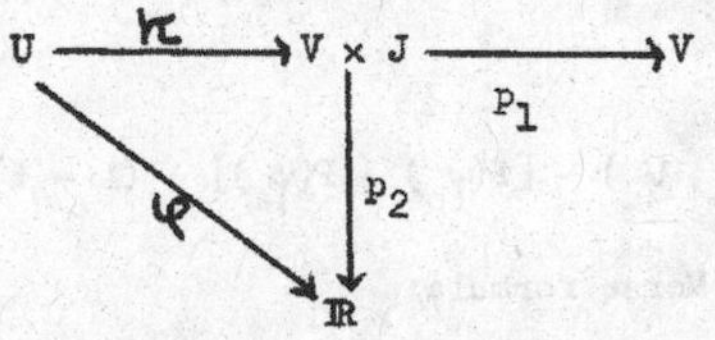

(1)

Otherwise the point x is called <u>critical</u> and $\varphi(x)$ is called a <u>critical value</u>. For example, the function $\varphi = z$ has on the sphere $\{(x, y, z) \in \mathbb{R}^3 : x^2 + y^2 + z^2 = 1\}$ and on the ball $\{(x, y, z) \in \mathbb{R}^3 : x^2 + y^2 + z^2 \leq 1\}$ only two critical points, namely $(0, 0, 1)$

This is a considerably extended version of one of a series of four lectures given at the Liverpool Summer School on Singularities, June 1970.

and $(0, 0, -1)$ in each case.

We write

$$\varphi_s = \{x \in X : \varphi(x) \leq s\}. \tag{2}$$

We assume moreover that any pair of spaces (φ_b, φ_a) for $a < b$, has finite dimensional singular homology groups over a field $\mathbb{F}$ to be kept fixed, with the usual exact split triangle in homology:

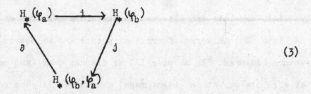

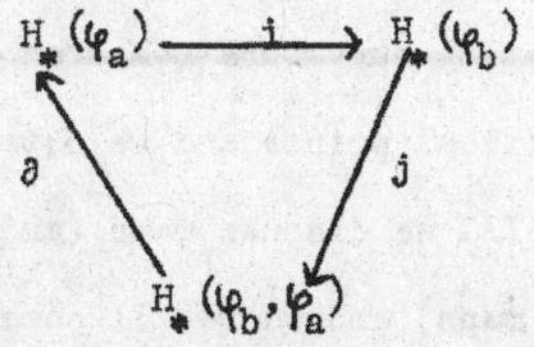

(3)

Recall that $\partial : H_k(\varphi_b, \varphi_a) \to H_{k-1}(\varphi_a)$ decreases the index. We have written $H_* = \underset{k}{\oplus} H_k$, etc.

Define the Poincare polynomials:

$$\begin{aligned} P(\varphi_a) &= \sum_k \dim H_k(\varphi_a).t^k \\ P(\varphi_b, \varphi_a) &= \sum_k \dim H_k(\varphi_b, \varphi_a).t^k \\ P(\mathrm{Im}\, \partial) &= \sum_k \dim (H_k(\varphi_a) \cap \mathrm{Im}\, \partial).t^k. \end{aligned}$$

Then (3) yields

$$P(\varphi_b, \varphi_a) - [P(\varphi_b) - P(\varphi_a)] = (1 + t)\, P(\mathrm{Im}\, \partial)$$

or we have the preliminary Morse formula:

$$\frac{P(\varphi_b, \varphi_a) - [P(\varphi_b) - P(\varphi_a)]}{1 + t} \succ 0. \tag{4}$$

Here $\succ$ is an order relation between polynomials such that $P_1 \succ P_2$ as well as $P_1 - P_2 \succ 0$ means that all coefficients of the polynomial $P_1 - P_2$ are non negative.

Observe that if two polynomials $\sum_{k>0} \mu_k t^k$ and $\sum_{k>0} \beta_k t^k$ obey the Morse relation

$$\frac{\Sigma\mu_k\, t^k - \Sigma\beta_k\, t^k}{1+t} \succ 0 \tag{5}$$

then we get the usual expression for the Morse inequalities:

$$t = -1 \text{ gives } \Sigma\,(-1)^k \mu_k = \Sigma\,(-1)^k \beta_k\,; \tag{6a}$$

$$\Sigma(\mu_k - \beta_k)\, t^k \succ 0 \text{ gives } \mu_k \geq \beta_k\,; \tag{6b}$$

and

$$\Sigma(\mu_k - \beta_k)\, t^k\, \Sigma_{i=0}^{\infty} (-t)^i \succ 0 \text{ gives}$$

$$\mu_j - \mu_{j-1} + \dots + (-1)^j \mu_0 \geq \beta_j - \beta_{j-1} + \dots + (-1)^j \beta_0. \tag{6c}$$

We now prove some lemmas.

Lemma 1. If φ_b is compact and there is no critical value in the closed interval $[a, b]$, then φ_a is a deformation retract of φ_b.

Proof. (Following M. Morse's methods) For any point $x \in \varphi_b \setminus \varphi_a$ we use a chart κ as in (1) for which $U \subset (X \setminus \varphi_a)$ and we define easily a continuous map $h_x : X \to X$ with the following properties:

$$\left.\begin{array}{ll} h_x(y) = y & \text{for } y \notin U \\ \left.\begin{array}{l} p_1\kappa(h_x(y)) = p_1\kappa(y) \\ p_2\kappa(h_x(y)) \leq p_2(\kappa y)) \\ p_2\kappa(h_x(x)) < p_2\kappa(x) \end{array}\right\} & \text{for } y \in U \end{array}\right\}. \tag{7}$$

A deformation $h_{x,t}$, $0 \leq t \leq 1$, leading from the identity to this map can be defined by the same equations as h_x, replacing the inequalities by

$$p_2\kappa(h_{x,t}(y)) = (1-t)\, p_2\kappa(y) + t p_2\kappa h_x(y). \tag{8}$$

For a point x with $\varphi(x) = a$ we use a chart κ as in (1) but we now define the continuous map $h_x : X \to X$ so that is has the properties:

$$\begin{array}{rcll} h_x(y) & = & y & \text{for } y \notin U \text{ and for } \varphi(y) \leq a \\ p_1\kappa h_x(y) & = & p_1\kappa(y) & \text{for } y \in U \\ a \leq p_2\kappa(h_x(y)) & \leq & p_2\kappa(y) & \text{for } y \in Y \text{ and } \varphi(y) \geq a \\ p_2\kappa h_x(y) & = & a & \text{for } y \in U \text{ and } \varphi(y) \geq a \\ & & & \text{and } y \text{ in some neighbourhood of } x. \end{array} \tag{9}$$

$h_{x,t}$ is again defined as above (see (3)).

There is a finite set of continuous maps (2) and (8) such that every point y in $\varphi_b \setminus \varphi_a$ is non-invariant under at least one of them. Let $h_1, h_2, \ldots, h_L$ be such a set. Then the composition $h_L h_{L-1} \ldots h_2 h_1 = h$ is a continuous map of X onto X. By compactness there exists $\epsilon > 0$ such that $h(\varphi_{a+\epsilon}) = \varphi_a$, and there exists $\delta > 0$ such that

$$\varphi(h(y)) \leq \varphi(y) - \delta \quad \text{for} \quad a + \epsilon \leq \varphi(y) \leq b.$$

Then it follows that

$$h^N(\varphi_b) = \varphi_a \qquad \text{for} \quad (N-1)\delta > b - a - \epsilon,$$

and

$$h^N(y) = y \qquad \text{for} \quad y \in \varphi_a.$$

h^N is a retraction of φ_b onto φ_a. By recalling that it is the composition of $N \times L$ elementary deformations (7) or (9) we see that h^N is a deformation retraction namely the composition (one after the other) of the corresponding deformations $h_{x,t}$.

<u>Lemma 2.</u> <u>If</u> φ_b <u>is compact and the only critical value in</u> $[c, b]$ <u>is</u> c, <u>and if all critical points with this critical value</u> c <u>are isolated, then</u> φ_c <u>is a deformation retract of</u> φ_b.

<u>Proof.</u> Let K be the (finite) set of critical points at level $= c$, and denote:

$$\overset{0}{\varphi}_c = \{x \in X : \varphi(x) < c\} \subset \varphi_c.$$

For any point x in $\varphi_b \setminus \{\overset{0}{\varphi}_c \cup K\}$ we can find a neighbourhood U not meeting K and a continuous map h_x as in either (7) or (9). Let $U_1 \supset \overline{U_2} \supset U_2 \supset \overline{U_3} \supset U_3 \ldots$ be a nested sequence of neighbourhoods of K in X with intersection $\bigcap_k U_k = K$. We may just as well assume that each U_i has one component for each point of K, namely one containing that point.

There exists a finite set of maps (7) and (9) such that every point y in the compact set $\varphi_b \setminus U_{i+1}$ is non-invariant under at least one of them. Let h'' be the composition of these maps. Of course $h''(y) = y$ for $y \in \varphi_a$. There is by the construction of h'' an $\epsilon'' > 0$ such that

$$h''(\varphi_{c+\epsilon''} \setminus U_{i+1}) \subset \varphi_a.$$

Let $\rho(\epsilon) = \sup_{y \in \varphi_{c+\epsilon}}$ distance $(h''(y), y)$ where the distance is with respect to any fixed metric on X. $\rho(\epsilon)$ is a continuous function of $\epsilon \geq 0$ and $\rho(0) = 0$. Hence we can find $0 < \epsilon' < \epsilon''$ such that

$$\rho(\epsilon') \leq \inf_{y \in U_{i+1}, z \notin U_i} \text{distance } (y, z).$$

Let h' be a deformation retraction of φ_b onto $\varphi_{c+\epsilon'}$ as constructed by lemma 1.

$$h^{(i)} = h'' \circ h'.$$

Then $h^{(i)}$ is a continuous map of φ_b into $\varphi_a \cup U_i$ (!) with

$$h^{(i)}(y) = y \quad \text{for} \quad y \in \varphi_a.$$

And $\lim_{k \to \infty} h^{(k)} h^{(k-1)} \ldots h^{(2)} h^{(1)}$ is the required retraction of φ_b onto φ_a, which by its construction as a composition of deformations, and the fact that U_i for $i \to \infty$ converges to the finite point set K, is a deformation retraction as required in the lemma.

Lemma 3. Let $c \in (a, b)$ be the only critical value of the function $\varphi: X \to \mathbb{R}$ in the closed interval $[a, b]$ and let the critical points be isolated. Hence, they form a finite set K. Then we have isomorphisms for the natural maps

$$H_k(\varphi_b, \varphi_a) \xleftarrow{\cong} H_k(\varphi_c, \varphi_a) \xrightarrow{\cong} H_k(\varphi_c, \varphi_c \setminus K) \xrightarrow{\cong} \bigoplus_{x \in K} H_k(\varphi_c, \varphi_c \setminus \{x\}). \tag{10}$$

Proof. The first isomorphism is a consequence of lemma 2.
The last isomorphism follows by the excision axiom. The second homomorphism is an isomorphism, because $H_k(\varphi_c \setminus K, \varphi_a) = 0$ for all k. This is the case because every (finite) singular chain in $\varphi_c \setminus K$ is contained in some compact part of $\varphi_c \setminus K$ and the union of φ_a and that part can be homotoped inside $\varphi_c \setminus K$, keeping fixed every point of φ_a, first into $\varphi_{c-\epsilon}$ for some $\epsilon > 0$ and then onto φ_a, with a construction as before involving maps (7) and (9). Therefore the homomorphism

$$H_*(\varphi_a) \longrightarrow H_*(\varphi_c \setminus K)$$

is injective (take a singular cycle representing any element in $H_*(\varphi_a)$ and suppose it bounds a chain in $\varphi_c \setminus K$), as well as surjective (take a singular cycle in $\varphi_c \setminus K$). Hence it is an isomorphism, and so $H_*(\varphi_c \setminus K, \varphi_a) = 0$.

If x is an isolated critical point of $\varphi : X \to \mathbb{R}$ with critical value c, then

$$M(\varphi, x) = \Sigma \mu_k(\varphi, x)\, t^k = \Sigma \dim H_k(\varphi_c, \varphi_c \setminus \{x\}) t^k \tag{11}$$

is called the Morse polynomial of x, and $\mu_k(\varphi, x)$ is the multiplicity of index k at the critical point.

$$\mu(\varphi, x) = \Sigma_k \mu_k(\varphi, x) \text{ is the (total) multiplicity at } x. \tag{12}$$

Examples (The field $\mathbb{F}$ is fixed)

1) The function $\varphi = -x_1^2 - x_2^2 - \dots - x_k^2 + x_{k+1}^2 + \dots + x_n^2 : \mathbb{R}^n \to \mathbb{R}$ has a critical point 0 with

$$M(\varphi, 0) = t^k,$$

and with total multiplicity $\mu(\varphi, 0) = \mu_k(\varphi, 0) = 1$.

2) The function $\varphi = z$ on the set

$$\{(x, y, z) \in \mathbb{R}^3 : x^2 + y^2 + z^2 \leq 1\}$$

has critical points

$(0, 0, -1)$ with total multiplicity (of index 0) one

and

$(0, 0, 1)$ with total multiplicity zero.

3) The function $\varphi = x_1^3 - 3x_1x_2^2$ (= real part of $(x_1 + ix_2)^3$; monkey saddle) on $\mathbb{R}^2$ has a critical point 0 with Morse polynomial

$$M(\varphi, 0) = 2t$$

and total multiplicity = multiplicity of index $1 = \mu(\ , 0) = 2$.

4) The function $\varphi = x_1^3 - 3x_1(x_2^2 + x_3^2)$ on $\mathbb{R}^3$ has a critical point 0 with Morse polynomial

$$M(\varphi, 0) = t + t^2,$$

with $\mu_1(\varphi, 0) = \mu_2(\varphi, 0) = 1$ and with total multiplicity 2.

5) Let X be the cone with top p_0 on a space Y obtained from $Y \times I$ by identifying $Y \times 1$ to one point, and let φ be the projection into $I = [0, 1] \subset \mathbb{R}$.

Then φ has the isolated critical point p_0 with value 1 (and of course also a critical value 0).

In view of the exact sequence

$$H_k(\text{point}) \cong H_k(\varphi_1) \to H_k(\varphi_1, \varphi_0) \to H_{k-1}(\varphi_0) \to H_{k-1}(\varphi_1)$$
$$\| $$
$$H_{k-1}(Y)$$

we obtain

$$M(\varphi, p_0) = t(P(Y) - 1).$$

Y can be for example a real projective plane. We see therefore that the choice of field has an influence on the Morse polynomials. This general model occurs in some neighbourhood of an isolated critical point p_0 for any piecewise linear function φ on a simplicial complex X', hence for any real analytic function φ on an analytic manifold X'. If c is the critical value then there is a closed conical neighbourhood $X = \text{Cone}\ (Y)$ of p_0 in φ_c. Compare with example 4).

6) For a non critical point x the definition of the Morse polynomial applies and gives the polynomial 0.

7) If we compose a piecewise linear embedding f of a compact piecewise linear manifold in some euclidean space E, with a linear function $z : E \to \mathbb{R}$ we get a function zf with in general isolated critical points with complicated Morse polynomials.

In the smooth case we get in general only Morse polynomials of the kind t^k!

For a function $\varphi : X \to \mathbb{R}$ on a compact metrisable space X with isolated critical points, we define the <u>Morse polynomial</u> of the function φ:

$$M(\varphi) = \Sigma \mu_k(\varphi) \cdot t^k \overset{\text{def}}{=} \sum_{x \in X} M(\varphi, x) = \sum_{x \in X} \sum_k \mu_k(\varphi, x) \cdot t^k. \tag{13}$$

<u>Theorem</u> (<u>Morse relations</u>). If $\varphi : X \to \mathbb{R}$ <u>is a continuous function on a compact metric space</u> X <u>with isolated critical points, then</u>

$$\frac{M(\varphi) - P(X)}{1 + t} \succ 0. \tag{14}$$

Proof. Let $c_1 < c_2 < \dots < c_L$ be the critical values of φ and let

$$-\infty = a_0 < c_1 < a_1 < c_2 < \dots < a_{L-1} < c_L < a_L = \infty.$$

By (4) we have

$$\sum_{i=0}^{L-1} \frac{P(\varphi_{a_{i+1}}, \varphi_{a_i}) - [P(\varphi_{a_{i+1}}) - P(\varphi_{a_i})]}{1 + t} \succ 0.$$

Substituting (10), (11), (13) we obtain (14).

From the interpretation (6a, b, c) of (14) we obtain in particular the Euler characteristic

$$\Sigma\, (-1)^k\, \mu_k(\varphi) = \Sigma\, (-1)^k\, \beta_k(X) = \chi(X) \tag{15}$$

and

$$\mu(\varphi) = \Sigma_k \mu_k(\varphi) \geq \beta(X) = \Sigma \beta_k(X). \tag{16}$$

If we define $\gamma(X) = \inf_\varphi \mu(\varphi)$ (17)

and

$$\bar{\gamma}(X) = \inf_\varphi \tfrac{1}{2}[\mu(\varphi) + \mu(-\ )], \tag{18}$$

with φ running over all functions with isolated critical points on X, then we have for any φ

$$\mu(\varphi) \geq \gamma(X) \geq \beta(X) \tag{19}$$

and

$$\tfrac{1}{2}[\mu(\varphi) + \mu(-\varphi)] \geq \bar{\gamma}(X) \geq \gamma(X) \geq \beta(X). \tag{20}$$

Examples. For a 3-manifold X which has the homology but not the homotopy type of S^3 one has

$$\gamma(X) \geq 4 > \beta(X) = 2.$$

For a space Y homeomorphic to the letter Y one has

$$\bar{\gamma}(Y) = 2 > \gamma(Y) = \beta(Y) = 1.$$

Problem. For a manifold X let $\gamma_{ND}(X)$ be equal to the minimal number of critical points a non degenerate function φ (that is one that gives rise to a handle decomposition in smooth, piecewise linear or topological topology) can have. $\gamma_{ND}(X)$ is independent of the field $\mathbb{F}$. By definition:

$$\gamma_{ND}(X) \geq \gamma(X). \tag{21}$$

Is there a smooth or piecewise linear manifold X for which $\gamma_{ND}(X) > \gamma(X, \mathbb{F})$ for some or every field $\mathbb{F}$?

It is known that not every topological manifold has a handle decomposition. Does every topological manifold have a function with isolated critical points?

A space X is called a <u>Morse</u> - <u>equality</u> space in case $\bar{\gamma}(X) = \beta(X)$ for some $\mathbb{F}$.

All closed surfaces are Morse - equality spaces, as can be seen by taking $\mathbb{F} = \mathbb{Z}_2$. Hence $\bar{\gamma}(X) = \gamma(X) = \beta(X) = 4 - \chi(X)$.

II. Polynomial curvature measures

Let $f : X \to E^N$ be a continuous map of a compact metric space X into a Euclidean vector space E^N. For any linear function $z : E^N \longrightarrow \mathbb{R}$ with gradient $z^* \in S^{N-1} \subset E^N$ a unit vector, the composition zf yields a function on X and on any open set $U \subset X$. Fix again a field $\mathbb{F}$ and suppose zf has isolated critical points. Then we have the Morse polynomial

$$M(zf, U) = \sum_{x \in U} M(zf, x).$$

We now make the assumption that zf has isolated critical points for almost every unit covector z on E^N. This assumption is fulfilled for smooth immersions of smooth manifolds and for PL-immersions of PL-manifolds.

Let α be the homogeneous measure on S^{N-1} with total value $\int_{S^{N-1}} \alpha = 1$. Assume moreover that the following integral converges for every open U in X:

$$T(f, U) = \Sigma_k \tau_k(f, U)\, t^k = \delta_{z^* \in S^{N-1}} M(zf, U) = \tag{22}$$

$$\underline{\underline{\text{def}}} \quad \int_{z^* \in S^{N-1}} M(zf, U) \,.\, \alpha.$$

Its value is a polynomial in t, called the <u>polynomial</u> <u>curvature</u> <u>measure</u> of f on U. In particular

$$T(f) = T(f, X) \tag{23}$$

is the polynomial curvature of f.

In particular, if $t = 1$, we get, by (22) and (20),

$$\tau(f) = \Sigma_k \tau_k(f) = \mathcal{E}_{z^* \in S^{N-1}} \mu(zf)$$

$$= \mathcal{E}_{z^* \in S^{N-1}} \tfrac{1}{2}[\mu(zf) + \mu(-zf)] \geq \bar{\gamma}(X). \tag{24}$$

We also have the Morse relations for curvature measures

$$\frac{T(f) - P(X)}{1+t} \succ 0. \tag{25}$$

By using the definition (22) and (23) of $T(f)$, and the inequality (14), we have

$$\frac{T(f) - P(X)}{1+t} = \mathcal{E}_z \left(\frac{M(zf) - P(X)}{1+t} \right) \succ 0.$$

A consequence of (25) obtained by substituting $t = -1$ in (25) is

$$\tau_{alt}(f) \overset{\text{def}}{=\!=} \Sigma_k (-1)^k \tau_k(f) = \chi(X). \tag{26}$$

This generalises the theorem of Gauss - Bonnet.

It applies for example to a piecewise differentiable manifold with boundary, embedded in E^N.

$\tau(f)$ is called the total absolute curvature of f.

$\tau_{alt}(f)$ is called the alternating curvature of f.

The real-valued alternating curvature measure $\tau_{alt}(f, U)$ for variable U is an intrinsic metric invariant for smooth immersions f of smooth manifolds in E^N (see [5]) and also for piecewise linear embeddings or immersions of simplicial complexes (T. Banchoff [2] proved this last fact and he has interesting interpretations of τ_{alt}, which measure in this case is concentrated in the 0-skeleton!).

Some well known interpretations for curves and surfaces are as follows. For 2-dimensional smooth surfaces in E^N one has

$$\tau_{alt}(f, U) = \int_u \frac{K|d\sigma|}{2\pi}.$$

For smooth surfaces immersed in E^3 one has:

$$\tau_0(f, U) = \tau_2(f, U) = \int_{U\cap(K>0)} \frac{|K d\sigma|}{4\pi} ;$$

$$\tau_1(f, U) = \int_{U\cap(\ <0)} \frac{|K d\sigma|}{2\pi} ;$$

$$\tau(f, U) = \int_U \frac{|K d\sigma|}{2\pi} .$$

For curves smoothly immersed in E^N one has

$$\tau(f, U) = \int_U \frac{|\rho d\sigma|}{\pi} .$$

III. Tight maps and sets

A map $f : X \to E^N$ is called tight in case it fulfills the assumptions mentioned in II and it has minimal total absolute curvature (compare (24)):

$$\tau(f) = \bar{\gamma}(X).$$

There are several results on tight smooth embeddings of smooth manifolds without boundary into Euclidean spaces, their existence or non-existence, and their forms in case of existence. Compare [5] and [8], where other references can be found.

T. Banchoff [1] has studied tight piecewise linear embeddings of piecewise linear surfaces in Euclidean spaces. For higher dimensional PL-manifolds nothing is known. Perhaps tight PL-embeddings are very rare?

In the topological category we see easily that for any convex compact set X in E^N with interior points and boundary ∂X, X and ∂X are tight sets with total absolute curvature 1 and 2 respectively. Vice versa any compact set X in E^N with total absolute curvature 1 is convex (essentially a proof is in [4]), and any tight topological immersion of S^n in E^N is an embedding onto the boundary of a convex (n+1)-dimensional body [8].

Tightness generalises convexity therefore, and several interesting problems remain open in particular in the topological category. I mention the problem of finding all topological tight embeddings of the real projective plane substantially into Euclidean 5-space, and the problem of finding all closed topological tight embeddings (suitably defined) of the open Moebius band into Euclidean spaces.

(An example in E^4 was found by S. Carter and A. West [3].) For the Moebius band with boundary all possible tight embeddings seem to be within reach [9]. There exists no smooth tight embedding. Tight embeddings of the figure Y into Euclidean space have an image which consists of three straight segments meeting in one end point, such that that end point is not a corner point for the convex hull of the figure. The image lies therefore in a plane. The property "tightness" is of course invariant under affine transformations of the target space E^N. For Morse equality spaces the property is also invariant under projective transformations of E^N (keeping the image inside E^N of course). For a proof see [3].

References

[1] Banchoff, T. F. Tightly embedded 2-dimensional polyhedral manifolds. Am. J. of Math. 87, 462-472, (1965).

[2] Banchoff, T. F. Critical points and curvature for embedded polyhedra. J. of Diff. Geom. 1, 245-256, (1967).

[3] Carter, S. and West, A. Tight and taut immersions. (forthcoming).

[4] Fary, I. A characterization of convex bodies. Am. Math. Monthly 69, 25-31, (1962).

[5] Ferus, D. Total Absolutkrümmung in Differentialgeometrie und -topologie. Lecture Notes Springer 66, (1968).

[6] Kuiper, N. H. La courbure d'indice k et les applications convexes. Sém. de top. et géom. diff. C. Ehresmann. Fac. des Sc. Paris 15 pages, (1960).

[7] Kuiper, N. H. Der Satz von Gauss-Bonnet für Abbildungen im E^N. Jahr. Ber. D.M.V. 69, 77-88, (1967).

[8] Kuiper, N. H. Minimal absolute total curvature for immersions. Inv. Math. 10, 209-238, (1970).

[9] Kuiper, N. H. Tight embeddings of the Moebius band. Forthcoming.

BLOWING UP SINGULARITIES

Harold I. Levine[1]

0. Let f be an immersion of X, a compact, connected, orientable, n-manifold into Y a connected, orientable, p-manifold $p > n$. Lashof and Smale have defined a generalized normal degree of f and have shown that this normal degreee generalizes the usual notion and determines the Euler characteristic of X, $\chi(X)$, if Y is not compact and determines $\chi(X)$ modulo $\chi(Y)$ if Y is compact. The results in case $Y = \mathbb{R}^p$, and general Y are discussed in paragraphs 1. and 2. respectively. In paragraph 3, the essential content of the Lashof-Smale theorem is recast in Theorem 5 which states that $\chi(X)$ can be computed as an integral over the unit normal bundle of f, $\mathbb{S}N$. The integrand here can be taken as the pull back of the (p-1)-form Π on $\mathbb{S}TY$ defined by Chern in his proof of the Gauss Bonnet formula. (The only properties of Π needed here are that it pulls back to a closed form and that it specializes to the volume element of a fibre of $\mathbb{S}TY$ when restricted to that fibre.) It is this integral formula of the theorem that we generalize.

In paragraph 4, the class $\mathcal{F}$ of mappings from X to Y is discussed. An element $f \in \mathcal{F}$ is a map whose first jet extension, j^1f, is transversal to the first order singularities. Let S(f) be the set of singular points of f. Over X-S(f), f is an immersion so it has a normal bundle, $N|$. Our theorem now states that if we pull back the same form and integrate it over $\mathbb{S}(N|)$, that the integral converges - even though the integration is over a non-compact manifold and in fact converges to $\chi(X)$ once again.

---

[1] This work was partially supported by NSF GP 9606 and 23117.
The author would also like to thank the Department of Pure Mathematics at Liverpool whose hospitality he enjoyed while working on this paper.

The convergence of the integral is proven by showing that its value can be computed by integrating the same differential form lifted to a sphere bundle over a compact manifold obtained from X by blowing X up repeatedly along centers which are submanifolds of S(f). To compute the actual value of this integral we apply a result of Lashof-Smale.

In paragraph 4, the theorem is given as the Corollary of Theorem 6, an integral theorem about T-homomorphisms between vector bundles whose proof is carried out along the lines outlined above. Paragraph 5 is devoted to a brief and simple description of blowing up, and paragraphs 6 and 7 deal with the proofs of convergence and evaluation of the integral of Theorem 6.

1. Given an immersion $f : X \to \mathbb{R}^p$, where X is a connected, compact oriented, n-manifold $(n < p)$, Chern [2] defines the normal map of f of the normal sphere bundle, ΣN, into Σ^{p-1}, the unit sphere of $\mathbb{R}^p$ by mapping a unit vector perpendicular to the image of $f(X)$ at $f(x)$, into its translate at the origin. The normal degree of f is defined to be the degree of the normal map of f.

Theorem 1 The normal degree of f is the Euler characteristic, $\chi(X)$.

In the special case of $p = n+1$, the normal bundle of the immersion is an oriented line bundle and so its associated sphere bundle is a trivial zero-sphere bundle. Thus the normal sphere bundle has two connected components each of which is diffeomorphic to X. The normal map consists then of two maps of X into Σ^n. If we let $\nu : X \to \Sigma^n$ map a point x to the normal* to $f(X)$ at $f(x)$ translated to the origin of $\mathbb{R}^{n+1}$, then the normal degree of f is the degree of ν plus the degree of $\alpha \circ \nu$ where α is the antipodal map of Σ^n. In case n is odd these two degrees cancel one another.

In case n is even we obtain the Hopf Index Theorem [3, p.116]

---

* We choose the normal $\nu(x)$ so that the image of the orientation of X at x followed by $\nu(x)$ orients $\mathbb{R}^{n+1}$.

Theorem 2 *If* $\dim X = 2k$ *and* f *is an immersion of* X *in* R^{2k+1}, *then the map* $\nu : X \longrightarrow \mathbb{S}^{2k}$ *which maps* x *to the translate at the origin of* R^{2k+1} *of the normal to* $f(X)$ *at* $f(x)$ *has degree* $\frac{1}{2}\chi(X)$.

2. Theorem 1 of the preceding section was generalized by Lashof and Smale [4] to the case of an immersion $f : X \longrightarrow Y$ where Y is a connected, oriented manifold when $p = \dim Y > \dim X = n$. Let $\pi_Y : TY \longrightarrow Y$ be the tangent bundle projection and let π and ϕ be the usual maps in the fibred-product diagram:

$$\begin{array}{ccc} E = f^*TY & \xrightarrow{\ \phi\ } & TY \\ \Big\downarrow{\scriptstyle \pi} & & \Big\downarrow{\scriptstyle \pi_Y} \\ X & \xrightarrow{\ f\ } & Y \end{array}$$

As usual define N, the normal bundle of the immersion f, by the exactness of:

$$0 \longrightarrow TX \longrightarrow E \longrightarrow N \longrightarrow 0$$

Using a metric on Y to obtain a splitting of this sequence we have an inclusion, $i : N \longrightarrow E$, which restricts to a mapping of sphere bundles $i : \mathbb{S}N \longrightarrow \mathbb{S}E$.

Theorem 3 *Let* s *be the class in* $H_{p-1}(\mathbb{S}E)$ *of an oriented fibre in* $\mathbb{S}E$ *and* $[\mathbb{S}N]$ *be the fundamental class of the appropriately oriented manifold* $\mathbb{S}N$. *Then*

$$i_*[\mathbb{S}N] = \chi(X).s$$

Note: Since we assumed that $p > n$, the portion of the Gysin sequence:

(1) $$0 \longrightarrow H_0(X) \xrightarrow{k_*} H_{p-1}(\mathbb{S}E) \xrightarrow{\pi_*} H_{p-1}(X) \longrightarrow 0$$

shows that $H_{p-1}(\mathbb{S}E)$ is torsion free. Hence $i_*[\mathbb{S}N]$ determines an integer, $\chi(X)$. Lashof and Smale consider instead of $i_*[\mathbb{S}N]$, the class $\phi_* \circ i_*[\mathbb{S}N]$ which is in $H_{p-1}(\mathbb{S}TY)$. Their result is

<u>Theorem 4</u> <u>Let</u> t <u>be the class in</u> $H_{p-1}(\mathbb{S}TY)$ <u>of an oriented fibre of</u> $\mathbb{S}TY$, <u>then</u> $\phi_* o i_*[\mathbb{S}N] = \chi(X)t$.

<u>Note</u>: If Y is compact, t is of order $\chi(Y)$ and so $\phi_* o i_*[\mathbb{S}N]$ determines an integer modulo $\chi(Y)$.

In the special case of $p = n+1$, let $\nu : X \longrightarrow \mathbb{S}E$ be the map analogous to the ν of the first paragraph. Since $i_*[\mathbb{S}N] = (1 + (-1)^n)\nu_*[X]$ we have the generalization of the Hopf Index Theorem:

<u>Theorem 5</u> <u>If</u> n <u>is even</u> a) $2\,\nu_*[X] = \chi(X)s$ <u>and</u> b) $2\,\phi_* o \nu_*[X] = \chi(X).t$.

3. It is <u>Theorem</u> 3 and its corollary <u>Theorem</u> 5a that we want to generalize to a wider class of maps $f : X \longrightarrow Y$. However the fact that $E = f^*TY$, for f an immersion of X in Y, plays no role in the statements and proofs of Theorems 3 and 5a. All we need is that E and N are vector bundles over X and are the terms of an exact sequence:

(2) $$0 \longrightarrow TX \longrightarrow E \longrightarrow N \longrightarrow 0$$

Thus this is all we assume about the E and N in Theorem 3 and 5a. The theorem that we actually generalize is an integral theorem equivalent to Theorem 3. To obtain this formulation, consider the parts of the Gysin cohomology and homology sequences:

(3) $$\begin{array}{ccccccccc} 0 & \longrightarrow & H^{p-1}(X) & \xrightarrow{\pi^*} & H^{p-1}(\mathbb{S}E) & \longrightarrow & H^0(X) & \longrightarrow & 0 \\ & & & & & \searrow^{\delta^*} & \downarrow{\cong\Phi^*} & & \\ & & & & & & H^p(\mathbb{B}E, \mathbb{S}E) & & \end{array}$$

(4) $$\begin{array}{ccccccccc} & & H_p(\mathbb{B}E,\mathbb{S}E) & & & & & & \\ & & \Phi_*\downarrow\cong & \searrow^{\partial_*} & & & & & \\ 0 & \longrightarrow & H_0(X) & \xrightarrow{k_*} & H_{p-1}(\mathbb{S}E) & \xrightarrow{\pi_*} & H_{p-1}(X) & \longrightarrow & 0 \end{array}$$

where $\mathbb{B}E$ is the unit ball bundle, and if $U \in H^p(\mathbb{B}E, \mathbb{S}E)$ is the orientation class of E over $\mathbb{Z}$ (see [7] p. 259), Φ^* and Φ_* are the Thom isomorphisms given by

$$\Phi_*(z) = \pi_*(U \cap z) \quad \text{and} \quad \Phi^*(v) = \pi^*(v) \cup U .$$

Let θ be any class in $H^{p-1}(\mathbb{S}E)$ such that $\delta^* \theta = U$, then if $<,>$ denotes the pairing of cohomology and homology:

$$(5) \qquad < \theta, i_*[\mathbb{S}N] > = \chi(X).$$

This follows from the fact that $s = k_*(1) = \partial_* U^*$, where $U^* = \Phi_*^{-1}(1)$ for 1 the generator of $H_0(X)$ corresponding to a point and as a consequence of the definition of Φ_*, $< U, U^* > = 1$.

If we knew that $i_*[\mathbb{S}N]$ were in the kernel of π_* then Theorem 3 would be equivalent to (5) for any $\theta \in \delta^{*^{-1}}(U)$. For $n < p - 1$ that fact is trivial; it is true also for $n = p-1$ using:

<u>Lemma</u> <u>Let</u> N <u>be an oriented Euclidean line bundle over a manifold</u> X. <u>Let</u> $\pi : N \longrightarrow X$ <u>be the bundle projection. Then</u>

$$\pi_*[\mathbb{S}N] = 0 .$$

<u>Proof</u> Since N is orientable it has a nowhere zero section, ν, which trivializes N:

$$i : X \times R \longrightarrow N : (x, t) \longrightarrow t \cdot \nu(x)$$

We assume that the length of ν is 1. Let $i_t(x) = i(x, t)$. Since $\mathbb{S}N$ is oriented as the boundary of $i : X \times I \longrightarrow N$, we have $[\mathbb{S}N] = (i_1)_*[X] - (i_{-1})_*[X]$. Since $\pi \circ i_t(x) = x$ for all t, we have $\pi_*[\mathbb{S}N] = 0$.

Thus we have essentially shown that the following theorem is equivalent to Theorem 3.

<u>Theorem 3'</u> <u>Let</u> ω <u>be any closed</u> $(p-1)$-<u>form on</u> $\mathbb{S}E$ <u>which when restricted to the fibres of</u> $\mathbb{S}E$ <u>gives the volume element. Then</u>

$$\int_{i_*[\mathbb{S}N]} \omega = \chi(X) ,$$

<u>where</u> $i_*[\mathbb{S}N]$ <u>means any singular chain representing the homology class</u> $i_*[\mathbb{S}N]$.

The only point left to check is that if θ is the cohomology class of ω, that $\delta^*\theta = U$. But that is easy, for if $j : (\mathbb{B}, \mathbb{S}) \longrightarrow (\mathbb{B}E, \mathbb{S}E)$ is the injection of a fibre, then we have the commutative diagram:

$$\begin{array}{ccc} H^{p-1}(\mathbb{S}E) & \xrightarrow{\delta^*} & H^p(\mathbb{B}E,\mathbb{S}E) \cong H^0(X) \\ \Big\downarrow j^* & & \Big\downarrow j^* \\ H^{p-1}(\mathbb{S}) & \xrightarrow[\delta^*]{} & H^p(\mathbb{B},\mathbb{S}) \cong H^0(pt) \end{array}$$

Since $H^0(X) = \mathbb{Z}$, $\delta^*\theta = k\cdot U$ for some integer k, so $j^*\delta^*\theta = k\cdot u$, where $u = j^*U$ is the generator of $H^p(B,\mathbb{S})$. However by assumption $j^*\theta$ is the orientation class of $\mathbb{S}$, so $\delta^* j^*\theta = u$. Thus $k = 1$.

In case $E = f^*TY$ and N is the normal bundle of f, an immersion of X into Y, we can take for ω the pull back of a form on $\mathbb{S}TY$. In fact, Chern in [C_1, p. 40] constructs a (p-1)-form, Π, on $\mathbb{S}TY$ which when restricted to each fibre of $\mathbb{S}TY$ gives the volume element of the fibre and such that $d\Pi = \pi_Y^*\Omega$, where Ω is a differential p-form on Y. For the ω in Theorem 3' we may take $\phi^*\Pi$, where $\phi : E \longrightarrow TY$ is the map covering f introduced in paragraph 2. We need merely check that $d\phi^*\Pi = 0$ but that is obvious since $n < p$.

4. In this paragraph we define the family, $\mathcal{F}$, for which we obtain the generalization of Theorem 3', and we state the theorem. We first recall the notion of a T-homomorphism between vector bundles A and B over a manifold, V. (See [6]). In $B \otimes A^*$, we have smooth sub-bundles $S_k(B \otimes A^*)$ consisting of those elements of $B \otimes A^*$ of rank $m-k$, where m is the minimum of the fibre dimension of A and B. Here to make sense of the notion of rank of a point of $B \otimes A^*$, we've identified $(B \otimes A^*)_v$ with $\mathrm{Hom}(A_v, B_v)$. For any bundle homomorphism $\alpha : A \longrightarrow B$ we have the associated section $Z_\alpha : V \longrightarrow B \otimes A^*$, where $Z_\alpha(v)$ is the element of

$\mathrm{Hom}(A_v, B_v)$ defined by $Z_\alpha(v)(a) = \alpha(a)$.

Definition A homomorphism α from A to B is called a T-homomorphism if Z_α is transversal to all of the $S_k(B \otimes A^*)$ for $k = 0,1,2,\dots$.

Definition If X and Y are manifolds, $\mathcal{T} = \mathcal{T}(X,Y)$ is the set of all maps $f : X \longrightarrow Y$ such that $Tf: TX \longrightarrow f^*TY$ is a T-homomorphism.

Remark: The condition that $f \in \mathcal{T}$ is equivalent to the condition that $J^1(f)$ is transversal to $S_k(X, Y)$ for all k. (See [5] for notation and definitions).

Lemma [6] If $\alpha: A \longrightarrow B$ is a T-homomorphism and $S_k(\alpha) = Z_\alpha^{-1}(S_k(B \otimes A^*))$ then if $\dim V = s$ and A and B are n-and p-plane bundles respectively then

1) If $k(|p-n| + k) > s$, then $S_k(\alpha) = \phi$

2) If $k(|p-n| + k) \leq s$, then either $S_k(\alpha) = \phi$ or $S_k(\alpha)$ is a submanifold of V of codimension $k(|p-n| + k)$ in V

3) $\overline{S_k}(\alpha) = \bigcup_{j \geq k} S_j(\alpha)$.

Let $\alpha: A \longrightarrow B$ be a T-homomorphism of an n-plane bundle into an oriented p-plane bundle over an n-manifold V, where $n < p$. Let $A|$ and $B|$ be the restriction of these bundles to $V - S(\alpha)$. Define $N|$ as the cokernel of the map $\alpha : A| \longrightarrow B|$. Take any metric on $B|$ and by means of it split the sequence

$$0 \longrightarrow A| \longrightarrow B| \longrightarrow N| \longrightarrow 0$$

so that we have an inclusion $i: N| \longrightarrow B|$ which restricts to an inclusion: $i : \mathbb{S}(N|) \longrightarrow \mathbb{S}(B|)$.

Theorem 6 Let ω be any closed (p-1)form on $\mathbb{S}B$ which when restricted to the fibres of $\mathbb{S}B$ gives the volume element, then

$$\int_{i_*[\mathbb{S}(N)]} \omega = \chi(A), \text{ the Euler number}$$

of the bundle A. Here the integral on the left is taken over an infinite singular chain representing the non-compact orientable submanifold $i(\mathbb{S}(N))$ of $\mathbb{S}B$.

Note: The differential form, ω, in the statement of Theorem 6 represents a real cohomology class $\theta \in H^{p-1}(\mathbb{S}B)$ such that $\delta^*\theta = U$, the orientation class of B over Z.

Applying Theorem 6 to the case $\alpha = Tf$ for $f \in \mathcal{F}$:

Corollary Let X and Y be connected, orientable manifolds, X compact. Let $f \in \mathcal{F}(X,Y)$. Let Π be any (p-1)-form on $\mathbb{S}TY$ which when restricted to the fibres of $\mathbb{S}TY$ gives the volume element. Let $N|$ be the normal bundle of $f|X\text{-}S(f)$ which we regard as a sub-bundle of f^*TY. Let $\phi : f^*TY \longrightarrow TY$ be the map covering f, namely $(x,v) \longrightarrow v$ for $v \in (TY)_{f(x)}$. Then

$$\int_{N|} \phi^*\Pi = \chi(X).$$

Such a form, Π, is explicitly constructed in [1, p.40]. We spend the next sections proving Theorem 6. We must prove two things: that the integral converges and that its value is the Euler number. The convergence is proven by showing this integral has the same value as one taken over a compact manifold. The compact manifold is a sphere bundle over a manifold obtained by blowing V up along its singular manifolds $S_k(\alpha)$. The evaluation of this new integral is done using a result of Lashof and Smale on the Gysin sequence of a Whitney sum. We begin by giving a simple description of the blowing up process.

5. Let V be a manifold and Z be a locally closed submanifold of codimension m in V. Blowing V up with center Z yields a new manifold $\hat{V}$ and a map $\sigma : \hat{V} \longrightarrow V$ such that σ is a diffeomorphism over V-Z and $\sigma : \sigma^{-1}Z \longrightarrow Z$ is the projectified normal bundle of Z in V. Thus in constructing $\hat{V}$ we do not alter V except at Z. Therefore we work inside a tubular neighbourhood of Z. We identify this neighbourhood of Z with the normal bundle of Z in V (or a neighbourhood of the zero section in this normal bundle), where Z is identified with the zero section of this vector bundle. Thus it suffices to define the blowing up of a vector bundle over Z with center the zero section.

Let E be an m-plane bundle over Z. Let PE be the projectification of E, the bundle associated to E with projective (m-1)-space as fibre. Let $\pi : PE \longrightarrow Z$ be the projection for this bundle. We have the fibered product $\pi^*E = \{([\omega],v) \in PE \times E | [\omega]$ and v lie over the same $z \in Z\}$.

<u>Define</u> $\hat{E} = \{([\omega],v) \in \pi^*E | v \in [\omega]$ (i.e. $v \wedge \omega = 0)\}$ and $\sigma : \hat{E} \longrightarrow E: ([\omega],v) \longrightarrow v$.

$\hat{E}$ <u>is the blow-up of</u> E <u>with center</u> Z.

Since a non-zero vector uniquely determines the line it is on, σ^{-1}(E-(zero section)) is obviously a diffeomorphism. Also the part of $\hat{E}$ which lies over the zero section of E is trivially identifiable with PE.

<u>Remark:</u> If we take the other projection $\hat{E} \longrightarrow PE$, we obtain a vector bundle which over each $z \in Z$ is the dual of the hyperplane section line bundle over projective (m-1)-space.

6. Let $\alpha : A \longrightarrow B$ be a T-homomorphism of n- and p-plane bundles over an n-manifold V. We assume that both A and B are oriented bundles. It is no restriction to assume that B is a Euclidean space bundle. As usual $n < p$. Over $V-S(\alpha)$ we have an exact sequence defining $N|$:

$$(6) \qquad 0 \longrightarrow A| \longrightarrow B| \longrightarrow N| \longrightarrow 0$$

where $|$ means restriction to $V-S(\alpha)$.

Let $\mathrm{cor}(\alpha)$ = maximal k such that $S_k(\alpha) \neq \phi$.

<u>Theorem 7</u> ([6] p.373) <u>Let</u> $\mathrm{cor}(\alpha) = k$. <u>Let</u> $V_k = S_k(\alpha)$, <u>and let</u> $\hat{V}$ <u>be the blow-up of</u> V <u>with center</u> V_k <u>and let</u> $\sigma_1 : \hat{V} \longrightarrow V$ <u>be the projection. Then there is a bundle</u> A_1 <u>over</u> $\hat{V}$, <u>and homomorphisms</u> α_1 <u>and</u> λ_1 <u>such that:</u>

1. *The following diagram commutes:*

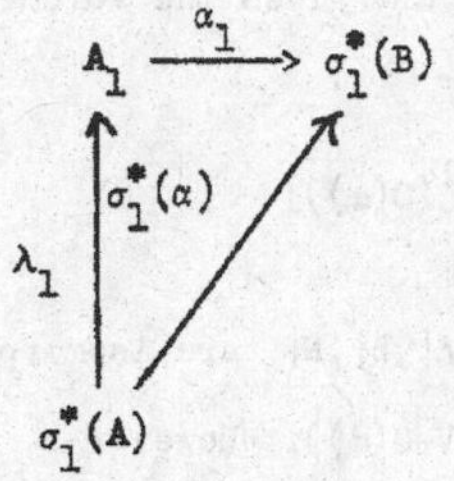

2. λ_1 *is an isomorphism except over* $\sigma_1^{-1}(V_k)$
3. α_1 *is a T-homomorphism with* $cor(\alpha_1) = cor(\alpha)-1$
4. *Restricted to* $\sigma_1^{-1}(V_k)$, $\ker \lambda_1 = \sigma_1^*(\ker \alpha)$.

We apply this theorem again to the T-homomorphism α, obtaining A_2, α_2, λ_2. Repeat this process k times, always blowing up with center the worst singularity that α_i presents. Finally we obtain a manifold $\tilde{V}$ and a mapping $\sigma : \tilde{V} \longrightarrow V$, the composition of all the blowing up projections, a bundle $\tilde{A}$ and two homomorphisms $\tilde{\alpha}, \tilde{\lambda}$ such that:

1. The diagram commutes

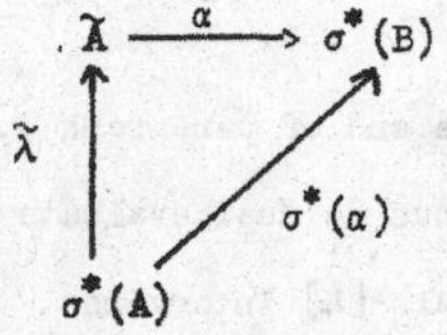

2. $\tilde{\lambda}$ is an isomorphism except over $\sigma^{-1}(S(\alpha))$
3. $\tilde{\alpha}$ has rank n everywhere
4. Restricted to each $\sigma^{-1}(V_j)$, $\ker \tilde{\lambda} = \sigma^*(\ker \alpha)$.

Define $\tilde{N}$ over $\tilde{V}$ by the exactness of:

(7) $$0 \longrightarrow \tilde{A} \longrightarrow \sigma^* B \longrightarrow \tilde{N} \longrightarrow 0$$

Restricted to $\tilde{V}-\sigma^{-1}(S(\alpha)) = \sigma^{-1}(V-S(\alpha))$, this sequence is the same as (6). Using the metric we've assumed to exist on B we can split both (6) and (7). Let $i: N| \longrightarrow B$ and $\tilde{i} : \tilde{N} \longrightarrow \sigma^* B$. Both of these inclusions restrict, of course, to the respective sphere bundles. Let ω be any closed differential (p-1)-form on $\mathbb{S}B$ which when restricted to the fibres of $\mathbb{S}B$ gives the volume element. Then

$$(8) \qquad \int\limits_{i_*[S(N|)]} \omega \;=\; \int\limits_{\tilde{i}_*[\mathbb{S}\tilde{N}|V-\sigma^{-1}(S(\alpha))]} \omega$$

since, if we avoid $S(\alpha)$ and $\sigma^{-1}(S(\alpha))$, $A|, B|, N|$ are isomorphic to $\tilde{A}|, \sigma^*B|, \tilde{N}|$ over diffeomorphic bases $V-S(\alpha)$ and $\sigma^{-1}(V-S(\alpha))$. However

$$\int\limits_{\tilde{i}_*[\mathbb{S}\tilde{N}|V-\sigma^{-1}(S(\alpha))]} \sigma^*\omega \;=\; \int\limits_{\tilde{i}_*[\mathbb{S}\tilde{N}]} \sigma^*\omega$$

since the sets we're integrating over differ by a set of measure zero. Thus we've proved the first part of the theorem 6 - namely that the integral converges.

7. To evaluate the integral $\int\limits_{\tilde{i}_*[\mathbb{S}\tilde{N}]} \omega$, note first that if θ is the cohomology class of $\sigma^*\omega$ we know by the argument at the end of paragraph 3, that $\delta^*\theta = U$, the orientation class of $\sigma^*(B)$ over Z. Thus we must evaluate $< \theta, \tilde{i}_*[\mathbb{S}\tilde{N}]>$. However here the general result of Lashof-Smale [4] intervenes. We give only that part that applies to our situation.

Theorem 8 Let $E = F \oplus G$ be vector bundles over a compact n-manifold M, where F is an n-plane bundle and E is a p-plane bundle and all the bundles are orientable. Then the following diagram commutes

$$\begin{array}{ccccccccc}
0 & \longrightarrow & H_0(M) & \xrightarrow{k_*} & H_{p-1}(\mathbb{S}E) & \longrightarrow & H_{p-1}(M) & \longrightarrow & 0 \\
 & & \uparrow{\scriptstyle \chi(F)} & & \uparrow{\scriptstyle i_*} & & \uparrow{\scriptstyle =} & & \\
0 & \longrightarrow & H_n(M) & \xrightarrow{k'_*} & H_{p-1}(\mathbb{S}G) & \xrightarrow{\pi_*} & H_{p-1}(M) & \longrightarrow & 0
\end{array}$$

where the vertical maps are as follows: $\cdot\chi(F) : H_n(M) \longrightarrow H_0(M)$ takes the fundamental class of $H_n(M)$ into $\chi(F)$ times the class of a point, and i is just the inclusion of G in E. (Here π is the bundle projection and k_* and k_*' the compositions of the inverse of the Thom isomorphisms and ∂_* as in (4))

In all cases $\pi_*[SG] = 0$, either because $p-1>n$ or by the lemma of paragraph 3. Thus we have

$$i_*[SG] = i_*k_*'[M] = k_*(\chi(F)\cdot[pt]) = \chi(F)s$$

where s is the class of a fibre of SE.

Thus applying this result to our computation we have

$$<\theta,\tilde{i}_*[S\tilde{N}]> = \chi(\tilde{A}) .$$

We have therefore:

$$\int_{i_*[SN|]}\omega = \int_{\tilde{i}_*[S\tilde{N}]}\sigma^*\omega = \chi(\tilde{A}) .$$

To finish the proof we need merely check that

Lemma $\chi(\tilde{A}) = \chi(A)$.

Proof We construct a section in $\tilde{A}$ which meets the zero section of $\tilde{A}$ transversally in the same number of points as section of A. It suffices to construct a section of A over V that meets the zero section transversally and which is never zero over $S(\alpha)$. Further the non-zero values of the section are never to be annihilated by α. If we had such a section, say τ, then it lifts to a section of $\sigma^*A, \sigma^*\tau$, which is again obviously transversal to the zero section. By the description of $\tilde{\lambda}$ in paragraph 6 $\tilde{\lambda}\circ\sigma^*\tau$ is a section of $\tilde{A}$ which is only zero where $\sigma^*\tau$ was zero - since the values of τ were never annihilated by α over $S(\alpha)$. Thus the number of zeros of $\tilde{\lambda}\circ\sigma^*\tau$ is the same as the number of zeros of τ so $\chi(\tilde{A})$ would be $\chi(A)$. The construction of such a τ is simple by the following transversality argument.

Let $V_k = S_k(\alpha)$, and $A|V_k = A_k = K \oplus C$ where $K = \ker \alpha|A_k$.

It suffices to show that a dense set of sections of C over V_k are never zero. But that is the case if

$$2(\dim C - \dim V_k) > \dim C, \text{ or}$$

$$\dim C > 2 \dim V_k .$$

But $\dim C = \dim V_k + (n-k)$ and by the Lemma cited in paragraph 4, $\dim V_k = n - k(p - n + k)$. Thus we must merely observe that

$$n - k > n - k(p - n + k).$$

References

[1] Chern, S. S. Topics in Differential Geometry, Institute for Advanced Study (1951).

[2] Chern, S. S. La geometrie des sous variétés d'un espace euclidéen à plusieurs dimensions, L'Enseignement Mathématique, 40, 26-46 (1955).

[3] Hicks, N. J. Notes on Differential Geometry, Van Nostrand Mathematical Studies, 3 (1965).

[4] Lashof, R. K. and Smale, S. S. On immersions of manifolds in Euclidean space, Annals of Math. 68, 562-583 (1958).

[5] Levine, H. I. Singularities of differentiable mappings, duplicated notes, Mathematics Institute, Bonn University (1959) and Vol. I of the Springer Lecture notes on Singularities 1970.

[6] Levine, H. I. A generalization of a formula of Todd, Anais da Acad. Bras. de Ciencias, 37, 369-384 (1965).

[7] Spanier, E. H. Algebraic Topology, McGraw-Hill Series in Higher Mathematics (1966).

SINGULARITIES IN THE DIFFERENTIAL GEOMETRY OF SUBMANIFOLDS

W.F. Pohl

Let N be a differentiable manifold provided with some geometric structure, such as a Riemannian metric, projective connection, or complex structure, and let $f : M \to N$ be a differentiable mapping of a differentiable manifold M. For simplicity let us assume that f is an immersion. By the singular points of f, or rather the one-point singularities of f with respect to the geometrical structure on N, we mean the points of M which satisfy some geometrical condition defined in terms of this structure.

A simple example of what we have in mind is an inflection point of a plane curve. Another simple example is a vertex of a plane curve, which is defined as a point at which the derivative of the curvature vanishes. It may be shown that this condition is equivalent to the condition that the curve have higher contact with its osculating circle at that point, so that a vertex of a plane curve is defined with respect to the conformal geometry, or Möbius geometry, of the plane. And on the other hand, an inflection point is defined with respect to the projective connection in the plane [18] . Or consider, for a surface in ordinary space, an umbilic, which is defined as a point at which the two principal curvatures are equal. Since such a point p is characterized by the property that any sphere tangent to the surface at p, except possibly for one, locally supports the surface, an umbilic is defined with respect to the conformal geometry of space.

By a two-point or a q-point singularity we mean a pair or a q-tuple of points of M satisfying a geometrical condition. For example, a double point, i.e. a pair of points $x, y \in M$ such that $f(x) = f(y)$, or a triple point. Double and triple points, of course, depend on no geometric structure on N. Another example is a double tangent of a plane curve, that is a pair of points for which the tangent lines coincide. This is, of course, an affine singularity. Similar examples, for f a curve in ordinary space, are a cross tangent, i.e. $x, y \in M$ such that the tangent line at $f(x)$ contains $f(y)$, and a trisecant, which is a triple of collinear points.

The notions of genericity, transversality, and the Thom polynomials, which are being considered so extensively for the usual singularities of differentiable mappings, also apply to the singularities we are considering here. For the case of one-point singularities this has been worked out with applications and examples by E.A. Feldman [2,3]. The basic consideration is this. The geometric structure on N gives rise to a stratification of the jet space $J^r(M, N)$, finer than the usual one, in such a way that each stratum corresponds to an r-th order singularity type. If $f : M \to N$ is as before and $G \subset M \times N$ is the graph of f then if $j_r(f) : G \to J^r(M,N)$ is transversal to a given stratum, we say that the corresponding singularities are generic. By the Thom transversality theorem, the set of mappings for which singularities of a given kind are generic is dense in the function space with a suitable topology. For a transversality theorem for q-point singularities, see Mather [17] (cf. also [7,8]).

It is, of course, useful to know the conditions on the derivatives of f which imply and are implied by genericity. But in some cases it is quite inconvenient to consider the whole r-jet of the map. J. Little [12] has developed a method for finding these conditions, in a wide range of cases, by examining only the differential geometric invariants.

The Thom polynomial theorem for one-point singularities may be stated as follows. Given the dimensions of M and N and a universal singularity type, there exists a polynomial depending only on these data such that if $f : M \to N$ has generic singularities of the given type, then the Poincaré dual of the homology class of the locus of singular points equals the polynomial evaluated on the characteristic classes of M and the "pull-backs" of the characteristic classes of N. For a similar statement in the case of q-point singularities, see Haefliger [7,8].

These Thom polynomials have been worked out extensively for inflection points by Feldman [2,3], Suzuki [27, 28,29], and Yoshioka [33]. A typical result is the following (Feldman). Let M be two-dimensional and let $f : M \to E^N$ be an immersion in an euclidean space, $N \geq 9$. Represent f locally by a vector-valued function of two variables, $X(u, v)$. We say that f is <u>third-order non-degenerate</u> at $p \in M$ if the vectors

$$\frac{\partial X}{\partial u}, \frac{\partial X}{\partial v}, \frac{\partial^2 X}{\partial u^2}, \frac{\partial^2 X}{\partial u \partial v}, \frac{\partial^2 X}{\partial v^2}, \frac{\partial^3 X}{\partial u^3}, \frac{\partial^3 X}{\partial u^2 \partial v}, \quad \frac{\partial^3 X}{\partial u \partial v^2}, \frac{\partial^3 X}{\partial v^3}$$

afe linearly independent at p. Otherwise we say that p is a <u>third-order inflection point</u>. Then any immersion of the real projective plane in E^{10} must have third-order inflection points. Feldman also shows, using the Thom transversality theorem, that the real projective plane may be immersed in E^{11} without third-order inflection points.

The Thom polynomials may also be applied to prove and generalise many of the classical enumerative formulas of algebraic geometry, in particular the Plücker formulas. (See Pohl [19]) However, the singularities involved are often not generic, so that auxiliary maps must be constructed and the formulas applied for these maps. A very general and complete treatment of the Plücker formulas for curves has been given by Macdonald [15, 16] and Schwarzenberger [23] using suitable product spaces.

Useful as the techniques of transversality and the Thom polynomials are, in lower dimensions in the real case they yield very little by themselves. In fact, what one can prove with them alone in this situation is usually either trivial, or proved already by a much simpler argument. For example, if one has a closed surface in ordinary space whose parabolic points are generic, then one can prove by a Thom-polynomial argument that the locus of parabolic points is homologous to zero in the surface. However, the parabolic points are just the points at which the Gauss curvature is zero, so that this theorem is really just a special case of the more general fact that the zero locus of a smooth real-valued function with generic zeros on a closed differentiable manifold is homologous to zero. And on the other hand these methods do not yield certain simple classical theorems, as one might expect. For example, a theorem of Möbius (see Sasaki [22]) states that a simple closed curve in the projective plane which is not contractible to a point must have at least three inflection points. The related four-vertex theorem states that a simple closed curve in the ordinary plane must have at least four vertices. A great many proofs of the four-vertex theorem are known (see Guggenheimer [6], Barner [1], Haupt and Künneth [10]), but neither the Möbius theorem nor the four-vertex theorem has been given a proof based on modern topological ideas. I think any such proof would have to be quite ingenious.

Our singularities may be studied both from a local and a global point of view, and, as usual in differential geometry, local theorems imply global theorems. For examples of recent work on the local theory I might refer to the work of Titus on the Caratheodory and Loewner conjectures, reported on by him in this symposium, and to the work of Little [12] on singularities of submanifolds of higher dimensional euclidean spaces, particularly surfaces in four-space. I wish to report here mainly on some recent global results on these singularities. The results fall in general into three categories :

1) formulas relating the numbers of singular points of various types with other intrinsic and extrinsic invariants;
2) theorems on the existence or non-existence of immersions, having given properties including existence or non-existence of singularities of a given type;
3) classification of submanifolds free from singularities of a given type up to homotopy or isotopy through submanifolds free from the singularities.

Results of the last sort may appear to be concerned with singularities only in a negative way, but in fact to prove such results usually requires a close look at the nature of the singularities involved, and in some cases one is called on to "cancel" pairs of singularities.

In what follows I will consider mainly curves in the plane and ordinary space, partly because of a predilection for lower dimensions and partly because of the ease in stating the results. Some of these things have higher dimensional generalizations. Throughout we will assume that all maps are differentiable of class C^∞ and that curves are immersed, unless the contrary is specifically indicated.

Some particularly fine examples of theorems of the first category are furnished by the results of Benjamin Halpern [9] on plane curves. A typical one is the following. Consider a closed immersed plane curve and consider a double tangent of the curve, that is a pair of points of the curve at which the tangent lines are the same. We assume for simplicity that the line supports a neighbourhood in the curve of each point of tangency. If these neighbourhoods lie on the same side of the line we say that the double tangent is "of the first kind"; if they lie on opposite sides we say that it is "of the second kind". Let I denote the number of double tangents of the first kind, II the number of double tangents of the second kind, C the number of self-intersections (we assume for simplicity that these are transversal crossings), and F the number of inflections. Then

$$I = II + C + \tfrac{1}{2}F.$$

The theorem is proved by defining a certain vector field on the space of unordered pairs of points of the curve and using the fact that the sum of the indices of the singularities of the vector field equals the Euler characteristic of the space. Concerning plane curves there are also results of James White [30,31,32] which relate the index of rotation, the winding number, and certain two-point singularities.

Let us now turn our attention to smooth immersed curves in ordinary euclidean space. Such curves have three basic local invariants, the arc length s, the curvature κ, and, provided $\kappa \neq 0$, the torsion τ. And these constitute a complete set of invariants in the sense that $\kappa(s)$, $\tau(s)$ determine the curve up to a rigid motion. Points where $\kappa = 0$, or where $\tau = 0$ are very natural to consider as singularities. It may be shown [2,3] that points with $\kappa = 0$ are not generic, but that τ vanishes generically at isolated points. And in a one-parameter family of space curves κ will vanish generically at isolated points of isolated curves.

The first natural global question that arises, therefore, is to give geometric conditions for a closed space curve with nowhere vanishing curvature which imply that there are actually points at which $\tau = 0$. Let us call such points inflection points. The first such theorem comes from the four-vertex theorem. As we have remarked, this theorem states that for a simple closed plane curve there are at least four points at which the osculating circle has higher contact with the curve. If we now project our curve stereographically onto the sphere, circles and orders of contact are preserved, so that vertices correspond to points of higher contact with the osculating plane. Thus we obtain the following theorem: a simple closed curve on the sphere has at least four inflection points. The conclusion remains true for simple closed curves on an arbitrary strictly convex surface (Barner [1]). Sharon Jones [11] has shown similarly that a simple closed space curve with nowhere vanishing curvature, with no cross tangents, and with no trisecant lines, must have at least two inflection points. (Note that a simple closed curve on a strictly convex surface has these properties.) B. Segre [24, 25, 26] has shown that a closed space curve having no pair of directly parallel tangent vectors must have at least four inflection points. For a similar theorem of Scherk and Segre see [20].

A second natural geometric question which arises from the considerations above is, given two space curves, whether any deformation of one to the other must necessarily have intermediate curves with points of vanishing curvature. In the Figure we give an example of two space curves with $\varkappa \neq 0$ such that any deformations of one to the other has such intermediate curves.

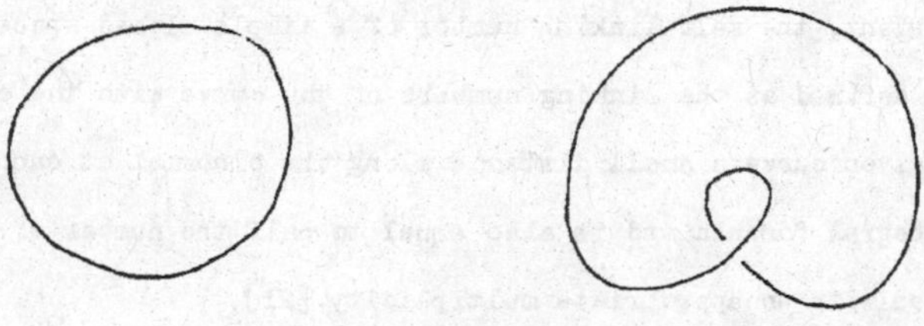

Figure

To see this we remark that the curvature of a space curve is non-zero if and only if the tangent indicatrix, (that is the curve on the unit sphere centred at the origin consisting of the unit tangent vectors to the space curve) is an immersion. Hence any deformation of one of the curves in the figure to the other through curves with $\varkappa \neq 0$ would give rise to a regular homotopy of their tangent indicatrices on the sphere. But such a regular homotopy is impossible, since the curve of unit tangent vectors of the tangent indicatrix of the left hand curve represents the generator of the fundamental group of the unit tangent bundle of the 2-sphere, while that of the right-hand one is contractible to a point.

The question can be put in a different way, namely, to classify closed space curves with $\kappa \neq 0$ up to deformation through curves with $\kappa \neq 0$. Or we can ask to classify closed space curves with $\kappa \neq 0$ and $\tau \neq 0$ up to deformation through such curves. And we can vary these questions by requiring in addition that the curves be embedded and that the deformations be isotopies. We are thus led to four classification problems. Three of them have been solved; the results are as follows.

I. Any closed space curve with $\kappa \neq 0$ is regularly homotopic to one of the curves in the Figure through curves with $\kappa \neq 0$. Thus there are two classes. (Feldman [4])

II. Two closed curves with $\kappa \neq 0$ and $\tau \neq 0$ are regularly homotopic through curves with $\kappa \neq 0$ and $\tau \neq 0$ if and only if they are regularly homotopic through curves with $\kappa \neq 0$ and the sign of the torsion is the same for each. Thus there are four classes (Little [14]).

III. Two simple closed space curves with $\kappa \neq 0$ are isotopic through such curves if and only if they are isotopic and have the same self-linking number. (J.F. Dillon and the author.)

This last invariant, the self-linking number of a simple closed space curve with $\kappa \neq 0$, may be defined as the linking numbers of the curve with the curve obtained by moving the given curve a small distance along the binormal at each point. It is given by an integral formula and is also equal to half the number of cross tangents, each counted with an appropriate multiplicity [21].

The theory of the self-linking number has been generalized to higher dimensions by White [30, 31, 32], who has also obtained a great many results belonging to category 1) mentioned above both in higher and lower dimensions. Some of these results are of a purely differential topological nature. White's results lead to simple and natural deformation questions in higher dimensions, generalizing I, II, and III above.

Feldman [5] has classified closed curves with nowhere vanishing geodesic curvature in a Riemannian 3-manifold up to deformation through such curves. Little [13] has done the same for curves on the ordinary round 2-sphere. But in the area of deformations free from geometric singularities there are a great many open questions.

REFERENCES

[1] M. Barner, Über die Mindestanzahl stationärer Schmiegebenen bei geschlossen strengkonvexen Raumkurven. Abh. Math. Sem. Univ. Hamburg, 20, 196-215 (1956).

[2] E.A. Feldman, The geometry of immersions, I, Trans. Amer. Math. Soc. 120, 185-224 (1965). II, same Trans. 125, 181-215 (1966).

[3] __________, On parabolic and umbilic points of immersed hypersurfaces. Trans. Amer. Math. Soc. 127, 1-28 (1967).

[4] __________, Deformations of closed space curves. Jour. Diff. Geom. 2, 67-75 (1968).

[5] __________, Non-degenerate curves on a Riemannian manifold. Mimeographed preprint.

[6] H. Guggenheimer, Differential Geometry, McGraw-Hill, New York, (1963).

[7] A. Haefliger, Sur les self-intersections des applications différentiables. Bull. Soc. Math. France. 87, 351-359. (1959).

[8] __________, Points multiples d'une application et produit cyclique reduit. Amer. J. Math. 83, 57-70,(1961)

[9] B. Halpern, Global theorems for closed plane curves. Bull. Amer. Math. Soc. 76, 96-100 (1970).

[10] O. Haupt and H.Künneth, Geometrische Ordnungen. Springer-Verlag, Berlin-Heidelberg - New York, 1967.

[11] E.S. Jones, A generalization of the two-vertex theorem for space curves. Thesis, University of Minnesota, 1970.

[12] J.A. Little, On singularities of submanifolds of higher dimensional Euclidean spaces. Annali di Matematica (4) 83, 261-336 (1969).

[13] __________, Non-degenerate homotopies of curves on the unit 2-sphere. J. Diff. Geom. 4 (1970)

[14] __________, Third order nondegenerate homotopies of space curves. J. Diff. Geom. to appear.

[15] I.G. Macdonald, Some enumerative formulas for algebraic curves. Proc.Cambridge Phil. Soc. 54, 399-416 (1958).

[16] I.G. Macdonald, Symmetric products of an algebraic curve, Topology 1, 319-343 (1962).

[17] J. Mather, Stability of C^∞ Mappings: V, Transversality, to appear.

[18] W.F. Pohl, Connexions in differential geometry of higher order. Trans. Amer. Math. Soc. 125, 310-325 (1966).

[19] ________, Extrinsic complex projective geometry. Proceedings of the Conference on Complex Analysis (Minneapolis, 1964). Springer-Verlag, Berlin-Heidelberg-New York, 18-29 (1965).

[20] ________, On a theorem related to the four-vertex theorem . Annals of Math. 84, 356-367 (1966).

[21] ________, The self-linking number of a closed space curve. J. Math. Mech. 17, 975-986 (1968).

[22] S. Sasaki, The minimum number of points of inflexion of closed curves in the projective plane. Tôhoku Math. J. (2) 9, 113-117 (1957).

[23] R.L.E. Schwarzenberger, The secant bundles of a projective variety. Proc. London Math Soc. (3) 14, 369-384 (1964).

[24] B. Segre, Sulle coppie di tangenti fra loro parallele relative ad una curva chuisa sghemba. Hommage au Professeur Lucien Godeaux, 141-167. Libraire Universitaire, Louvain, 1968.

[25] ________, Alcune proprietà differenziali in grande della curve chuise sghemba. Rend. Mat. (6) 1, 237-297 (1968).

[26] ________, Global differential properties of closed twisted curves. Rend. Sem. Mat. Fie. Milano 38, 256-263 (1968).

[27] H. Suzuki, Characteristic classes of some higher order tangent bundles of complex projective spaces. J. Math. Soc. Japan, 18,386-393 (1966).

[28] ________, Bounds for dimensions of odd order nonsingular immersions of RP^n. Trans. Amer. Math. Soc. 121, 269-275 (1966).

[29] ________, Higher order non-singular immersions in projective spaces. Quart. J. Math. Oxford 2nd Series 20, 33-44 (1969).

[30] J.H. White, Self-linking and the Gauss integral in higher dimensions. Amer. J. Math. 91, 693-728 (1969).

[31] J.H. White, Some differential invariants of submanifolds of euclidean space. J. Diff. Geom. 4, 207-224 (1970).

[32] ________, Self-linking and the directed secant span of a differentiable manifold, to appear.

[33] C. Yoshioka, On the higher order non-singular immersions. Sci. Rep. Niigata Univ. Ser. A No.5, 23-30 (1967).

A PROOF OF A CONJECTURE OF LOEWNER AND OF THE CARATHEODORY CONJECTURE CONCERNING UMBILIC POINTS

C.J. Titus

The following conjecture of Loewner was motivated by the study of umbilic points on surfaces and by various other geometrical investigations in differential equations, see especially Loewner [7]. Let u be a real analytic function on the closed unit disc $D \subset \mathbb{R}^2, x^2 + y^2 \leq 1$. Let $\nabla = \partial_x + i\partial_y$ represent the gradient operator in complex form (also $\nabla = \frac{1}{2}\frac{\partial}{\partial \bar{z}}$) and consider iterates of the operator ∇ acting on u, $\nabla^n u$, as vector fields on the disc D.

The Loewner Conjecture (circa 1950)

If $u \in C^{\omega}(D, \mathbb{R})$ is such that the vector field $\nabla^n u$ has a singularity only at the origin then the index of $\nabla^n u$ is no greater than n.

For $n = 1$ the Conjecture is a classical fact, for $n = 2$ it is the key statement required for a proof of the Caratheodory conjecture on umbilic points, see Hamburger [2, 3, 4], Bol[1], Klotz [5]. For $n \geq 3$, Little [6] has conjectured geometric implications for immersions of the sphere in Euclidean spaces of higher dimension.

The main result of this paper is a proof for $n \geq 1$ of a sharpened form of the Loewner Conjecture. For $u \in C^{\omega}(D, \mathbb{R})$ write $u = u_p + \dots + u_q + \dots$ where each u_q is a form (= homogeneous polynomial) of degree q. Let $\#_k u_p$ be the number of distinct linear factors of multiplicity at least k (note $\#_k u_p \geq \#_{k+1} u_p$).

Theorem

If $u \in C^{\omega}(D, \mathbb{R})$ is such that $\nabla^n u$, $n \geq 1$, has a singular point only at the origin, then with $u = u_p + \dots$ and u_p the form of lowest degree so that $\nabla^n u_p \neq 0$, the index of $\nabla^n u$ satisfies the inequalities (1) and (2).

$$(1)\begin{cases} \operatorname{Ind} \nabla^n u_p \leq n - (\#_{2n-p} u_p + \dots + \#_n u_p) + (\#_{n+1} u_p + \dots + \#_p u_p), & p \leq 2n-1, \\ \operatorname{Ind} \nabla^n u_p \leq n - (\#_1 u_p + \dots + \#_n u_p) + (\#_{n+1} u_p + \dots + \#_{2n} u_p), & 2n \leq p. \end{cases}$$

(2) $n - p \leq \mathrm{Ind}\, \nabla^n u \leq p - n$. (By choice of u_p, $n \leq p$.)

Examples

A. Let Q_p be a strictly positive definite form of degree p and let $u = Q_p + \ldots$, then the Theorem gives $-(p-n) \leq \mathrm{Ind}\, \nabla^n u \leq n$ and there exist such Q_p and higher order terms so that either bound is achieved.

B. Let $L_1, \ldots, L_r$ be r pairwise independent linear forms, let $\alpha_1 + \ldots + \alpha_r = p \geq 2n$ be a partitioning of p with $1 \leq \alpha_i \leq n$ and let $u = L_1^{\alpha_1} \ldots L_r^{\alpha_r} + \ldots$, then the Theorem computes $\mathrm{Ind}\, \nabla^n u = n - p$.

C. Let $k \geq 0$ and $u = x^{n+k} + \ldots$, then the Theorem gives $-k \leq \mathrm{Ind}\, \nabla^n u \leq \min(k,n)$ and there exist higher order terms so that either bound is achieved.

Sketch of Proof

The main idea of the proof is, first, to make a detailed geometrical study of the action of the operator ∇^n on forms u_q. This is shown to be contained in the study of the action of a certain semi-group $\mathfrak{b}$ on $C^\omega(S, \mathbb{C})$.

The semi-group $\mathfrak{b}$ is defined in the following way. For $c \in \mathbb{C}$, $a \in \mathbb{R}$ define the operators (c, a): $C^\omega(S, \mathbb{C}) \to C^\omega(S, C)$ by sending $\zeta \in C^\omega(S, \mathbb{C})$ to $(c, a)\zeta = \zeta + a(\mathrm{Im}\, \bar{c}\zeta')c$, $(\zeta' = \frac{d\zeta}{d\theta})$.

Under the composition law, naturally induced by the action, these operators satisfy the following identities for $c \in C$ and $\forall a, b \in \mathbb{R}$:

1. $(b, c)(a, c) = (a + b, c)$;
2. $(0, c) = (a, 0) = \text{identity}$;
3. $(a, bc) = (ab^2, c)$.

Identities (1, 2) show that each operator is bijective, for $(a, c)^{-1} = (-a, c)$. So the algebraic closure, under compositions and inverses, of the generating set $\{(a, c) | a \in \mathbb{R}, c \in \mathbb{C}\}$ forms a group $\mathcal{G}$. (Algebraically $\mathcal{G}$ is characterized as a free product of the one-parameter groups with the relations (1, 2, 3) and the action of this group on polynomials has been studied by Norton [8].) The closure, under composition alone, of the generating set $\{(a, c) | a \in \mathbb{R}^+, c \in \mathbb{C}\}$ forms a semigroup $\mathcal{S} \subset \mathcal{G}$. Note that $S \in \mathcal{S}$ may involve differentations of arbitrarily high order; in fact, using the identities (1, 2, 3), one has

<u>Lemma 1</u> (Unique factorization)

<u>Given</u> $S \in \mathcal{S}$ $\exists$ <u>a sequence</u> $c_1, \ldots, c_m \in \mathbb{C}$, $\operatorname{Im} \bar{c}_k c_{k+1} > 0$, <u>such that</u> $S = (1, c_m) \ldots (1, c_1)$.

The length m of this canonical word is called the order of S.

Given $\zeta \in C^\omega(S^1, \mathbb{C})$ the integral $\frac{1}{2\pi}\int_0^{2\pi} \frac{d}{d\theta} \arg \zeta' \, d\theta$, $\arg \zeta'$ taken mod π (instead of mod 2π), is seen always to exist and is simply the number of times the (non-oriented) line from 0 to $\zeta(\theta)$ turns around (counter clockwise = positive). This half-integer, $\omega\zeta$, exists even if ζ sometimes takes the value 0; when ζ is never 0 it gives the ordinary winding number about 0. Let $\#_k\zeta$ denote the number of distinct zeros of ζ of multiplicity at least k.

<u>Lemma 2</u> (Monotonicity of ω under $\mathcal{S}$) <u>For almost all</u> $S \in \mathcal{S}$.

$\omega S \zeta \geq \omega \zeta + \frac{1}{2}(\#_1\zeta + \ldots + \#_m\zeta)$, m = order of S.

In [6] Loewner proved $\omega S\zeta \geq \omega\zeta$.

<u>Lemma 3</u> (Lowering of multiplicities)

<u>Given</u> ζ, $\theta \in S^1$, <u>the multiplicity of the zero of ζ at θ is reduced to zero or to</u> m <u>by almost all operators</u> $S \in \mathcal{S}$ <u>of order</u> m.

<u>Lemma 4</u> (Alignment of tangents at 0)

<u>Given</u> $S \in \mathcal{S}$, $S = (1, c_m) \ldots (1, c_1)$, <u>and</u> ζ <u>with a zero of multiplicity</u> $\geq m$ <u>at</u> θ, <u>the tangent line to</u> ζ <u>at</u> θ <u>contains the point</u> c_m.

Finally it turns out that the rate of change of $\arg \zeta$, (mod π) can be computed at 0, in a simple way, when in application to the Theorem, and has a monotonicity depending on the degree of the form u_q and the order of the zero of u_q on S. The above Lemmas and this fact are the essential tools in the proof of the Theorem. The complete proof will appear in 1971.

REFERENCES

[1] Bol, G., Über Nabelpunkte auf einer Eifläche, Math. Zeit. 49, (1944), 389-410.

[2] Hamburger, H., Beweis einer Caratheodoryschen Vermutung I, Ann. of Math (2) (1940) 63-86.

[3,4] Hamburger, H., ______, II, III, Acta Math., (1941), 175-228, 229-332.

[5] Klotz, T., On G. Bol's Proof of the Caratheodory Conjecture, Comm. Pure Appl. Math. 12, (1959), 277-311

[6] Little, J., Geometric Singularities, This volume, pp.

[7] Loewner, C., A Topological Characterization of a Class of Integral Operators, Ann. of Math. (2), 41 (1940), 63-86.

[8] Norton, V.T. Differential and Polynomial Transvections in the Plane, Thesis, Univ. of Mich. (1970), Ann Arbor.

[9] Titus, C.J., The Combinatorial Topology of Analytic Functions on the Boundary of a disc, Acta Math. 106 (1961), 45-64.

[10] Titus, C.J., Characterizations of the Restriction of a Holomorphic Function to the Boundary of a disc.

[11] Titus, C.J. & Young, G.S. An Extension Theorem for a Class of Differential Operators, Mich. Math. J. 6 (1959), 195-204.

GEOMETRIC SINGULARITIES

J.A. Little

Let $\nabla = \frac{\partial}{\partial x} + i\frac{\partial}{\partial y}$ be the gradiant operator on real valued functions of two real variables. The iterated gradiant operator ∇^n is defined by composition ; for example ∇^2 is $\frac{\partial^2}{\partial x^2} - \frac{\partial^2}{\partial y^2} + 2i\frac{\partial^2}{\partial x \partial y}$. Charles Titus has proved [2] a conjecture of Loewner; namely that for any real analytic function such that $\nabla^n u$ has an isolated singular point the index of the vector field $\nabla^n u$ is less than or equal to n . This result implies the Caratheordory conjecture; namely any real analytic immersion of a two sphere in E^3 has at least two umbilics.

The purpose of this paper is to find geometric problems for which the result of Titus is the solution. Unfortunately I have not yet been able to show that the Theorem of Titus gives the solution of the proposed geometric problem.

Consider a second order nondegenerate immersion of a surface in E^6, $X : M^2 \to E^6$. Second order nondegenerate means that $X_u, X_v, X_{uu}, X_{uv}, X_{vv}$ span a 5-dimensional linear space at each point, where u, v are local coordinates. Let e_6 be the unit normal to this space. We may define a cubic form $\eta_p : T_p \to \mathbb{R}$, where T_p is the tangent 2-plane to M at p as follows. For any tangent vector v at p , let X(s) be a curve through p such that $\frac{dX}{ds}(p) = v$. Then $\eta_p(v) = \frac{d^3X}{ds^3}(p) \cdot e_6$. One verifies that η_p depends only on the tangent vector and not on the curve. Since η_p is a cubic form on T_p. the restriction of η_p to the unit tangent circle may therefore be written

$$\eta_p(e) = a\cos^3\theta + 3b\cos^2\theta\sin\theta + 3c\cos\theta\sin^2\theta + d\sin^3\theta \quad ,$$

where $e = \cos\theta\, e_1 + \sin\theta\, e_2 : e_1, e_2$ are fixed orthonormal frames in T_p. We may write this cubic polynomial using the trigonometric identities in the form

$$\eta_p(e) = A\cos 3\theta + B\sin 3\theta + C\cos\theta + D\sin\theta \ .$$

The 3θ and θ part of $\eta_p(e)$ are individually invariant under a change of the defining frame so that we may define a field of tangent 3 crosses and a tangent vector field as the tangent directions whose arguments respectively satisfy

$$A \cos 3\theta + B \sin 3\theta = 0 , \quad C \cos \theta + D \sin \theta = 0 ,$$

Let us call a point where $A = B = 0$ an S point.

<u>Theorem</u> <u>Every second order nondegenerate surface in E^6 has a field of 3-crosses with singularities at the S points. Generically the S points are isolated with index $\pm {}^{1}/_{3}$. There are at least $3\chi(M)$ S points on any generic surface.</u>

We remark that it is possible to immerse the 2-sphere in E^6 second order nondegenerately; namely as the Veronese surface in E^5.

A tangent r cross is a set of r unit tangent vectors whose tips form a regular polygon. A singular point is a point where the cross field is not defined. In [1] we defined the index of a singular point of a field of r-crosses and showed that the sum of the indices of a field of tangent r-crosses is $\chi(M)$, the Euler characteristic of M.

Suppose p_0 is a singular point of a field of r-crosses given locally by $A \cos r\theta + B \sin r\theta = 0$. Then if the map $\phi : M^2 \to \mathbb{R}^2$ given by $\phi(p) = (A(p), B(p))$, where $\phi(p_0) = (0, 0)$ has nonzero Jacobian at p_0 the index of the field of r-crosses at p_0 is $\pm {}^{1}/_{r}$. By using the methods of [1] we may verify that the S points defined above are generically isolated points with index $\pm {}^{1}/_{3}$. Generic means that the maps $X : M^2 \to E^6$ with this property are dense in the space $C^\infty(M^2, E^6)$.

<u>Conjecture</u> <u>Let $X : M^2 \to E^6$ be a real analytic second order nondegenerate immersion. The index of an isolated S point is ≤ 1. Thus every second order nondegenerate real analytic immersion of the 2-sphere in E^6 must have at least two S points.</u>

References

[1] J.A. Little, On singularities of submanifolds of higher dimensional Euclidean spaces. Ann. Mat. Pur. Appl . (IV) 83 (1969) 261-336.

[2] C.J. Titus, A proof of a conjecture of Loewner and of the Caratheodory conjecture concerning umbilic points. This volume, pp.114-117.

REMARK ON GEOMETRIC SINGULARITIES

C.T.C. Wall

The decomposition of $\eta_p(e)$ in the preceding paper into a 3θ and a θ part can be generalised. Note that any manifold M has higher order tangent bundles $T_r M$, a point of which over $x \in M$ is a differential operator of order $\leq r$ acting on germs at x of functions on M. There are exact sequences

$$0 \to T_{r-1}M \to T_r M \to 0^r TM \to 0 ,$$

where 0^r denotes the rth symmetric power. Given a connection on M, we have natural splittings of these sequences.

An immersion $f : M \to V$, where V has a metric, hence a connection, is <u>r^{th} order nondegenerate</u> if the composite map

$$T_r M \to T_r V \to TV$$

is injective on fibres. But we can split $T_r M$ up as follows. First, the metric on V induces one on M and hence splittings of $T_r M$ as a sum of symmetric powers of TM. Next, we can split each of these, corresponding to irreducible representations of the orthogonal group, as described on p.157 of H. Weyl's book 'The Classical Groups', as follows. In fact $0^r T$ is isomorphic to the direct sum of an irreducible module and $0^{r-2}T$, where the latter is embedded by multiplying by the canonical element (corresponding to the scalar product) in $0^2 T$. In terms of differential operators, this means composing with the Laplacian.

We can now define new classes of singularities by insisting that the above map $T_r M \to TV$ be injective merely on a sum of a chosen set of the subbundles of $T_r M$ constructed above.

GEOMETRIC DIFFERENTIATION - A THOMIST VIEW OF DIFFERENTIAL GEOMETRY

I.R. Porteous

0. The normal singularities of a submanifold

Let M be an m-dimensional smooth $(= C^\infty)$ submanifold of $\underset{\sim}{R}^n$. Locally M is the image of a smooth embedding $g : A\ (\subset \underset{\sim}{R}^m) \to \underset{\sim}{R}^n$, say. In this note (for fuller details see [1]) we consider the generic singularities $\Sigma^I f$ [2] (see also [3]) of the map

$$f : A \times \underset{\sim}{R}^n \to \underset{\sim}{R} \times \underset{\sim}{R}^n \ ; \ (t, x) \rightsquigarrow (|x - g(t)|^2, x)$$

and hence the generic singularities $\Sigma^I \phi$ of the map

$$\phi : M \times \underset{\sim}{R}^n \to \underset{\sim}{R} \times \underset{\sim}{R}^n \ ; \ (w, x) \rightsquigarrow (|x - w|^2, x)$$

for small values of m and n. The inspiration for studying ϕ in this way is a remark of R. Thom in his book [4], where he justifies the use of the word 'umbilic' to describe certain of the elementary catastrophes.

Differentiation of f shows at once that $\Sigma^m f$, the subset of $A \times \underset{\sim}{R}^n$ on which df has kernel rank m, is the set

$$\{(t, x) \in A \times \underset{\sim}{R}^n : \big(x - g(t)\big)\cdot dgt = 0\} \ ,$$

so that $\Sigma^m \phi$ may be regarded as the normal bundle NM of M in $\underset{\sim}{R}^n$. It is a smooth m-dimensional submanifold of $M \times \underset{\sim}{R}^n$. The sets $\Sigma^{m'} f$, with $m' > m$, are clearly null.

The higher-order singularities of f are the singularities of $f|\Sigma^m f$. It is readily verified that the kernel rank of $f|\Sigma^m f$ is everywhere equal to the kernel rank of its second component, the map

$$\Sigma^m f \to \underset{\sim}{R}^n \ ; \ (t, x) \rightsquigarrow x \ .$$

The higher-order singularities of ϕ are therefore the <u>normal singularities</u> of M in $\underset{\sim}{R}^n$, namely the singularities of the map

$$\psi : NM \to \underset{\sim}{R}^n ; \ (w, x) \rightsquigarrow x \ .$$

Now the map ψ is not generic, at least when $\dim M > 1$, nor is it generic in the sense of Thom's paper on envelopes [5]. Here we have a possibly generic setting for it. Indeed John Mather has verified that the map f is, generically, transverse to all κ -singularity types, in the terminology of his paper [6]. The κ -singularity types include the Boardman types Σ^I.

In the following examples the bundle projection map $NM \to M$; $(w, x) \leadsto w$ will be denoted by π.

1. Curves in $\underset{\sim}{R}^2$.

Let $m = 1, n = 2$. Then $NM = \Sigma^1\phi$, while $\Sigma^{1,1}\phi$ is a smooth curve in NM whose image in $\underset{\sim}{R}^2$ by ψ is the evolute of M in $\underset{\sim}{R}^2$. Generically $\Sigma^{1,1,1}\phi$ is a discrete set of points of $\Sigma^{1,1}\phi$ whose images in $\underset{\sim}{R}^2$ are cusps on the evolute. These points are the centres of curvature of the vertices of M.

2. Curves in $\underset{\sim}{R}^3$.

Let $m = 1, n = 3$. Then $\Sigma^1\phi = NM$ and, for any $w = g(t) \in M$, $\Sigma^{1,1}\phi \cap NM_w$ is the line with equation

$$dgt \cdot dgt \; - \; \big(x - g(t)\big) \cdot d^2gt \; = \; 0 \; ,$$

where $\big(x - g(t)\big) \cdot dgt = 0$, the $\cdot$ being given the obvious interpretation in each case. (Notice the appearance of the first and second fundamental forms here!) This line, the focal line at w of M, is the polar with respect to the unit circle, with centre w, of the end $w + \kappa(w)$ of the curvature vector $\kappa(w)$ at w. The union of the set of focal lines of M is the focal set in NM of M. In the generic case its image in $\underset{\sim}{R}^3$ is the focal developable of M, with the image of $\Sigma^{1,1,1}\phi$ as cuspidal edge. The set $\Sigma^{1,1,1}\phi$ is a smooth curve in NM. Its intersection with the focal line at w is the centre of spherical curvature of M at w. Generically also $\Sigma^{1,1,1,1}\phi$ is a discrete set of points of $\Sigma^{1,1,1}\phi$, whose images in $\underset{\sim}{R}^3$ by ψ are cusps on the edge of regression of the focal developable and whose images on M by π are the vertices of M.

3. Surfaces in $\underset{\sim}{R}^3$.

Let $m = 2, n = 3$. Then $\Sigma^2\phi = NM$. The focal set consists of the centres of principal curvature of M, at most two on each fibre of the normal bundle, namely those points $x \in NM_w$, with $w = g(t)$, where the kernel rank of the map

$$dgt \cdot dgt \; - \; \big(x - g(t)\big) \cdot d^2gt : \underset{\sim}{R}^2 \to L(\underset{\sim}{R}^2, \underset{\sim}{R})$$

is > 0, that is, = 1 or 2 . (Notice the first and second fundamental forms again! The image by ψ of the focal set is the centro-surface of M. Generically the subset $\Sigma^{2,1}\phi$ is the non-singular part of the focal set, a smooth submanifold of NM of dimension 2, while $\Sigma^{2,2}\phi$ is its set of singularities, the centres of curvature of the umbilics of M.

Further singularity sets on NM in the generic case are $\Sigma^{2,1,1}\phi$, a smooth curve on $\Sigma^{2,1}\phi$, and $\Sigma^{2,1,1,1}\phi$, a discrete set of points on $\Sigma^{2,1,1}\phi$. The image of $\Sigma^{2,1,1}\phi$ on M by π is then a smooth immersed curve on M. We call the components of $\Sigma^{2,1,1}\phi$ the <u>ribs</u> of M and the images by π of the ribs the <u>base-ribs</u> or <u>ridges</u> of M. These are curves on M. A given base-rib is usually transverse to the members of one of the two families of curves of curvature on M, those points where it fails to be transverse, that is, where it touches a member of the appropriate family of curves of curvature, being the projections on M by π of points of $\Sigma^{2,1,1,1}\phi$.

An ellipsoid with distinct semi-axes has six base-ribs [7], namely the major-mean and mean-minor principal sections and the components of the complement in the major-minor principal section of the set of umbilics, which all lie on this section and are four in number. Each base-rib is a plane curve of curvature and in this case the set $\Sigma^{2,1,1,1}\phi$ is null.

4. <u>What happens at an umbilic</u>

Let M be a generic surface in $\underset{\sim}{R}^3$, and suppose that K is the tangent space to M at an umbilic of M . Choose g so that g(0) is the umbilic. Then we may identify K with $\underset{\sim}{R}^2$ by the differential of g . On K we have the quadratic first fundamental form

$$u \rightsquigarrow dg0u \cdot dg0u$$

and also a cubic form, the third derivative of f w.r.t. t, namely the map

$$u \rightsquigarrow (x - g(t)) \cdot d^3gt(u)(u)(u) \; - \; 3\, dgt(u) \cdot d^2gt(u)(u)$$

(apart from a factor -2) . We call the associated symmetric trilinear form P_3.

The cubic form determines three lines in the tangent spaces at the umbilic, with representative tangent vectors u_1, u_2, u_3 say. In general either one or three of these is real. Let α, β be representatives of the Hessian lines of the cubic. These are real if only one of the u_i is real and they are complex if the three u_i are real and distinct. Let $\overline{u_i}$ be a representative of the harmonic conjugate of the line $[u_i]$ w.r.t. the lines $[\alpha]$ and $[\beta]$.

Theorem (i) Let C be a base-rib of M passing through the umbilic. Then the limiting value at the umbilic of the kernel along C is $[u_1]$, $[u_2]$ or $[u_3]$, and if $[u_i]$ is this limiting value then the tangent direction $[t]$ of C at the umbilic is the harmonic conjugate of the direction $[v_i]$, orthogonal to $[u_i]$, with respect to $[u_i]$ and $[\overline{u_i}]$.

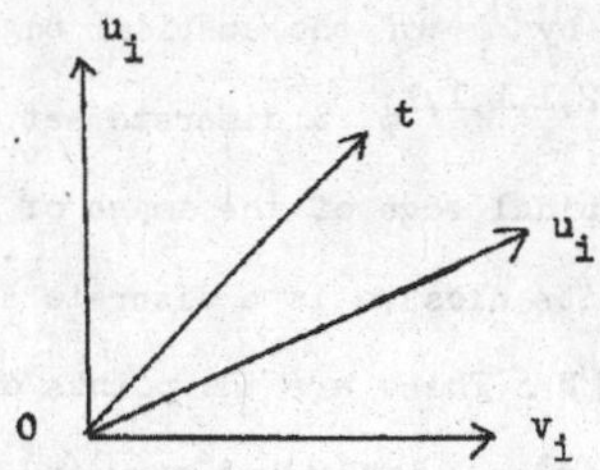

(ii) Let C be a curve through the umbilic lifting to $\Sigma^{2,1}f$ with horizontal tangent at the umbilical centre. Then the limiting value at the umbilic of the kernel along C is either $[\alpha]$ or $[\beta]$ and the corresponding tangent direction at the umbilic is $[\beta]$ or $[\alpha]$.

(iii) Let C be a line of curvature through the umbilic. Then the limiting value of the kernel $[u]$ at the umbilic is also the tangent direction there, and

$$P_3(u)(u)(v) = 0$$

where v is orthogonal to u.

Note that if $\overline{u_i}$ is orthogonal to u_i in (i) then $t = \overline{u_i} = v$, and also $P_3(\overline{u_i})(\overline{u_i})(u_i) = 0$. In general the reality conditions for the possible base-rib directions are not in accord with the reality conditions for the possible directions of lines of curvature at the umbilic.

5. Surfaces in $\mathbb{R}^4$

Let $m = 2$, $n = 4$. Again $\Sigma^2\phi = NM$. On any normal plane NM_w the focal set is a conic that has no tangents passing through w [8]. Generically there is a curve, the curve of umbilics [9], in M over each point of which the focal conic is a pair of intersecting lines. The set $\Sigma^{2,1}\phi$ is the non-singular part of the focal set, while $\Sigma^{2,2}\phi$ consists of the nodes of the focal line-pairs. The reciprocal of the focal conic with respect to the unit circle is an ellipse, the ellipse of curvature. This degenerates to a line segment in the case that the focal conic is a line-pair. (Cf. also [10], [11], [12] and [13]. Other earlier references are to be found in [14]).

Further singularities in the generic case are $\Sigma^{2,1,1}\phi$, a smooth surface in NM, $\Sigma^{2,1,1,1}\phi$, the inverse image by ψ of the cuspidal edge of the image of $\Sigma^{2,1,1}\phi$ in $\underset{\sim}{R}^4$, and, finally, $\Sigma^{2,1,1,1,1}\phi$, a discrete set of points on $\Sigma^{2,1,1,1}\phi$, mapped by ψ to cusps on the cuspidal edge of the image of $\Sigma^{2,1,1}\phi$. Generically the complement of $\Sigma^{2,1,1,1}\phi$ in its closure is a discrete set of points of $\Sigma^{2,2}\phi$, the set of parabolic umbilics of M. These are the points of M where the cubic form $u \rightsquigarrow (x - g(t)) \cdot d^3gt(u)(u)(u) - 3\,dgt(u) \cdot d^2gt(u)(u)$ on the tangent space to M has two coincident roots.

REFERENCES

[1] I.R. Porteous, The normal singularities of a submanifold. To appear in Jour. Diff. Geom.

[2] J.M. Boardman, Singularities of differentiable maps. Inst. Hautes Études Sci. Publ. Math. 33 (1967), 21-57.

[3] I.R. Porteous, Simple singularities of maps. Columbia Notes 1962, reprinted with slight revision in Vol.1 of these Proceedings. Springer Lecture notes no. 192 (1971).

[4] R. Thom, Stabilité structurelle et morphogenèse. Benjamin (to appear)

[5] R. Thom, Sur la théorie des enveloppes. J. Mat. Pur. Appl. 41 (1962) 177-192.

[6] J. Mather, Stability of C^∞ mappings III. Finitely determined map-germs. Inst.Hautes Études Sci. Publ. Math. 35 (1968) 279-308.

[7] A. Cayley, On the centro-surface of an ellipsoid. Trans. Camb. Phil. Soc. 12 (1873) 319-365 (Collected Works Vol.VIII, paper 520)

[8] K. Kommerell, Riemanschen Flächen im ebenen Raum von vier Dimensionen. Mat. Ann. 60 (1905) 548-596.

[9] D. Perepelkine, Sur la courbure et les espaces normaux d'une V_m dans R_n. Rec. Math. (Mat. Sbornik) N.S. 42 (1935) 81-100.

[10] Y-C. Wong, A new curvature theory for surfaces in a Euclidean 4-space. Comm. Mat. Helv. 26 (1952) 152-170.

[11] K.S. Ramazanova, On the theory of two-dimensional surfaces in E_4. (Russian) Volzh. Mat. Sb. Vyp. 3 (1965) 296-311 M.R. 34 # 693.

[12] K.S. Ramazanova, The theory of curvature of X_2 in E_4. (Russian) Izv. Vyssh. Uchebn. Zaved. Mat. No. 6 (55) (1966) 137-143 M.R. 35 # 888.

[13] J.A. Little, On singularities of submanifolds of higher-dimensional Euclidean spaces. Thesis. U. of Minnesota (1968).

[14] J.A. Schouten, und D.J. Struik, Einführung in die neueren Methoden der Differential geometrie. Vol.2. Groningen-Batavia 1938.

LECTURES ON THE THEOREM OF GROMOV

A. Haefliger

In his thesis [2], Gromov proves a very general theorem which contains as particular cases the Smale-Hirsch theorem on immersions [7] and [4] , and Phillips' theorem on submersions [5], as well as many other remarkable new theorems in differential topology. The idea of the proof is essentially the one Smale used to prove his immersion theorem, and which has been clarified successively by many people (see Thom [8], Hirsch [4], Hirsch-Palais (unpublished seminar), Haefliger-Poenaru [3], Phillips [5], etc.)

In part I, we state the main result of Gromov's thesis and discuss some of the particular cases considered by Gromov. In part II we give the proof, which follows exactly the same pattern as the one given in Phillips [5] . (Our treatment is maybe not as general as in Gromov's thesis.)

I. The main theorem and some particular cases

Given a C^∞ m-manifold M , we consider a differentiable bundle E(M) "naturally" associated to the differentiable structure of M.

"Naturally" means the following. To any m-manifold M is associated a differentiable bundle E(M), so that if U is open in M, then E(U) is the restriction of E(M) to U. Moreover to any diffeomorphism f of an open set U on an open set V of a manifold N is associated a diffeomorphism $\bar{f} : E(U) \to E(V)$ covering f, such that $\bar{g} \circ \bar{f} = \overline{g \circ f}$ and $\overline{\text{identity of } U}$ = identity of E(U). Also $\bar{f}$ depends continuously on f .

It is then clear that the pseudogroup $\mathcal{D}(M)$ of local diffeomorphism of M acts on E(M) and also on the space $\Gamma E(M)$ of sections of E(M); namely if $f : U \to V$

is a diffeomorphism and if $\sigma : V \to E(V)$ is a section, then the section $\bar{f}\sigma : U \to E(U)$ is defined by $\bar{f}^{-1} \circ \sigma \circ f$.

For instance $E(M)$ could be the trivial bundle $M \times X$, where X is a fixed smooth manifold, with the trivial action of $\mathfrak{D}(M)$. Another example is $E(M) = TM$ the tangent bundle of M (with the usual action induced by the differential).

Denote by $E^r(M)$ the bundle whose fiber above $x \in M$ is the space of r-jets of local C^r-sections at x. It is also naturally associated to the differentiable structure of M.

Let $E^r_0(M)$ be an open subbundle of $E^r(M)$ invariant by $\mathfrak{D}(M)$.

Let $\Gamma E^r_0(M)$ be the space of continuous sections of $E^r_0(M)$ with the compact-open topology.

Let $\Gamma_0 E(M)$ be the space of C^r-sections σ of $E(M)$ whose r-jet $j^r\sigma$ is a section of $E^r_0(M)$; one puts on $\Gamma_0 E(M)$ the C^r-topology.

The manifold M is open if $M - \partial M$ has no compact component.

<u>Main theorem.</u> <u>If M is open, the map</u>

$$j^r : \Gamma_0 E(M) \to \Gamma E^r_0(M)$$

<u>is a weak homotopy equivalence</u> (abbreviated w.h.e.)

There is also a relative version. Let U be a closed n-submanifold with boundary in M such that $\operatorname{int} M - \operatorname{int} U$ has no compact component. Let $g \in \Gamma_0 E(U)$. Then j^r is a w.h.e. on the subspaces of $\Gamma_0 E(M)$ and $\Gamma E^r_0(M)$ of sections whose restrictions to U are g and $j^r g$ respectively.

This theorem translates the problem of classifying the C^r-sections of $E(M)$ whose r-jet verifies an intrinsic differential inequality into a classical problem in algebraic topology namely the classification of continuous sections of a bundle.

One can also express the theorem as follows. Let us call "holonomic section" of E^r a section which is the r-jet of a C^r-section of E. Then the inclusion of the space of holonomic sections of $E^r_0(M)$ in the space of all continuous sections of $E^r_0(M)$ is a w.h.e. . Hence up to homotopy, the integrability conditions are irrelevant.

Note that the theorem is not in general true for a <u>closed</u> manifold M.

We now list a few examples.

Example 1 : k-mersions

Let N be a C^∞ n-manifold. Take for $E(M)$ the trivial bundle $M \times N$. The sections of $E(M)$ correspond to the maps of M to N; $E^r(M)$ is the bundle $J^r(M,N)$ of r-jets. For $r = 1$, $E^1(M)$ is simply the bundle over $M \times N$ whose fiber over (x, y) is the space of all linear maps of the tangent space T_xM into the tangent space T_yN.

Define $E^1_0(M)$ to be the subspace of those linear maps of rank $\geq k$. Then $\Gamma_0 E(M)$ is the space $\mathrm{Imm}_k(M, N)$ of k-mersions, namely the space of C^1-maps $f : M \to N$ whose differential df is of rank $\geq k$; $\Gamma E^1_0(M)$ is the space $\mathrm{Hom}_k(TM, TN)$ of bundle homomorphisms $\phi : TM \to TN$ whose restriction to each fiber is linear of rank $\geq k$.

The theorem says that, if M is open, the differential

$$\mathrm{Imm}_k(M, N) \to \mathrm{Hom}_k(TM, TN)$$

induces a weak homotopy equivalence.

In fact the theorem is also true for M closed if $k < \dim N$ (cf. Feit [1]). In the case of immersions ($k = \dim M < \dim N$), it is easy to prove the theorem in the closed case, replacing M by the total space of a suitable normal vector bundle over M(that being an open manifold).

More generally (as in the next example), one can take for E^r_0 any subbundle of $J^r(M, N)$ whose typical fiber is an open set of $J^r_0(\mathbb{R}^m, N)$ (r-jets at $0 \in \mathbb{R}^m$) invariant by the action of the group of r-jets at 0 of diffeomorphisms of $\mathbb{R}^m$ leaving 0 fixed.

It would be very interesting to get some information on the following vague question.

Problem : Single out a class of invariant open sets in $J^r_0(\mathbb{R}^m, \mathbb{R}^n)$ for which the theorem holds when M is a closed manifold.

Example 2. Maps transverse to a field of k-planes

Let us consider on N a field of k-planes, i.e. a subbundle η of rank k in TN. Let ν be the quotient bundle TN/η and let $\pi : TN \to \nu$ be the natural projection.

Let $E_0^1(M) \subset J^1(M, N)$ be the subbundle of 1-jets of maps which are transverse to the given field η. Then $\Gamma_0 E(M)$ is the space $Tr(M, \eta)$ of C^1-maps $f : M \to N$ such that $\pi \circ df$ belongs to the space Epi (TM, ν) of epimorphism of TM on ν. $\Gamma E_0^1(M)$ has the same homotopy type as the space Epi (TM, ν). Hence if M is open, the map $\pi \circ d$

$$Tr(M, \eta) \to \text{Epi}(TM, \nu)$$

is a w.h.e.

This has also been proved by Phillips when the field η is integrable [cf. 6]. Note that if η is integrable (and f is transverse to η), then so is the subbundle $df^{-1}(\eta) \subset TM$.

Let me give an application of this theorem which is a particular case of a general result (cf. Haefliger, Topology 9 (1970), 183-194).

Let G be a subgroup of the group of diffeomorphisms of a m-manifold A. Suppose that G is strongly effective in the following sense: if $g \in G$ is the identity on some open set of A, then g is the identity everywhere. By definition, a G-structure on M is given by an open covering $\{U_i\}$ of M, and a family $f_i : U_i \to A$ of diffeomorphisms on open subsets of A such that each change of charts $f_j f_i^{-1}$ is the restriction toan open set of an element $g_{ji} \in G$.

For instance if G is the group of affine transformations of $\mathbb{R}^m$, one gets the classical notion of an affine structure on M.

To such a structure is associated a fiber bundle E over M with fiber A and structural group G (with the discrete topology) defined by the constant transition functions $g_{ij} : U_i \cap U_j \to G$; the bundle E is the quotient of the disjoint union of the $U_i \times A$ by the equivalence relation : $(x_i, a_i) \in U_i \times A$ is equivalent to $(x_j, a_j) \in U_j \times A$ iff $x_i = x_j$ and $a_j = g_{ji} a_i$. On E there is a foliation, complementary to the fibers, whose restriction to $U_i \times A$ is given by the slices $U_i \times a$, $a \in A$. The local sections $U_i \to U_i \times A$ defined by $x \mapsto (x, f_i x)$ define a global section f of E which is transversal to that foliation.

Conversely if $p : E \to M$ is a bundle with fiber A and discrete structural group G, one has on E a foliation transverse to the fibers. If $f : M \to E$ is a section transverse to that foliation, then it defines on M a G-structure. If M is open, such a map exists iff there is an epimorphism of TM on the normal bundle ν to that foliation, which is the same as the tangent bundle along the fibers of E.

Consider the particular case where G is the affine group acting on $\mathbb{R}^m$. Let ν_0 be the vector bundle associated to E by the natural homomorphism $G \to GL(m, \mathbb{R})$. The vector bundle ν is the pullback of ν_0 by the projection $p : E \to M$. So we get the following.

Theorem : The open manifold M admits an affine structure iff the structural group of its tangent vector bundle can be reduced to a discrete group.

Of course, this affine structure is not complete in general. Also the theorem is not true in general if M is closed. For instance the tangent bundle of S^3 is trivial, but S^3 does not admit an affine structure because S^3 cannot be immersed in $\mathbb{R}^3$.

If we come back to the general case, we have to consider the universal bundle A_G with fiber A over the classifying space BG for the discrete group G. Let ν_G be the vector bundle over A_G tangent to the fibers. Then the open manifold M admits a G-structure iff there is an epimorphism of the tangent bundle TM on ν_G.

Example 3. Symplectic structures

Let M be a manifold of dimension 2n. A symplectic structure on M is given by a differential 2-form ω such that $d\omega = 0$ and $\omega^n \neq 0$. Such a form is locally the exterior derivative of a 1-form α.

Consider the bundle $E(M) = T^*M$ dual to TM and let $E_0^1(M)$ be the bundle of 1-jets of 1-forms α such that $(d\alpha)^n \neq 0$. Then $\Gamma_0 E(M)$ is the space of 1-forms α such that $(d\alpha)^n \neq 0$. The space $\Gamma E_0^1(M)$ has the same homotopy type as the space of 2-forms β on M such that $\beta^n \neq 0$. Indeed the exterior derivative defines a vector bundle epimorphism of $E^1(M)$ onto the bundle of antisymmetric bilinear forms on TM. It is also well known that the space of 2-forms β such that $\beta^n \neq 0$ has the same homotopy type as the space of almost complex structures on M.

Hence the main theorem implies that if M is open, there is a symplectic structure ω on M (in fact $\omega = d\alpha$) iff M has an almost complex structure.

In fact the de Rham cohomology class of ω can be chosen arbitrarily in $H^2(M, \mathbb{R})$. Indeed given such a class h, it is represented by a 2-form ω' with $d\omega' = 0$. Let α be a 1-form such that $(d\alpha)^n \neq 0$ and $d\alpha$ defines an almost complex structure homotopic to the given one. It ϵ is a small enough positive number, then $\epsilon\omega' + d\alpha$ is again a symplectic form(at least on a relatively compact open set of M) which defines an almost complex structure homotopic to the given one and which represents the cohomology class ϵh. Then $\omega = \omega' + \epsilon^{-1} d\alpha$ is a symplectic form whose cohomology class is h. So in the case M open, the cohomology class of ω and the homotopy class of the almost complex structure defined by ω can be chosen independently.

This is no longer true if M is closed because if ω is a symplectic form on M, its cohomology class h must satisfy $h^n \neq 0$, so is never 0. For instance S^6 admits an almost complex structure, but no symplectic structure.

Example 4. Contact structure

Let M be a manifold of dimension $2n + 1$. A contact form ω on M is a 1-form such that $\omega \wedge (d\omega)^n \neq 0$. The theorem of Gromov implies that, if M is open, there is a one to one correspondence between the homotopy classes of contact forms on M and the homotopy classes of reductions of TM to the group $U(n) \subset GL(2n + 1)$. Lutz and Martinet have proved that this is also true for closed 3-manifolds.

II. Proof of the main theorem

First remark that the m-manifold M is open iff there is on M a proper Morse function $f : M \to [0, \infty)$ with all critical points of index $< m$ (cf. Phillips [5]). We can order the critical points $a_1, a_2, \ldots$ of f such that the critical values $c_i = f(a_i)$ are increasing.

Around each a_i, there are local coordinates $(x_1, \ldots, x_m)$ such that f is of the form

$$f = c_i - x_1^2 - \dots - x_k^2 + x_{k+1}^2 + \dots + x_m^2 ,$$

the index k being less than m.

For each i, let $M_i = f^{-1}[0, c_i + \epsilon_i]$ where $c_{i+1} - c_i > \epsilon_i > 0$ and ϵ_i is small. Let $M_{i-1}^{\urcorner}$ be M_i minus the set of points $x = (x_1, \dots, x_m)$ in the neighbourhood of a_i such that $x_1^2 + \dots + x_k^2 < \delta_i/2$, where δ_i is a very small positive number.

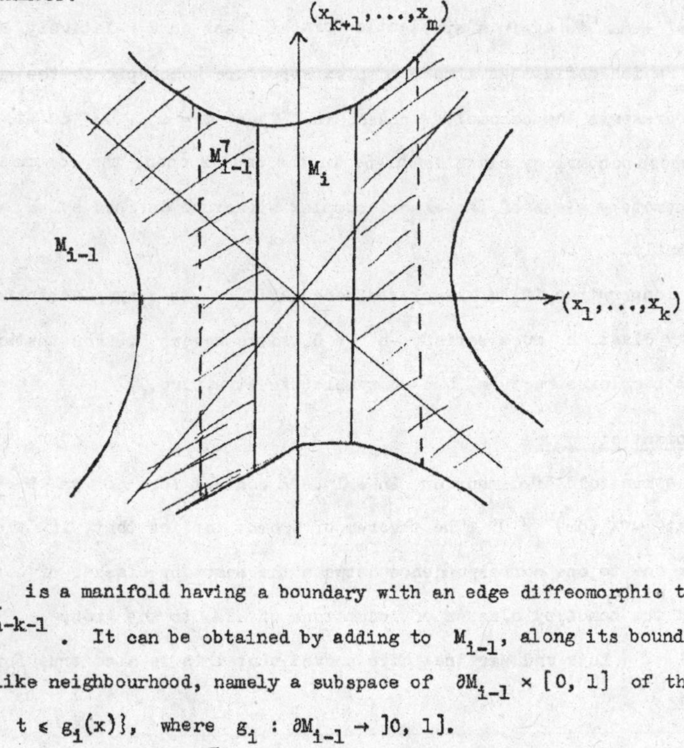

$M_{i-1}^{\urcorner}$ is a manifold having a boundary with an edge diffeomorphic to $S^{k-1} \times S^{n-k-1}$. It can be obtained by adding to M_{i-1}, along its boundary ∂M_{i-1}, a collarlike neighbourhood, namely a subspace of $\partial M_{i-1} \times [0, 1]$ of the form $\{(x, t),\ t \leq g_i(x)\}$, where $g_i : \partial M_{i-1} \to]0, 1]$.

M_i is the union of $M_{i-1}^{\urcorner}$ and A^k which is diffeomorphic to $D^k \times D^{m-k}$, $M_{i-1}^{\urcorner} \cap A^k$ being diffeomorphic to a collar neighbourhood B of $\partial D^k \times D^{m-k}$.

We shall say that M_i is M_{i-1} with a k-handle attached.

M is represented as the union of an increasing sequence

$$M_1 \subset \dots \subset M_{i-1} \subset M_{i-1}^{\urcorner} \subset M_i \subset M_i^{\urcorner} \subset \dots$$

of compact manifolds with boundary.

The theorem follows from the three following propositions.

Write $\Gamma_0(M)$ for $\Gamma_0 E(M)$ and $\Gamma(M)$ for $\Gamma E_0^r(M)$.

<u>Proposition 1. The theorem is true if M is the m-disc D^m, namely</u>

$$j^r : \Gamma_0(D^m) \to \Gamma(D^m)$$

<u>is a w.h.e.</u>

<u>Proposition 2. Suppose that $M^\urcorner$ is obtained from M by glueing along its boundary a collarlike neighbourhood.</u>

<u>Then the restriction maps</u>

$$\Gamma_0(M^\urcorner) \to \Gamma_0(M)$$
$$\Gamma(M^\urcorner) \to \Gamma(M)$$

<u>are w.h.e. and Serre fibrations.</u>

<u>Proposition 3. Let</u> $A = D^k \times D^{m-k}$ <u>and</u> $B = D^k_{\frac{1}{2}} \times D^{m-k}$, <u>where</u> $D^k_{\frac{1}{2}} = \{x \in D^k ; \frac{1}{2} \leq |x| \leq 1\}$. <u>Then the restriction maps</u>

$$\rho_0 : \Gamma_0(A) \to \Gamma_0(B)$$
$$\rho : \Gamma(A) \to \Gamma(B)$$

<u>are Serre fibrations.</u>

Assuming these three propositions, the theorem is proved by induction (as in the immersion case) using the following lemma.

<u>Lemma</u> : Let $p : E \to B$ and $p' : E' \to B'$ be Serre fibrations. Let $\bar{g} : E \to E'$ be a fibre map with projection $g : B \to B'$. Assume g is a w.h.e. Then

a) If $\bar{g}$ is a w.h.e., then so is its restriction $\bar{g}_x : E_x \to E'_{gx}$ to each fibre E_x of E.

b) Conversely, if $\bar{g}_x : E_x \to E'_{gx}$ is a w.h.e. for each fibre, then so is $\bar{g}$.

This follows immediately from the homotopy exact sequence of the fibrations and the five lemma.

Assume by induction that the theorem is true for compact manifolds which are unions of handles of index $< k$. Proposition 1 is the start of the induction. We want to prove that if the theorem is true for the m-manifold M, it is also true for the manifold M' obtained from M by attaching a handle of index $k < m$. As before $M^\urcorner$ is the manifold M with a collarlike neighbourhood added along its boundary and $M' = M^\urcorner \cup A$, $M^\urcorner \cap A = B$.

Consider the following commutative diagram

$$(1)\qquad \begin{array}{ccc} \Gamma_0(A) & \xrightarrow{j^r} & \Gamma(A) \\ \rho_0\downarrow & & \rho\downarrow \\ \Gamma_0(B) & \xrightarrow{j^r} & \Gamma(B) \end{array}$$

By proposition 3, the maps ρ_0 and ρ are Serre fibrations. By proposition 1 (and proposition 2), $j^r : \Gamma_0(A) \to \Gamma(A)$ is a w.h.e., and also $j^r : \Gamma_0(B) \to \Gamma(B)$ because B is the union of a 0-handle and a (k-1)-handle. The lemma implies that j^r is a w.h.e. on each fibre.

Consider now the corresponding diagram

$$(2)\qquad \begin{array}{ccc} \Gamma_0(M') & \xrightarrow{j^r} & \Gamma(M') \\ \rho_0\downarrow & & \downarrow\rho \\ \Gamma_0(M^{\urcorner}) & \to & \Gamma(M^{\urcorner}) \end{array}$$

By restriction to A and B, the diagram (2) is mapped into the diagram (1) and the vertical maps of (2) are the pull back fibrations of the fibrations ρ_0 and ρ in (1). So they also are Serre fibrations and the restriction of j^r to each fiber also is a w.h.e. By assumption and proposition 2, $j^r : \Gamma_0(M^{\urcorner}) \to \Gamma(M^{\urcorner})$ is a w.h.e., hence the lemma implies that $j^r : \Gamma_0(M') \to \Gamma(M')$ is also a w.h.e.

To get the theorem for manifolds which are unions of an infinite number of handles, one remarks that $\Gamma_0 M$ and ΓM are the inverse limits of the $\Gamma_0 M_i$ and ΓM_i, and one applies the following lemma (cf. Phillips [5]).

Lemma : Consider the commutative diagram

$$\begin{array}{ccccccccc} \cdots \to & A_{i+1} & \to & A_i & \to & A_{i-1} & \to \cdots \to & A_1 \\ & \downarrow j_{i+1} & & \downarrow j_i & & \downarrow j_{i-1} & & \downarrow j_1 \\ \cdots \to & B_{i+1} & \to & B_i & \to & B_{i-1} & \to \cdots \to & B_1 \end{array}$$

where all the horizontal maps are Serre fibrations and the maps j_i are w.h.e.

Then $j = \varprojlim j_i : \varprojlim A_i \to \varprojlim B_i$ is also a w.h.e.

Proof of proposition 1.

The fibre bundle $E(D^m)$ is a product bundle $D^m \times F$. Sections of this bundle are identified with maps of D^m in F, and the fibre of $E^r(D^m)$ above $0 \in D^m$ with $J^r_0(D^m, F)$.

It is clear that the restriction map

$$\rho \quad : \quad \Gamma(D^m) \longrightarrow \Gamma(0)$$

where $\Gamma(0)$ is the fiber of $E^r_0(D^m)$ above $0 \subset D^m$, is a homotopy equivalence. Hence it is sufficient to prove that

$$\rho \circ j^r : \Gamma_0(D^m) \to \Gamma(0)$$

is a w.h.e.

Let us prove that $\pi_i(\Gamma_0(D^m)) \to \pi_i(\Gamma(0))$ is surjective. We can consider F as a closed submanifold in some euclidean space $\mathbb{R}^N$ and let $\pi : W \to F$ be a differentiable retraction, where W is an open tubular neighbourhood of F in $\mathbb{R}^N$.

Let $f : S^i \to \Gamma(0) \subset J^r_0 \ (D^m, F) \subset J^r_0(D^m, \mathbb{R}^N)$. Replacing each jet of $J^r_0(D^m, \mathbb{R}^N)$ by its polynomial representative of degree r, we get a map $F : S^i \times D^m \to \mathbb{R}^N$ such that, for each $s \in S^i$, the map $F_s : D^m \to \mathbb{R}^N$ defined by $F_s(x) = F(s, x)$ is of class C^r, $j^r F_s$ is continuous in s and $j^r F_s(0) = f(s)$.

There is a neighbourhood of V of 0 in D^m such that $F(S^i \times V) \subset W$. Let $F'_s = \pi \circ F_s | V$.

As $E^r_0(D^m)$ is an open subbundle, there is a neighbourhood U of 0 in V such that $F'_s | U \in \Gamma_0(U)$. Let h be an embedding of D^m in U which is the identity on a neighbourhood of 0. Then $g(s) = \bar{h}^{-1} \circ F_s \circ h$ defines a map $g : S^i \to \Gamma_0(D^m)$ whose r-jet at 0 is f.

This proves surjectivity. Injectivity is proved in a similar way.

Proof of proposition 2. Left to the reader. One uses the fact that sections can be extended to a neighbourhood U of M in $M^{\urcorner}$, and that there is an isotopy, fixed on a neighbourhood of M, deforming the identity of $M^{\urcorner}$ into an embedding of $M^{\urcorner}$ in U.

Proof of proposition 3.

Denote by $D^k_{[a,b]}$ the annulus $\{x \in \mathbb{R}^k, a \leq |x| \leq b\}$, by D^k_a the disc

$\{x \in \mathbb{R}^k, |x| \leq a\}$ and by S_a the sphere $\{x \in \mathbb{R}^k, |x| = a\}$. We shall denote by A the product $D_2^k \times D^{m-k}$ and B will be $D^k_{[1,\,2]} \times D^{m-k}$.

The bundle $E(\mathbb{R}^m)$ is a product bundle $\mathbb{R}^m \times F$ (but the action of the diffeomorphism group of $\mathbb{R}^m$ may not be trivial). Sections of this bundle will be identified with maps of $\mathbb{R}^m$ to F.

Let P be a polyhedron (considered as a parameter space). If U is a subspace of $\mathbb{R}^m \times P$, all the maps $f : U \to F$ we shall consider will be assumed to be of class C^r on the slices $U \cap (\mathbb{R}^m \times \{p\})$ and their r-jets $j^r f$ (on such slices) to be continuous on $\mathbb{R}^m \times P$. If this r-jet belongs to the given subbundle E^r_0, we shall say that f is <u>admissible</u>.

The fact that $\Gamma(A) \to \Gamma(B)$ is a Serre fibration follows immediately from the usual covering homotopy property for locally trivial bundles.

Saying that $\Gamma_0(A) \to \Gamma_0(B)$ is a Serre fibration means the following : given a polyhedron P, a continuous map $f : P \times I \to \Gamma_0(D^k_{[1,2]} \times D^{m-k})$ and a continuous map $g_0 : P \times 0 \to \Gamma_0(D_2^k \times D^{m-k})$, there is a continuous map $g : P \times I \to \Gamma_0(D_2^k \times D^{m-k})$ extending g_0 and such that $\rho g = f$.

$$\begin{array}{ccc} P \times 0 & \xrightarrow{\quad g_0 \quad} & \Gamma_0(D_2^k \times D^{m-k}) \\ \downarrow & \nearrow^{g} & \downarrow \rho \\ P \times I & \xrightarrow{\quad f \quad} & \Gamma_0(D^k_{[1,2]} \times D^{m-k}) \end{array}$$

Equivalently, we are given admissible maps $f : D^k_{[1,2]} \times D^{m-k} \times P \times [0,1] \to F$ and $g_0 : D_2^k \times D^{m-k} \times P \times 0 \to F$ which coincide on the common part. We have to find an admissible extension g to the whole of $D_2^k \times D^{m-k} \times P \times [0,1]$.

The construction will be done in three steps.

a) Extend f to an admissible map f' of a neighbourhood, $D^k_{[\alpha,2]} \times D^{m-k} \times P \times [0,1]$ where $\alpha < 1$, f' being equal to g_0 where both are defined.

b) We claim there is an increasing sequence $0 = t_0 < t_1 < \dots < t_s = 1$ and for each n, $0 \leq n < s$, there is an admissible map μ_n defined on a neighbourhood of $D^k_{[\alpha,2]} \times D^{m-k} \times P \times [t_n, t_{n+1}]$ such that :

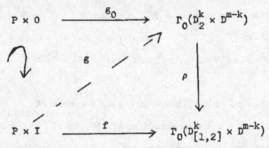

$\mu_n(x, y, p, t) = f'(x, y, p, t)$ for $t = t_n$ or

for x in a neighbourhood of $D^k_{[1,2]}$

$\mu_n(x, y, p, t) = \mu_n(x, y, p, t_n)$ for x in some neighbourhood of S_α.

This is because such a map μ (not necessarily admissible) can be constructed _uniformly_ for each value of t; as admissibility is an open condition, μ will be admissible for small variations of t. We can construct the t_i using the compactness of the interval $[0,1]$.

Using μ_0, we can construct the extension g on the strip $D^k_2 \times D^{m-k} \times P \times [0, t_1]$ by the formula

$$g(x, y, p, t) = \begin{cases} \mu(x, y, p, t) & \text{for } x \in D^k_{[\alpha,2]} \\ g_0(x, y, p, 0) & \text{for } x \notin D^k_{[\alpha,2]} \end{cases}.$$

c) Suppose inductively that we have already constructed an admissible map $g_n : D^k_2 \times D^{m-k} \times P \times [0, t_n] \to F$ extending g_0 and equal to f' on a neighbourhood of $D^k_{[\beta,2]} \times D^{m-k} \times P \times [0, t_n]$, where $\alpha < \beta < 1$.

Let $U \subset D^k_2 \times D^{m-k}$ be some neighbourhood of $D^k_{[\alpha,\beta]} \times 0$ on which μ_n and f' are both defined, and such that $U \cap (D^k_{[1,2]} \times D^{m-k}) = \emptyset$.

As $k < m$, there is an isotopy Δ_t of $D^k_2 \times D^{m-k}$, $0 \leq t \leq t_n$, such that

1) Δ_t is the identity outside of U and on a neighbourhood of $S_\beta \times 0$ and $S_1 \times 0$. Also Δ_t is the identity for $t \leq t_n/2$.

2) Δ_{t_n} maps $S_\gamma \times 0$ on $S_\alpha \times 0$, where γ is such that $\beta < \gamma < 1$.

One first defines g_{n+1} on a small enough neighbourhood of the core $C = [(D^k_2 \times 0) \cup (D^k_{[1,2]} \times D^{m-k})] \times P \times [0, t_{n+1}]$ by the following formulas :

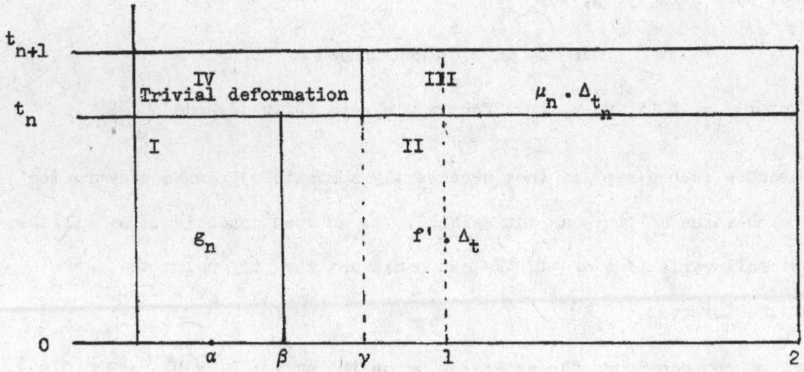

On part I (defined by $\|x\| \leq \beta$, $0 \leq t \leq t_n$)

$$g_{n+1} = g_n$$

On part II (defined by $\beta \leq \|x\| \leq 2$, $0 \leq t \leq t_n$)

$$g_{n+1}(x, y, p, t) = \bar{\Delta}_t^{-1} f'(\Delta_t(x, y), p, t)$$

On part III (defined by $\gamma \leq \|x\| \leq 2$, $t_n \leq t \leq t_{n+1}$)

$$g_{n+1}(x, y, p, t) = \bar{\Delta}_{t_n}^{-1} \mu_n (\Delta_{t_n}(x, y), p, t)$$

On part IV (defined by $\|x\| \leq \gamma$, $t_n \leq t \leq t_{n+1}$)

$$g_{n+1}(x, y, p, t) = g_{n+1}(x, y, p, t_n) .$$

By construction, all these maps coincide on their common domains of definition, if (x, y) belongs to a suitable neighbourhood V of C.

Let h_t be an isotopy of embeddings of $D_2^k \times D^{m-k}$ into itself such that h_0 is the identity, h_t is the identity on a neighbourhood of $D_{[1,2]}^k \times D^{m-k}$, and $h_t(D_2^k \times D^{m-k}) \subset V$ for $t \geq t_n/2$. Then g_{n+1} will be defined on the whole product $D_2^k \times D^{m-k} \times P \times [0, t_{n+1}]$ by

$$g_{n+1}(x, y, p, t) = \bar{h}_t^{-1} g_{n+1}(h_t(x, y), p, t) .$$

<u>Remarks on the topological case</u>

One can prove the analogous theorems for immersions or submersions of topological manifolds (cf. Lees, Bulletin AMS, 1969 and Gauld, thesis 1969), as well as the Gromov-Phillips theorem on maps transverse to a topological foliation (cf. a forthcoming paper of J.Cl. Hausmann).

One uses semi-simplicial complexes instead of topologies, the reason being that there is for instance no topology which makes the set of topological immersions open in the space of maps. The scheme of the proof remains essentially the same. The main difficulty is to lift small homotopies (to construct the maps μ_i in the proof of proposition 3, part 6). This is achieved using the Černowski - Kirby - Edwards theorem on local contractibility of spaces of homeomorphisms.

References

[1] Feit, S.D. k-mersions of manifolds, Acta Mathematica (1969).

[2] Gromov, M.I. Izv. Akad. Nauk SSSR 33 (1969), 707-734.

[3] Haefliger, A. and Poenaru, V. La classification des immersions combinatoires, Publications I.H.E.S.

[4] Hirsch, M.W. Immersions of manifolds, Trans. AMS 93 (1959), 242-276.

[5] Phillips, A. Submersions of open manifolds, Topology 6 (1967), 171-206.

[6] Phillips, A. Smooth maps transverse to a foliation, Bull. AMS (to appear).

[7] Smale, S. The classification of immersions of spheres in Euclidean spaces, Ann. of Math. 69 (1959), 327-344.

[8] Thom, R. La classification des immersions d'après Smale, Séminaire Bourbaki, Dec. 1957.

FORMES DE CONTACT SUR LES VARIÉTÉS DE DIMENSION 3

J. Martinet

Introduction

Le but de ce papier est de démontrer le

Théorème. Soit M une variété compacte et orientable de dimension 3. Il existe sur M une forme de contact, c'est à dire une forme différentielle ω telle que

$$\omega \wedge d\omega \neq 0 \text{ en tout point de } M.$$

Ce théorème répond à une question de S.S. Chern [1].

Le problème de l'existence d'une forme de contact sur une variété de dimension impaire et ouverte a été récemment (presque) résolu par Gromov [8]; elle équivaut dans ce cas à l'existence d'une "presque-structure de contact", c'est à dire un couple (α, β) d'une 1-forme différentielle α et d'une 2-forme β telles que $\alpha \wedge \beta^p \neq 0$ en tout point de M (où β^p désigne la puissance extérieure p-ème de β)

Par contre, à peu près rien n'est connu dans le cas des variétés compactes (voir cependant Gray [4] , qui donne les premières obstructions à l'existence d'une presque structure de contact). Le théorème annoncé résoud donc le premier problème qui se pose dans cette direction.

Mentionnons d'autre part un remarquable théorème de R. Lutz [5] , indiquant en particulier que, sur la sphère S^3, il existe une forme de contact dans toute classe d'homotopie de formes sans zéros; la généralisation de ce résultat aux variétés compactes orientables de dimension 3 parait facile.

La démonstration du théorème annoncé repose de façon essentielle sur la propriété de stabilité des structures de contact; cette propriété a été etablie par Gray [4], dans le cadre de la théorie des déformations de Kodaira-Spencer; j'en redonne ici une démonstration élémentaire et géométrique, qui se prête particulièrement bien aux applications.

Les deux autres clés de la démonstration sont le théorème de structure des variétés compactes orientables de dimension 3 du à Lickorish [6], et l'existence d'un grand nombre de formes de contact sur le tore plein $S^1 \times D^2$ (cf le §4).

1. Préliminaires

Tous les objets considérés sont de classe C^∞.

1.1 Soit M une variété de dimension n. On désigne par $T^*M \to M$ le fibré cotangent à M, et par $P \to M$ le fibré projectif associé à T^*M ; la fibre P_x de P en $x \in M$ est l'espace projectif des droites de la fibre T^*_xM de T^*M en x.

Toute section σ (C^∞) de P sera appelée une équation de Pfaff sur M ([2]); c'est donc un sous-fibré en droites de T^*M ; par passage à l'orthogonal, σ définit un sous fibré $\bar{\sigma}$ de codimension un du fibré tangent TM.

Désignons par T^*_o le complémentaire de la section nulle dans T^*M, par $\pi : T^*_o \to P$ la projection canonique.

Soit σ une équation de Pfaff sur M, et U un ouvert de M ; une 1-forme différentielle ω définie et non nulle en tout point de U est dite un relèvement de σ sur U si $\sigma = \pi \circ \omega$ sur U; en d'autres termes, ω est une section sans zéro de σ sur U; ou encore, l'équation $\omega = 0$ définit le fibré $\bar{\sigma}$ sur U. Si ω et ω' sont deux relèvements de σ sur U, alors $\omega' = f.\omega$ où f est une fonction ne s'annulant pas sur U.

L'existence d'un relèvement global ω d'une équation de Pfaff σ équivaut à la trivialité du fibré en droites σ, c'est à dire au fait que le sous fibré $\bar{\sigma}$ de TM est transversalement orientable.

1.2 Soit $V \to P$ le fibré vectoriel des vecteurs tangents à P, verticaux par rapport à la projection $P \to M$; si $p \in P$, la fibre V_p de V est l'espace tangent en p à l'espace projectif de T_xM, où x est la projection de p sur M.

Désignons par Σ l'ensemble des sections (C^∞) de P.

Pour tout $\sigma \in \Sigma$, on note $\sigma^* V$ le fibré vectoriel sur M , image réciproque de V par $\sigma : M \to P$; une section de $\sigma^* V$ attache à chaque point x de M un vecteur vertical tangent à P en $\sigma(x)$: c'est une "déformation infinitésimale" de σ. L'espace des sections (C^∞) de $\sigma^* V$ sera naturellement noté $T_\sigma \Sigma$ (espace tangent à Σ en σ).

Soient $\omega \in T_o^*$, et $p = \pi(\omega) \in P$; soit x l'origine de ω dans M. Il est clair que la dérivée de π en ω définit une application linéaire <u>surjective</u>

$$T_\omega \pi \; : \; T_x^* M \to V_p \quad .$$

Le noyau de $T_\omega \pi$ est la droite définie par le vecteur $\omega \in T_x^* M$.

Soit maintenant ω une <u>section</u> de T_o^* , et $\sigma = \pi \circ \omega$ l'équation de Pfaff définie par ω ; on notera encore

$$T_\omega \pi \; : \; T^* M \to \sigma^* V \quad .$$

le morphisme défini en chaque point comme précédemment.

1.3 Le groupe Diff(M) des difféomorphismes de M sur elle-même agit de façon naturelle sur Σ . En effet, soit ϕ un automorphisme de M ; on a le diagramme :

$$\begin{array}{ccc} T^*M & \overset{T^*\phi}{\longleftarrow} & T^*M \\ \downarrow & & \downarrow \uparrow \omega \\ M & \overset{\phi}{\longrightarrow} & M \end{array}$$

où $T^*\phi$ désigne l'automorphisme de T^*M déduit de ϕ par dérivation; $T^*\phi$ laisse invariant l'ouvert T_o^* , et commute avec la multiplication par un scalaire dans T^*M , donc passe au quotient par π, et définit un automorphisme de P, encore noté $T^*\phi$, tel que le diagramme suivant commute :

$$\begin{array}{ccc} P & \overset{T^*\phi}{\longleftarrow} & P \\ \downarrow & & \downarrow \uparrow \sigma \\ M & \overset{\phi}{\longrightarrow} & M \end{array} \quad .$$

Si ω (resp. σ) est une section de T^*M (resp. P), on pose $\phi^* \omega = T^*\phi \circ \omega \circ \phi$ (resp. $\phi^* \sigma = T^*\phi \circ \sigma \circ \phi$); ainsi, si ω est un relèvement de σ , on a :

(1) $$\phi^*\sigma = \pi \circ \phi^*\omega .$$

Si maintenant X est un champ de vecteurs sur M et σ une équation de Pfaff, la <u>dérivée de Lie</u> $\theta(X)\sigma$ de σ par rapport à X est une section de σ^*V définie de la manière suivante: soit ϕ_t le groupe (local) à un paramètre obtenu par intégration de X ; au voisinage de chaque point de M, on a une famille à un paramètre $\sigma_t = \phi_t^*\sigma$ d'équations de Pfaff ; on pose alors $\theta(X)\sigma = \frac{\partial\sigma_t}{\partial t}\Big|_{t=0}$.

Il est clair que si ω est un relèvement (local) de σ, et X un champ de vecteurs quelconque, on a, d'après 1.2 et la formule (1) :

$$\theta(X)\sigma = T_\omega\pi \circ \theta(X)\,\omega$$

en tout point de l'ouvert de définition de ω, où $T_\omega\pi$ est le morphisme défini en 1.2.

<u>Lemme</u>. <u>Soit</u> σ_t $(t \in [0,1])$ <u>une famille à un paramètre d'équations de Pfaff sur</u> M. <u>Soit</u> ϕ_t <u>une famille à un paramètre de difféomorphismes de</u> M. <u>Posons</u> $\dot\sigma_t = \frac{\partial\sigma_t}{\partial t}$, <u>et soit</u> X_t <u>la famille à un paramètre de champs de vecteurs</u> $X_t = \frac{\partial\phi_t}{\partial t} \circ \phi_t^{-1}$. <u>Alors, les conditions suivantes sont équivalentes</u> :

a) $\phi_t^*\sigma_o = \sigma_t$ pour tout $t \in [0,1]$

b) $\theta(X_t)\sigma_t = \dot\sigma_t$ " " .

<u>Démonstration</u>. Posons $\mu_t = \left(\phi_t^{-1}\right)^*\sigma_t$; un calcul standard et facile montre que :

$$\frac{\partial\mu_t}{\partial t} = (\phi_t^{-1})^*\,[\dot\sigma_t - \theta(X_t)\,\sigma_t]$$

où, cette fois, $(\phi_t^{-1})^*$ représente l'action de ϕ_t^{-1} sur le fibré V. L'équivalence entre a) et b) en résulte immédiatement.

1.4 Dorénavant, la variété M sera toujours supposée de dimension impaire $2p+1$.

Une forme de Pfaff ω définie sur un ouvert U de M est appelée une <u>forme de contact</u> sur U si $\omega \wedge d\omega^p \neq 0$ en tout point de U (où $d\omega^p$ désigne la puissance extérieure p-ème de la différentielle extérieure $d\omega$ de ω).

Une équation de Pfaff σ sur M est appelée une <u>structure de contact</u> sur M si, pour tout ouvert U de M et tout relèvement ω de σ sur U, ω est une forme de contact sur U.

Remarquons que si ω est une forme différentielle et f une fonction à valeurs numériques, on a :

$$\omega' \wedge d\omega'^p = f^{p+1} . \omega \wedge d\omega^p \quad \text{où} \quad \omega' = f . \omega .$$

On en déduit qu'une équation de Pfaff σ est une structure de contact si et seulement si il existe une famille $(U_i, \omega_i)_{i \in I}$ où les U_i constituent un recouvrement ouvert de M , et ω_i est une forme de contact relevant σ sur U_i pour chaque i .

<u>Une structure de contact est donc une équation de Pfaff de classe maximale en tout point</u> (cf. [3]), <u>c'est à dire sans singularités au sens de</u> [2].

Compte tenu du <u>Théorème de Darboux</u> (voir par exemple [2] ou [3]) la donnée d'une structure de contact sur M équivaut à la donnée d'un atlas de M tel que les changements de cartes appartiennent au <u>pseudo-groupe des transformations de contact</u> (cf [4]).

Plus géométriquement, si σ est une structure de contact, le sous fibré $\bar{\sigma}$ de TM (voir 1.1), considéré comme champ d'éléments de contact de codimension un, est aussi éloigné que possible de la complète intégrabilité (qui équivaut à la condition de Frobenius $\omega \wedge d\omega = 0$, ou ω est un relèvement local quelconque de σ) ; ceci sera précisé plus loin (Remarque 2.1).

La condition, pour une équation de Pfaff σ , d'être une structure de contact est, de façon évidente, une condition <u>ouverte</u> sur le 1-jet de σ en tout point; ainsi, si l'on munit l'espace Σ de la C^1-topologie de Whitney ([2]), l'ensemble des structures de contact est un C^1-ouvert de Σ .

Remarquons encore qu'une structure de contact σ se laisse définir par une forme de contact globale ω si et seulement si $\bar{\sigma}$ est transversalement orientable dans TM.

Rappelons enfin que si la variété M est de dimension $2p + 1$, p <u>impair</u>, toute structure de contact sur M définit canoniquement une orientation de M (GRAY [4], Prop. 2.2.1).

Ainsi, en dimension 3, le problème de l'existence d'une structure de contact ne se pose que dans le cas des variétés orientables.

2. Stabilité des structures de contact

2.1 Soit toujours M une variété de dimension impaire $2p + 1$. Soit σ une équation de Pfaff sur M. On se propose d'étudier l'équation :

$$\theta(X)\sigma = \tau$$

où τ est une section donnée de σ^*V, et où l'on cherche le champ de vecteurs X.

Remarquons d'abord que $X \mapsto \theta(X)\sigma$ est un opérateur différentiel d'ordre un de l'espace des champs de vecteurs sur M dans l'espace $T_\sigma\Sigma$ des sections de σ^*V.

On va considérer la restriction de cet opérateur au sous-espace des champs de vecteurs sections du fibré $\bar{\sigma} \subset TM$.

<u>Proposition.</u> <u>L'opérateur</u> $X \mapsto \theta(X)\sigma$, <u>restreint à l'espace des sections de</u> $\bar{\sigma}$, <u>est d'ordre</u> 0, <u>c'est à dire qu'il est défini par un morphisme</u> $u : \bar{\sigma} \to \sigma^*V$. <u>De plus,</u> u <u>est un isomorphisme de</u> $\bar{\sigma}$ <u>sur</u> σ^*V <u>si et seulement si</u> σ <u>est une structure de contact.</u>

<u>Démonstration.</u> La proposition est de nature locale. On peut donc utiliser un relèvement local ω de σ. On a alors, d'après 1.3, pour tout champ de vecteurs X:

$$\theta(X)\sigma = T_\omega\pi \circ \theta(X)\omega$$

et $$\theta(X)\omega = d(X \lrcorner \omega) + X \lrcorner d\omega$$

où $\lrcorner$ représente le produit intérieur.

Mais, si X est une section de $\bar{\sigma}$, on a :

$$X \lrcorner \omega = 0$$

par définition de ω.

Ainsi $$\theta(X)\sigma = T_\omega\pi \circ (X \lrcorner d\omega)$$

et la première partie de la proposition est démontrée, le morphisme u étant défini par :

$$u(\xi) = T_\omega\pi\,(\xi \lrcorner d\omega) .$$

Maintenant, remarquons que les fibrés $\bar{\sigma}$ et σ^*V sont de même dimension 2p.

D'autre part, le noyau du morphisme

$$T_\omega \pi \ : \ T^*M \ \to \ \sigma^* V$$

est, en chaque point $x \in M$, la droite définie par $\omega(x) \in T^*_x M$, d'après 1.2.

Enfin, pour chaque $x \in M$, le sous-espace $S_x = \{\xi \lrcorner\, d\omega \ ; \ \xi \in \bar{\sigma}(x)\}$ est inclus dans le <u>support</u> de $d\omega$ en x ([2], I.1.3).

Donc, si u est surjectif, S_x est transverse à $\omega(x)$ pour tout $x \in M$. Il en résulte que $d\omega$ est de rang 2p en chaque point, et que son support est exactement S_x ; donc ([2], Prop. I.4.1) $d\omega^p \neq 0$ en tout point, et le support de $d\omega^p$ en x est S_x ; d'après ([2], Prop. I.1.4.5), $\omega \wedge d\omega^p \neq 0$ en tout point, et σ est une structure de contact.

La réciproque est immédiate.

D'après la proposition précédente, si σ est une structure de contact, et si τ est une section quelconque de $\sigma^* V$, il existe un champ de vecteurs X et un seul, section de $\bar{\sigma}$, tel que $\theta(X)\,\sigma = \tau$.

Ceci exprime la "stabilité infinitésimale" de σ. Il est remarquable qu'elle s'établisse simplement par l'intermédiaire d'un opérateur linéaire.

<u>Remarque</u>. La proposition précédente admet aussi l'interprétation suivante, qui représente une propriété géométrique caractéristique des structures de contact: soient σ une structure de contact et X un champ de vecteurs <u>non nul</u> sur un voisinage d'un point de M, inclus dans le champ d'éléments de contact $\bar{\sigma}$; alors $\theta(X)\sigma$ est non nul; Soit S un élément d'hypersurface transvers à X, et soit π la projection sur S dont les fibres sont les courbes intégrales de X; pour tout x, la projection de $\bar{\sigma}(x)$ par π est un hyperplan tangent à S en $\pi(x)$; le fait que $\theta(X)\sigma \neq 0$ signifie que, lorsque x parcourt une trajectoire de X, l'hyperplan projection de $\bar{\sigma}(x)$ <u>"pivote"</u> autour de $\pi(x)$ (qui est fixe).

2.2 <u>Théorème</u>. ([4], Th.5.2.1) <u>Soit</u> M <u>une variété compacte, et</u> σ_t, $t \in [0,1]$, <u>une famille à un parametre de structures de contact sur</u> M. <u>Alors, il existe un isotopie</u> ϕ_t <u>de</u> M <u>telle que</u> :

$$\phi_t^* \sigma_o \ = \ \sigma_t \qquad \underline{\text{pour tout}} \quad t \in [0,1]$$

<u>Demonstration</u>. Posons

$$\dot{\sigma}_t \ = \ \frac{\partial \sigma_t}{\partial t}$$

Pour chaque t, d'après la proposition 2.1, il existe un champ de vecteurs X_t et un seul, section de $\bar{\sigma}_t$, tel que

$$\theta(X_t)\sigma_t = \dot{\sigma}_t$$

Il est clair que X_t dépend différentiablement de t ; l'intégration de l'équation différentielle

$$\frac{dx}{dt} = X_t(x)$$

fournit l'isotopie cherchée, d'après le lemme 1.3.

2.3 Corollaire (stabilité des structures de contact). Soit M une variété compacte; soit $\sigma \in \Sigma$ une structure de contact sur M . Alors il existe un voisinage U de σ dans Σ , pour la C^1-topologie, tel que, pour tout $\sigma' \in U$, σ' est une structure de contact et il existe un difféomorphisme ϕ de M tel que $\phi^*\sigma = \sigma'$ (i.e. σ' est isomorphe à σ).

En effet, d'après 1.4, les structures de contact forment un C^1-ouvert de Σ ; il suffit alors de choisir un voisinage U de σ tel que, pour tout $\sigma' \in U$, il existe une famille à un paramètre σ_t de structures de contact, avec $\sigma_o = \sigma$ et $\sigma_1 = \sigma'$, et d'appliquer le théorème précédent.

2.4 Remarques

1. La démonstration du théorème 2.2 ne suppose pas, au contraire de celle de Gray ([4]), les structures de contact représentables par des formes de contact globales, et, en particulier, ne suppose pas la variété M orientable.

2. Si ω_t désigne une famille à un paramètre de formes de contact, on obtient, en appliquant le théorème 2.2 à la famille σ_t définie par ω_t , une isotopie ϕ_t de M telle que

$$\phi_t^* \omega_o = f_t \cdot \omega_t$$

où f_t désigne pour tout t une fonction sans zéro sur M. Il est en général impossible de trouver des ϕ_t tels que $f_t = 1$, c'est à dire qu'une forme de contact n'est pas stable.

3. L'argument employé au théorème 2.2 fournit une démonstration particulièrement simple et naturelle du Théorème de Darboux :

Si ω est un germe de forme de contact à l'origine de R^{2p+1}, il existe des coordonnées locales $z, x_1, \ldots, x_p, y_1, \ldots, y_p$ telles que :

$$\omega = f.[dz + \sum_{i=1}^{p} x_i dy_i] \ , \quad f \neq 0 .$$

En effet, écrivons le développement de Taylor d'ordre 1 de à l'origine :

$$\omega = \omega^o + \omega^1 + \omega'$$

où ω^o est une forme linéaire sur R^{2p+1}, ω^1 une forme différentielle à coefficients linéaires, ω' une forme à coefficients d'ordre supérieur ou égal à deux.

Posons $\omega^o + \omega^1 = \omega_o$; ω_o est une forme à coefficients affines, qui s'identifie au 1-jets de ω à l'origine; ω_o est évidemment de contact, et, d'après ([2], Prop. I 4.3.2), il existe des coordonnées locales pour lesquelles

$$\omega_o = dz + \sum_{i=1}^{p} x_i dy_i \ .$$

Posons alors $\omega_t = \omega_o + t\omega'$ $\quad t \in [0,1]$,

En appliquant l'argument de 2.2 et la remarque 2 précédente, on obtient une famille à un paramètre ϕ_t de germes de difféomorphismes <u>conservant l'origine</u>, tels que

$$\phi_t^* \omega_o = f_t . \omega_t \ .$$

Donc $\phi_1^* \omega_o = f_1 . \omega$ et $f_1 \neq 0$.

Ainsi, via ϕ_1, ω s'écrit sous la forme requise.

3. <u>Un modèle pour les structures de contact transverses à une courbe fermée</u>

3.1 Considérons la variété $S^1 \times R^{2p}$; posons $\Gamma = S^1 \times \{0\}$; soit la forme différentielle

$$\omega_o = d\theta + \sum_{i=1}^{p} (x_i dy_i - y_i dx_i)$$

où $d\theta$ représente la forme fondamentale de S^1, et $(x_1, \ldots, x_p, y_1, \ldots, y_p)$ les coordonnées naturelles dans R^{2p}.

On a : $\omega_o \wedge d\omega_o^p = p!\ 2^p\ d\theta \wedge dx_1 \wedge dy_1 \wedge \ldots \wedge dx_p \wedge dy_p$;

donc ω_o est une forme de contact.

Désignons par σ_o la structure de contact définie par ω_o. L'expression de ω_o en un point quelconque $m \in \Gamma$ ($x_i = 0$, $y_i = 0$) est $\omega_o(m) = d\theta$; en ce point, l'hyperplan $\bar{\sigma}_o(m)$, défini par l'équation $d\theta = 0$, est donc <u>transverse à</u> Γ.

En général, on dira qu'une équation de Pfaff σ, définie sur un voisinage de Γ dans $S^1 \times R^{2p}$, est <u>transverse à</u> Γ si $\bar{\sigma}(m)$ est transverse à Γ en tout point $m \in \Gamma$.

3.2 <u>Proposition</u>. <u>Soit σ une structure de contact définie sur un voisinage de Γ dans $S^1 \times R^{2p}$, et transverse à Γ. Il existe alors un difféomorphisme ϕ d'un voisinage de Γ sur un voisinage de Γ, laissant fixe chaque point de Γ, tel que</u>

$$\phi^* \sigma = \sigma_o .$$

En d'autres termes, l'équation de Pfaff

$$\omega_o = d\theta + \sum_{i=1}^{p} (x_i dy_i - y_i dx_i) = 0$$

représente un <u>modèle</u> pour les <u>germes</u> de structures de contact transverses à Γ.

<u>Démonstration</u>

1) Désignons par X le champ de vecteurs unitaires tangent à Γ. Ainsi

$$X \lrcorner\, \omega_o = \omega_o(X) = 1 \qquad \text{sur } \Gamma .$$

D'autre part, il est clair que : $X \lrcorner\, d\omega_o = 0$

puisque $$d\omega_o = 2 \sum_{i=1}^{p} dx_i \wedge dy_i$$

On va d'abord montrer que la structure σ admet sur un voisinage de Γ un relèvement ω tel que

a) $X \lrcorner\, \omega = 1$

b) $X \lrcorner\, d\omega = 0$.

Comme σ est transverse à Γ, $\bar{\sigma}$ est transversalement orientable sur un voisinage de Γ ; σ y admet donc des relèvements; quel que soit le relèvement ω'

considéré, on a $X \lrcorner \omega' \neq 0$ sur Γ, toujours par l'hypothèse de transversalité; on peut donc supposer la condition a) réalisée pour un relèvement ω' fixé.

On va ensuite chercher une fonction f, définie sur un voisinage de Γ, égale à 1 sur Γ, telle que la condition b) soit réalisée pour $\omega = f.\omega'$. Ceci équivaut à $X \lrcorner d\omega = X \lrcorner (f.d\omega' + df \wedge \omega') = f \circ X \lrcorner d\omega' + (X \lrcorner df).\omega' - \omega'(X)df = 0$ en tout point de Γ.

Mais $\quad X \lrcorner df = 0$ puisque $f = 1$ sur Γ

et $\quad \omega'(X) = 1$ par construction.

On obtient donc

$$df = X \lrcorner d\omega' = \alpha \quad \text{en tout point de } \Gamma .$$

Comme $X \lrcorner \alpha = 0$, on vérifie trivialement l'existence d'une fonction f répondant à la condition ci-dessus en tout point de Γ.

2) Considérons maintenant le fibré $\bar{\sigma}$, restreint à Γ. Comme ω est une forme de contact, la forme bilinéaire alternée $d\omega$ définit, pour tout $m \in \Gamma$, une structure symplectique sur la fibre $\bar{\sigma}(m)$.

De même, le fibré $\bar{\sigma}_o = S^1 \times R^{2p}$ est muni d'une structure symplectique via $d\omega_o$.

Les fibrés $\bar{\sigma}$ et $\bar{\sigma}_o$ sont triviaux sur Γ. D'autre part, on sait que les formes symplectiques sur R^{2p} constituent une orbite de l'action canonique du groupe $Gl(2p, R)$ sur $\overset{2}{\wedge}(R^{2p})^*$ (cf. [2], Prop. I.4.2).

On en déduit V: le groupe symplectique etant connexe, qu'il existe isomorphisme

$$h : \bar{\sigma}_o \to \bar{\sigma}$$

qui, en chaque point de Γ, échange les structures symplectiques.

3) Munissons maintenant la variété $S^1 \times R^{2p}$ d'une métrique riemannienne quelconque, et soit exp l'application exponentielle correspondante.

La restriction

$\phi_o : \bar{\sigma}_o \to S^1 \times R^{2p}$ (resp. $\phi : \bar{\sigma} \to S^1 \times R^{2p}$)

de exp à $\bar{\sigma}_o$ (resp. $\bar{\sigma}$), où $\bar{\sigma}_o$ et $\bar{\sigma}$ désignent toujours les fibrés de base Γ, est un difféomorphisme d'un voisinage de la section nulle de $\bar{\sigma}_o$ (resp. $\bar{\sigma}$) sur un voisinage de Γ.

Alors $\quad \psi = \phi \circ h \circ \phi_o^{-1}$

définit un difféomorphisme local laissant fixe chaque point de Γ.

Soit $m \in \Gamma$; l'espace tangent T_m à $S^1 \times R^{2p}$ en m s'écrit

$$T_m = D_m \oplus \bar{\sigma}_o(m) \quad \text{d'une part}$$

$$T_m = D_m \oplus \bar{\sigma}(m) \quad \text{d'autre part}$$

où D_m désigne la tangente en m à Γ.

On vérifie immédiatement que la dérivée $T\psi(m)$ de ψ en m a pour expression

$$T\psi(m) = 1_{D_m} \oplus h \quad .$$

Considérons alors la structure de contact

$$\sigma_1 = \psi^* \sigma \quad .$$

Elle admet pour relèvement la forme $\omega_1 = \psi^* \omega$, et il résulte des propriétés de ω , $d\omega$, h et $T\psi$ aux points de Γ que :

$$\omega_1 = \omega_o$$

$$\text{et} \qquad d\omega_1 = d\omega_o$$

en tout point de Γ.

4) Considérons enfin la famille à un paramètre

$$\omega_t = \omega_o + t(\omega_1 - \omega_o) \qquad t \in [0,1] \quad .$$

En chaque point $m \in \Gamma$, on a, d'après 3) :

$$\omega_t(m) = \omega_o(m)$$

$$d\omega_t(m) = d\omega_o(m) \quad .$$

Il en résulte que la famille d'équations de Pfaff σ_t définie par $\omega_t = 0$ est une famille de structures de contact sur un voisinage de Γ , et que de plus

$$\dot{\sigma}_t = \frac{\partial \sigma_t}{\partial t} = 0 \text{ sur } \Gamma \text{, puisque } \frac{\partial \omega_t}{\partial t} \text{ y est nul.}$$

En appliquant la Proposition 2.1, on obtient donc une famille à un paramètre de champs de vecteurs X_t définis sur un voisinage de Γ dans $S^1 \times R^{2p}$, tels que :

$$X_t(m) = 0 \text{ pour tout } m \in \Gamma \text{ et tout } t \in [0,1]$$

$$\theta(X_t) \cdot \sigma_t = \dot{\sigma}_t \quad .$$

Par intégration de l'équation différentielle $\frac{dx}{dt} = X_t(x)$, on obtient une famille à un paramètre de germes de difféomorphismes ϕ_t laissant fixes les points de Γ , tels que, d'après le Lemme 1.3 :

$$\phi_t^* \sigma_o = \sigma_t \quad .$$

Ainsi, en posant $\phi = \psi \circ \phi_1$

on a $\phi^* \sigma = \sigma_o$

et la proposition est démontrée.

<u>Remarque</u>. La classification des germes de structures de contact transverses à Γ à <u>isomorphisme positif</u> près (i.e. via des difféomorphismes à jacobien positif, c'est à dire conservant l'orientation) est la suivante :

a) Si p est pair, tout tel germe est positivement isomorphe à σ_o .

b) Si p est impair, on a deux classes d'isomorphisme, représentées par les structures :

$$d\theta \pm \sum_{i=1}^{p} (x_i dy_i - y_i dx_i) = 0 .$$

4. <u>Formes de contact remarquables sur</u> $S^1 \times D^2$

4.1 Soit D^2 le disque unité dans le plan R^2 .

Soient $T = S^1 \times D^2 \subset S^1 \times R^2$, et ∂T le bord du tore plein T.

Soit ω (resp. $\tilde{\omega}$) une forme différentielle définie sur un voisinage de ∂T (resp. de T); on dira que $\tilde{\omega}$ est un <u>prolongement</u> de ω si ω et $\tilde{\omega}$ coïncident sur un voisinage de ∂T.

On se propose d'établir un critère pour qu'une forme ω <u>de contact</u> sur un voisinage de ∂T soit prolongeable en une forme $\tilde{\omega}$ <u>de contact</u> sur T.

Soit X le champ de vecteurs unitaire sur $S^1 \times R^2$, tangent en chaque point (a, b) à la courbe $S^1 \times \{b\}$.

Soit ω une 1-forme différentielle sur un ouvert de $S^1 \times R^2$; la forme ω est dite <u>invariante</u> si $\theta(X)\omega = 0$.

Cela signifie que ω est invariante par le groupe des <u>rotations</u>

$(\theta, x, y) \mapsto (\theta + \alpha, x, y)$ de $S^1 \times R^2$; en d'autres termes, si $d\theta$ designe la forme fondamentale sur S^1, une forme invariante s'exprime de la manière suivante :

$$\omega = \eta + f \cdot d\theta$$

où η est une forme differentielle sur un ouvert Ω de R^2, et f une fonction à valeurs numériques sur Ω.

En particulier, si ω est une forme invariante sur un voisinage de ∂T, η (resp. f) est une forme différentielle (resp. une fonction) définie sur un voisinage du cercle $C = \partial D^2$ dans R^2.

<u>Proposition</u>. <u>Soit</u> ω <u>une forme différentielle invariante sur un voisinage de</u> ∂T, <u>telle que</u> f <u>ne s'annule pas sur</u> $C = \partial D^2$. <u>Alors, si</u> ω <u>est de contact, elle admet un prolongement</u> $\tilde{\omega}$ <u>sur</u> T, <u>invariant et de contact</u>.

4.2 Pour démontrer cette proposition, (essentiellement due à R. Lutz) nous utiliserons le resultat auxiliaire suivant:

Soit $\Delta \subset R^2$ un compact, connexe, dont la frontière est réunion disjointe d'un nombre fini de courbes simples fermées sans points doubles γ_i, $i = 1, \ldots, p$.

Le domaine Δ est orienté par la 2-forme $\Omega = dx \wedge dy$ (x, y coordonnées dans R^2); $\partial\Delta = \sum_{i=1}^{p} \gamma_i$ désignera le bord <u>orienté</u> de Δ.

Soit maintenant μ une 1-forme différentielle, définie sur un voisinage de $\partial\Delta$ dans R^2, telle que, si $d\mu = h \cdot dx \wedge dy$, on ait $h > 0$ sur $\partial\Delta$.

<u>Lemme</u>. <u>Les conditions suivantes sont équivalentes</u> :

a) $\int_{\partial\Delta} \mu > 0$.

b) <u>Il existe une forme</u> $\tilde{\mu}$ <u>définie sur un voisinage de</u> Δ, <u>prolongeant</u> μ (i.e. $\tilde{\mu} = \mu$ <u>sur un voisinage de</u> $\partial\Delta$), <u>telle que, si</u> $d\tilde{\mu} = \tilde{h} \cdot dx \wedge dy$, <u>on ait</u> $\tilde{h} > 0$ <u>sur</u> Δ.

<u>Démonstration</u>.

b => a. Par la formule de Stokes, on a

$$\int_{\partial\Delta} \mu = \int_{\partial\Delta} \tilde{\mu} = \int_{\Delta} d\tilde{\mu} > 0 .$$

a => b On montre facilement l'existence d'une fonction $\bar{h}$, C^∞ et strictement positive sur Δ , telle que :

1. $\bar{h} = h$ sur un voisinage de $\partial\Delta$.

2. $\int_\Delta \bar{h}\, dx \wedge dy = \int_{\partial\Delta} \mu$.

Alors, la forme différentielle $\beta = \bar{h} . dx \wedge dy$ est exacte, car $H^2(\Delta, R) = 0$. Il existe donc, d'après le Théorème de De Rham et l'égalité 2, une 1-forme différentielle α sur Δ telle que :

3. $d\alpha = \beta$

4. $\int_{\gamma_i} \alpha = \int_{\gamma_i} \mu$ pour tout $i = 1, \ldots, p$.

D'après 1 et 3, on a

$$d(\mu - \alpha) = d\mu - \beta = 0 \text{ au voisinage de } \partial\Delta .$$

D'après 4, $\mu - \alpha$ est exacte sur un voisinage de γ_i , pour tout i ; on a donc

$$\mu = \alpha + df_i$$

où f_i est une fonction définie sur un voisinage de γ_i .

Il existe alors une fonction f sur Δ telle que $f = f_i$ au voisinage de chaque γ_i . Il suffit alors de poser

$$\tilde{\mu} = \alpha + df$$

et le lemme est démontré .

4.3 Démonstration de la proposition 4.1

4.3.1. Considérons d'abord une forme invariante quelconque

$$\omega = \eta + f . d\theta .$$

Donc $\qquad d\omega = d\eta + df \wedge d\theta$

et $\qquad \omega \wedge d\omega = (\eta \wedge df + f . d\eta) \wedge d\theta$

puisque $\qquad \eta \wedge d\eta = 0$, η étant une forme différentielle dans le plan .

La forme ω est donc de contact si et seulement si :

(1) $\eta \wedge df + f . d\eta \neq 0$ en tout point.

En un point où $f = 0$, (1) équivaut à

$$\eta \wedge df \neq 0$$

Ceci implique $df \neq 0$, donc l'ensemble des zéros de f est une <u>courbe</u> ; de plus, la restricition de η à cette courbe est sans zéro; en particulier, si f est nulle sur une courbe fermée γ, on a $\int_\gamma \eta \neq 0$.

Remarquons d'autre part que, si $f \neq 0$, on a l'identité

$$(2) \qquad \eta \wedge df + f \cdot d\eta = f^2 \cdot d\left(\frac{\eta}{f}\right) .$$

4.3.2 Revenons maintenant à la forme ω donnée au voisinage de ∂T dans $S^1 \times R^2$, invariante et de contact.

On peut supposer $f > 0$ sur C, quitte à changer ω en $-\omega$. Par hypothèse, et compte tenu de (1) et (2), on a

$$\eta \wedge df + f \cdot d\eta = f^2 \cdot d\left(\frac{\eta}{f}\right) \neq 0 \text{ en tout point de } C.$$

Soient (x, y) des coordonnées dans R^2 telles que

$$d\left(\frac{\eta}{f}\right) = \frac{1}{f^2}(\eta \wedge df + f \cdot d\eta) = h \cdot dx \wedge dy \text{ avec } h > 0 \text{ sur } C .$$

Le plan R^2 est alors orienté par la forme $dx \wedge dy$.

Deux cas sont à considérer :

a) $\int_C \frac{\eta}{f} > 0$ où $C = \partial D^2$ est le bord <u>orienté</u> de D^2.

Dans ce cas, on prolonge f en une fonction $\tilde{f}$ <u>strictement</u> <u>positive</u> sur D^2. Puis, posant $\mu = \frac{\eta}{f}$, on remarque que $d\mu$ est positive sur un voisinage de C par hypothèse, et que $\int_{\partial D^2} \mu > 0$.

D'après le Lemme 4.2, il existe une forme $\tilde{\mu}$ sur D^2 prolongeant μ, telle que $d\tilde{\mu}$ est positive sur D^2. On pose $\tilde{\eta} = \tilde{f} \cdot \tilde{\mu}$; il est clair, compte tenu de (2), que la forme

$$\tilde{\omega} = \tilde{\eta} + \tilde{f} \cdot d\theta$$

répond à la question.

b) $\int_C \frac{\eta}{f} \leq 0$.

Considérons la forme

$$\omega_1 = \eta_1 + f_1 \cdot d\theta = -xdy + ydx + \left(r^2 - \frac{1}{4}\right)d\theta \quad \text{ou} \quad r^2 = x^2 + y^2 .$$

On a $d\omega_1 = -2dx \wedge dy + 2rdr \wedge d\theta$;

donc $\omega_1 \wedge d\omega_1 = (-r^2 + \frac{1}{2})\, dx\, dy\, d\theta$

Soit γ_ϵ le cercle de centre 0 , de rayon $\frac{1}{2} + \epsilon$ $(0 < \epsilon < \frac{1}{\sqrt{2}})$ dans le plan (x, y). On vérifie immédiatement que

$$\int_{\gamma_\epsilon} \frac{\eta_1}{f_1} \to -\infty \text{ quand } \epsilon \to 0.$$

Fixons ϵ assez petit pour que

$$\int_C \frac{\eta}{f} - \int_{\gamma_\epsilon} \frac{\eta_1}{f_1} > 0 \quad .$$

Considérons alors la couronne Δ_ϵ située entre C et γ_ϵ. Comme f (resp. f_1) est strictement positive sur C (resp. sur γ_ϵ), il existe une fonction $\bar{f}$ sur Δ_ϵ , <u>strictement positive</u> et prolongeant f et f_1. On pose

$$\mu = \frac{\eta}{\bar{f}} \quad \text{au voisinage de } C$$

$$= \frac{\eta_1}{\bar{f}} \quad \text{"} \quad \text{"} \quad \gamma_\epsilon .$$

On a par hypothèse $d\mu > 0$ au voisinage de $\partial\Delta_\epsilon$, et $\int_{\partial\Delta_\epsilon} \mu > 0$. D'après le Lemme 4.2, il existe une forme $\tilde{\mu}$, telle que $d\tilde{\mu} > 0$ sur Δ_ϵ , et prolongeant μ. La forme différentielle

$$\tilde{\omega} = \bar{f}.\tilde{\mu} + \bar{f}\, d\theta \qquad \text{sur } S^1 \times \Delta_\epsilon$$

$$= \omega_1 \qquad \text{sur } S^1 \times (D^2 - \Delta_\epsilon)$$

répond alors à la question, et la proposition 4.1 est démontrée.

<u>Remarque</u>. La Proposition 4.1 represente un cas particulier d'un théorème utilisé par Lutz pour établir les résultats de [5] .

4.4 Considérons la variété à bord

$N = S^1 \times S^1 \times [1,+\infty[= \{(e^{i\theta}, e^{i\theta'}, r);\ \theta, \theta' \in \mathbb{R};\ 1 \leq r < +\infty\}$.

On se donne sur N la structure de contact définie par l' équation

$$\omega_0 = d\theta + r^2 d\theta' = 0 .$$

Remarquons qu'il s'agit de l'expression de la forme "canonique" $\omega_0 = d\theta + xdy - ydx$ introduite en 3.1, lorsqu'on utilise les coordonnées polaires définies par $x + iy = r.e^{i\theta'}$.

Soit maintenant

$$\bar{\phi}_A : S^1 \times S^1 \to S^1 \times S^1$$

un <u>automorphisme</u> <u>unimodulaire</u> du tore à deux dimensions, défini par

$$\theta = a\,\bar{\theta} + b\,\bar{\theta}'$$

$$\theta' = c\,\bar{\theta} + d\,\bar{\theta}'$$

où la matrice $A = \begin{pmatrix} a & b \\ c & d \end{pmatrix}$ est une matrice unimodulaire, c'est à dire que a, b, c, d $\in \mathbb{Z}$ et $ad-bc = \pm 1$.

Considérons enfin la variété $T \underset{\bar{\phi}_A}{\cup} N$ obtenue en recollant le tore plein $T = S^1 \times D^2$ et la variété N le long des bords, via le difféomorphisme

$$\bar{\phi}_A : \partial T = S^1 \times S^1 \to \partial N = S^1 \times S^1$$

<u>Proposition:</u> <u>Il existe sur</u> $T \underset{\bar{\phi}_A}{\cup} N$ <u>une forme de contact</u> $\tilde{\omega}_o$ <u>égale à</u> ω_o <u>sur</u> N.

<u>Démonstration.</u> C'est un corollaire de la Proposition 4.1. Il est clair en effet que la forme

$$\omega_A = d(a\,\bar{\theta} + b\,\bar{\theta}') + r^2 d(c\,\bar{\theta} + d\,\bar{\theta}')$$

définie et de contact sur

$$\dot{T} = S^1 \times (D^2 - \{0\}) = \{(e^{i\bar{\theta}}, \bar{x} + i\bar{y} = r\,e^{i\bar{\theta}'}),\ \bar{\theta}, \bar{\theta}' \in \mathbb{R},\ 0 < r \leq 1\}$$

prolonge la forme ω_o donnée sur N.

Mais $\omega_A = (b + dr^2)\,d\bar{\theta}' + (a + cr^2)\,d\bar{\theta} = \eta + f\,.\,d\bar{\theta}$ est <u>invariante</u> par les rotations en $\bar{\theta}$. De plus, le polynôme $f = a + cr^2$ n'est pas identiquement nul; soit donc r_o tel que $0 < r_o < 1$ et $a + cr_0^2 \neq 0$; le résultat est établi en appliquant la proposition 4.1 à ω_A et au tore plein $S^1 \times D^2_{r_o}$, ou $D^2_{r_o}$ désigne le disque de centre 0 et de rayon r_o.

5. <u>Théorème.</u> <u>Soit</u> M <u>une variété connexe, compacte, orientable de dimension</u> 3. <u>Il existe sur</u> M <u>une forme de contact</u>.

5.1 On va d'abord reformuler un résultat de Lickorish [6].

Soit toujours $T = S^1 \times D^2$ le tore plein à trois dimensions. Un <u>grand</u> <u>cercle</u> $\gamma \subset T$ est, par définition, le graphe dans T d'une application de S^1 dans

$\mathring{B}^2$ = intérieur de D^2 ; ainsi, tout voisinage tubulaire de γ est difféomorphe au tore plein.

Soit $\Gamma = (\gamma_i)$, $i = 1, \ldots, p$, une famille finie de grands cercles disjoints dans T, et soit $\mathfrak{T} = (T_i)$, $i = 1, \ldots, p$, une famille de voisinages tubulaires compacts des γ_i , telle que $T_i \cap T_j = \emptyset$ si $i \neq j$. Alors $T - \bigcup_{i=1}^{p} \mathring{T}_i$ (où $\mathring{T}_i$ = intérieur de T_i) est une variété à bord, et

$$\partial(T - \bigcup_{i=1}^{p} \mathring{T}_i) = (\bigcup_{i=1}^{p} \partial T_i) \cup \partial T$$

est réunion disjointe de $p + 1$ tores à deux dimensions.

Soient enfin $\tilde{T}_0, \tilde{T}_1, \ldots, \tilde{T}_p$ $p + 1$ exemplaires de T. Soient

$$\phi_o : \partial \tilde{T}_0 \to \partial T, \quad \phi_i : \partial \tilde{T}_i \to \partial T_i \quad (i = 1, \ldots, p)$$

des difféomorphismes; on désignera par $\emptyset$ la famille $\phi_o, \ldots, \phi_p$.

Posons $$M_{T, \mathfrak{T}, \emptyset} = (T - \bigcup_{i=1}^{p} \mathring{T}_i) \underset{\phi_0}{\cup} \tilde{T}_0 \underset{\phi_1}{\cup} \tilde{T}_1 \cup \ldots \underset{\phi_p}{\cup} \tilde{T}_p .$$

Alors $M_{T, \mathfrak{T}, \emptyset}$ est une variété compacte, connexe et orientable de dimension 3.

Théorème (Lickorish [6]). Pour toute variété M compacte, connexe et orientable de dimension 3, il existe un triplet $\Gamma, \mathfrak{T}, \emptyset$ tel que M soit difféomorphe à $M_{T, \mathfrak{T}, \emptyset}$.

Remarques

1. La famille Γ étant fixée, on peut choisir les voisinages tubulaires T_i aussi petits que l'on veut, dans le sens suivant : étant donnés un couple $(\mathfrak{T}, \emptyset)$ quelconque, associé à Γ , et des voisinages U_i des cercles γ_i , il existe un couple $(\mathfrak{T}', \emptyset')$ avec $T_i' \subset U_i$ pour tout i , tel que $M_{T, \mathfrak{T}', \emptyset'}$ soit difféomorphe à $M_{T, \mathfrak{T}, \emptyset}$.

2. Soit V un C^1-voisinage de l'âme $S^1 \times \{0\}$ du tore plein T; j'entends par là un ensemble de grands cercles, graphes d'applications de S^1 dans D^2 appartenant à un C^1-voisinage donné de l'application nulle. On montre facilement que le théorème précédent reste vrai si l'on impose aux familles Γ d'être constituées de grands cercles appartenant à V .

5.2 Supposons maintenant les familles Γ et $\mathscr{C}$ fixées. Supposons de plus donnés des difféomorphismes

$$\psi_i : T_i \to T \quad \text{pour tout } i = 1, \ldots, p .$$

Pour toute famille ϕ, posons

$$\bar{\phi}_i = \psi_i \circ \phi_i : \partial \tilde{T}_i \to \partial T \qquad i = 1, \ldots, p$$

$$\bar{\phi}_o = \phi_o : \partial \tilde{T}_0 \to \partial T \quad .$$

La famille ϕ est bien déterminée par la famille $\bar{\phi} = \bar{\phi}_o, \ldots, \bar{\phi}_p$, qui est une famille d'automorphismes du tore à deux dimensions.

On sait que, si $\bar{\phi}$ et $\bar{\phi}'$ sont isotopes, c'est à dire que, pour tout $i = 0, 1, \ldots, p$, $\bar{\phi}_i$ et $\bar{\phi}'_i$ sont des automorphismes isotopes de $S^1 \times S^1$, alors $M_{\Gamma, \mathscr{C}, \phi}$ et $M_{\Gamma, \mathscr{C}, \phi'}$ sont difféomorphes.

Rappelons enfin que tout automorphisme $\bar{\phi}$ de $S^1 \times S^1$ est isotope à une transformation unimodulaire et une seule (pour la définition et les notations, voir 4.4); soit en effet A la matrice représentant l'automorphisme

$$\bar{\phi}_* : \pi_1 (S^1 \times S^1) \to \pi_1 (S^1 \times S^1)$$

induit par $\bar{\phi}$ sur le groupe fondamental du tore; on a $\pi_1(S^1 \times S^1) = \mathbb{Z}^2$, et A est une matrice unimodulaire; alors l'automorphisme $\bar{\phi}_A^{-1} \circ \phi$ est homotope à l'identité, donc lui est isotope (on sait en effet que, si S est une surface compacte orientable, la composante connexe de l'identité dans Diff(S) est constituée des difféomorphismes homotopes à l'identité : voir par exemple [7]) .

5.3 <u>Démonstration du théorème 5</u>

1) Considérons sur le tore plein $T = S^1 \times D^2$ la structure de contact "canonique" σ_o (cf 3.1) définie par

$$\omega_o = d\theta + xdy - ydx = 0$$

Comme σ_o est transverse à l'âme $S^1 \times \{0\}$ de T , il existe un C^1-voisinage V de $S^1 \times \{0\}$ tel que σ_o est transverse à tout grand cercle appartenant à V.

2) Il existe alors des familles Γ, $\mathscr{C}$, ϕ et ψ telles que

a. M est difféomorphe à $M_{\Gamma, \mathscr{C}, \phi}$ (d'après le théorème 5.1)

b. Les grands cercles γ_i appartiennent à V (5.1 Remarque 2)

c. $\psi_i : T_i \to T$, $i = 1, \ldots, p$, est un difféomorphisme tel que $\psi_i^* \sigma_o = \sigma_{o,i}$, où $\sigma_{o,i}$ désigne la restriction de σ_o à T_i. Ceci se déduit de la Remarque 1 de 5.1 et de la Proposition 3.2, compte tenu de b.

d. Les difféomorphismes $\bar{\phi}_i$, $i = 0, 1, \ldots, p$, sont des automorphismes unimodulaires du tore $S^1 \times S^1$ (d'après 5.2).

3) Considérons maintenant

$$M \simeq M_{T, \tau, \phi} = (T - \cup \overset{\circ}{T}_i) \underset{\bar{\phi}_o}{\cup} \tilde{T}_0 \underset{\bar{\phi}_1}{\cup} \tilde{T}_1 \ldots \underset{\bar{\phi}_p}{\cup} \tilde{T}_p .$$

D'après c., d., et la Proposition 4.4, la structure de contact σ_o sur $T - \cup \overset{\circ}{T}_i$ se prolonge en une structure de contact sur chaque tore plein T_i, et nous avons ainsi construit une structure de contact σ sur M.

4) On vérifie facilement que la structure de contact σ ainsi construite est <u>transversalement orientable</u> ; elle se laisse donc définir globalement par une forme de contact ω sur M (voir 1.4) et le théorème est démontré.

5.4 <u>Remarques</u>

1. On sait, d'après un théorème classique de Haefliger, qu'il ne peut en général exister de forme différentielle <u>complètement intégrable</u> (i.e. $\omega \wedge d\omega = 0$ en tout point) et <u>analytique</u> sur une variété analytique réelle compacte de dimension 3.

Par contre, l'existence d'une forme de contact analytique est évidente : nous venons de construire une forme de contact ω au moins C^∞ ; les formes analytiques étant denses dans l'ensemble des formes C^∞ (muni de la C^1-topologie), il existe une forme analytique assez voisine de ω pour être encore de contact.

2. Si l'on effectue la construction indiquée en 5.3 à partir de la forme

$$\omega_o' = d\theta - xdy + ydx$$

on obtient sur M une forme de contact ω' telle que $\omega' \wedge d\omega'$ définisse l'orientation opposée a $\omega \wedge d\omega$ (voir 3.2 Remarque).

BIBLIOGRAPHIE

[1] S.S. Chern, The geometry of G-structures. Bull. Amer. Math. Soc., 72, 1966, 167-219.

[2] J. Martinet, Sur les singularités des formes différentielles extérieures. Ann. Inst. Fourier, Grenoble, 20, 1970, 95-178.

[3] E. Cartan, Les systèmes différentiels extérieurs et leurs applications géométriques. Hermann, Paris, 1945.

[4] J.W. Gray, Some global properties of contact structures. Ann. Math. 69, 1959, 421-450.

[5] R. Lutz, Sur l'existence de certaines formes différentielles remarquables sur la sphere S^3 . C.R.A.S., Paris.

[6] W.B. Lickorish, A foliation for 3-manifolds. Ann. Math. 82, 1965, 414-420.

[7] C.J. Earle and J. Eells, A fibre bundle description of Teichmüller theory. J. Diff. Geom., 3, 1969, 19-43.

[8] M.L. Gromov, Stable maps of foliations in manifolds. Izv. Akad. Nauk. S.S.S.R. 33 (1969) 707-734.

Algebraic Invariants for Pseudo-Isotopies*

J.B. Wagoner

We shall discuss how some groups arising in algebraic K -Theory are related to pseudo-isotopies of non-simply connected smooth manifolds. Our methods are an extension of the techniques used by Cerf in [1] to show that "pseudo-isotopy implies isotopy" in the simply connected case: we study one parameter families of "gradient-like" vector fields associated to certain one parameter families of real valued functions. This paper is semi-expository in nature. A more detailed account will appear in the future.

§1. Results and questions

Let M^n be a smooth n-manifold with possibly ∂M not empty. If $f, g \in \mathrm{Diff}(M, \partial M)$, a pseudo-isotopy from f to g is a diffeomorphism H of $(M \times I; \partial M \times I, M \times 0, M \times 1)$ to itself such that $H|M \times 0 = f$, $H|M \times 1 = g$, and $H|\partial M \times I$ is an isotopy. The pseudo-isotopy H will also be called a "pseudo-isotopy of f". Let $\mathcal{S}$ denote the group (under composition) of all pseudo-isotopies between pairs of diffeomorphisms of $(M, \partial M)$. For a fixed $f \in \mathrm{Diff}(M,\partial M)$, let $\mathcal{S}_f$ be the subset of $\mathcal{S}$ consisting of pseudo-isotopies of f. Note that $\mathcal{S}_{id}$ is a normal subgroup of $\mathcal{S}$. Let $\alpha : \mathrm{Diff}(M, \partial M) \to \mathcal{S}$ be the homomorphism which takes f to the constant isotopy $\alpha_f(x, t) = (f(x), t)$. Then $\mathcal{S}$ is the semi-direct product $\mathcal{S}_{id} \times_\alpha \mathrm{Diff}$. The pseudo-isotopy problem is to compute $\pi_0(\mathcal{S}_{id})$. Using the fact that pseudo-isotopy implies isotopy in codimensions greater than two it is possible to show by a direct geometric argument that $\pi_0(\mathcal{S}_{id})$ is abelian when $n \geq 6$.

Recall the definition of the algebraic K -theory functor K_2 given in [4].

---

* Thanks to Liverpool Symposium on Singularities and to I.H.E.S. for partial support of this paper.

Let Λ be any associative ring with unit. Let $3 \leq \ell \leq \infty$ and let F_Λ be the free group generated by symbols x_{ij}^λ where $i \neq j$, $1 \leq i, j \leq \ell$, and $\lambda \in \Lambda$. Let $N_\Lambda \subset F_\Lambda$ be the smallest normal subgroup generated by the

<u>Steinberg relations</u>

(i) $x_{ij}^\lambda \cdot x_{ij}^\mu = x_{ij}^{\lambda+\mu}$

(ii) $[x_{ij}^\lambda, x_{k\ell}^\mu] = 1$ for $i \neq \ell$, $j \neq k$

(iii) $[x_{ij}^\lambda, x_{jk}^\mu] = x_{ik}^{\lambda\mu}$ for $i \neq k$.

Sometimes x_{ij}^λ will be denoted by $x_{ij}(\lambda)$. Define the <u>Steinberg group</u> as $St(\ell, \Lambda) = F_\Lambda/N_\Lambda$. We will usually write $St(\Lambda)$ for $St(\infty, \Lambda)$. Let $E\ell(\Lambda) \subset GL(\Lambda)$ be the subgroup generated by the elementary matrices $e_{ij}^\lambda (i \neq j)$, where e_{ij}^λ is the identity along the diagonal and has exactly one possibly non-zero off diagonal entry - namely λ in the (i, j) spot. The correspondence $x_{ij}^\lambda \to e_{ij}^\lambda$ defines a homomorphism $\pi : St(\Lambda) \to E\ell(\Lambda)$ and by definition

$$K_2(\Lambda) = \ker \pi .$$

Now let $W(\pm \pi_1 M) \subset St(Z[\pi_1 M])$ denote the subgroup generated by the words $w_{ij}(\pm g)$ where $w_{ij}(\pm g) = x_{ij}(\pm g) \cdot x_{ji}(\mp g^{-1}) \cdot x_{ij}(\pm g)$ for $g \in \pi_1 M$. Let $W_0(\pm \pi_1 M) = W(\pm \pi_1 M) \cap K_2(Z[\pi_1 M])$. Let $\mathfrak{D} \subset \pi_0(\mathcal{P}_{id})$ denote the "unicity of death" subgroup (cf. §5). Our main result is

<u>Theorem A.</u> $(n \geq 8)$ <u>There is a surjective homomorphism</u>

$$\rho : K_2(Z[\pi_1 M]) \to \pi_0(\mathcal{P}_{id}) \bmod \mathfrak{D}$$

<u>such that</u> $W_0(\pm \pi_1 M) \subset \ker \rho$.

<u>Conjecture</u>. $\ker \rho = W_0(\pm \pi_1 M)$.

Since $\mathfrak{D} = 0$ when $\pi_1 M = 0$ (cf. [1]) and $K_2(Z) \cong Z_2$ has $(x_{12}^1 x_{21}^{-1} x_{12}^1)^4$ as a generator, the theorem implies

<u>Corollary</u> (Cerf [1]) $\pi_0(\mathcal{P}_{id}) = 0$ <u>whenever</u> $n \geq 8$ <u>and</u> $\pi_1 M = 0$.

The corollary is actually true whenever $n \geq 5$ (cf. [1]). The condition "$n \geq 8$" in the theorem can probably be improved to "$n \geq 5$".

It was originally announced by the author that $\mathcal{D} = 0$ for arbitrary $\pi_1 M$. However, the argument was mistaken and I wish to thank A. Chenciner for several discussions which led to discovering the error. As of now the best known estimate is that $\mathcal{D}$ is some quotient group of $(Z_2 \times \pi_2 M)^{\pi_1 M - 1}$, the group of functions from $\pi_1 M - 1$ into $Z_2 \times \pi_2 M$ with finite support (cf. [2]).

Non-trivial pseudo-isotopies were first found by L. Siebenmann on manifolds M which can be written as $M = W \times S^1$ for some smooth manifold W with $Wh(\pi_1 W) \neq 0$. Here "Wh" denotes the Whitehead group (see [5]). In fact for $n \geq 6$ there is a surjective homomorphism $\pi_0(\mathcal{P}_{id}) \to Wh(\pi_1 W)$ (see [7]). This homomorphism probably contains $\mathcal{D}$ in its kernel. If so, the groups $Wh(\pi_1 W)$ and $K_2(Z[\pi_1 W \times Z])$ are lower and upper bounds for $\pi_0(\mathcal{P}_{id})$ mod $\mathcal{D}$ when $M = W \times S^1$. A closely related algebraic fact is that for any associative ring with unit Λ there is a natural direct sum decomposition $K_2(\Lambda[t, t^{-1}]) = K_2(\Lambda) \oplus K_1(\Lambda) \oplus \{?\}$. (see [8]). Computing the unknown summand is probably closely connected to finding the obstruction for the following codimension one pseudo-isotopy problem: given a pseudo-isotopy $Q : W \times S^1 \times I \to W \times S^1 \times I$, when is Q isotopic to a pseudo-isotopy Q' such that $Q'(W \times 1 \times I) = W \times 1 \times I$? The summand $\{?\}$ is not even known when $\Lambda = Z$. Another interesting codimension one problem is when M is a connected sum $M = A \# B$ along $\partial A = \partial B = S^{n-1}$ and the codimension one submanifold is S^{n-1}. This is related to computing $K_2(Z[\pi_1 A * \pi_1 B])$ in terms of $K_2(Z[\pi_1 A])$ and $K_2(Z[\pi_1 B])$. For example, is $\bar{K}_2(Z[\pi_1 A * \pi_1 B]) = \bar{K}_2(Z[\pi_1 A]) \oplus \bar{K}_2(Z[\pi_1 B])$? Here $\bar{K}_2$ denotes K_2 modulo the image of $K_2(Z)$. What is $\bar{K}_2(Z[Z * Z])$?

Added in proof. The conjecture on ker ρ has recently been verified by the author and also by A. Hatcher of Stanford University, who has independently studied pseudo-isotopies and has obtained similar results. The present situation is the following :

For any group G , let $Wh_2(G)$ be defined by $Wh_2(G) = K_2(Z[G])/K_2(Z[G]) \cap W(\pm G)$.

Theorem B. ($n \geq 8$) *There is a homomorphism* $\sigma : \pi_0(\mathcal{P}_{id}) \to Wh_2(\pi_1 M)$ *such that* $\mathcal{D} \subset \ker \sigma$ *and such that the homomorphisms*

$$\rho : Wh_2(\pi_1 M) \to \pi_0(\mathcal{P}_{id}) \bmod \mathcal{D}$$

and

$$\sigma : \pi_0(\mathcal{P}_{id}) \bmod \mathcal{D} \rightarrow Wh_2(\pi_1 M)$$

<u>are inverses of one another.</u>

The condition "$n \geq 8$" can probably be improved to "$n \geq 5$". It has recently been shown by Milnor that $Wh_2(Z_{20})$ has at least five elements.

The σ-invariant satisfies product and duality formulae similar to those for the torsion invariant in Whitehead torsion theory.

<u>Questions</u>

1. Is $\pi_0(\mathcal{P}_{id}) \cong Wh_2(\pi_1 M) \oplus \mathcal{D}$?
2. What is $\mathcal{D}$?
3. For any group G is $K_2(Z[G]) \cap W(\pm G) = \operatorname{Im} h_*$, where $h_* : K_2(Z[F]) \rightarrow K_2(Z[G])$ is induced by any homomorphism $h : F \rightarrow G$ of a free group F onto G ? In particular is $Wh_2(F) = 0$?

§2. <u>Cerf's "functional approach" to pseudo-isotopies</u> (see [1])

Let $\mathcal{F}$ be the space of C^∞-functions $f : M \times I \rightarrow I$ such that $f(x, 0) = 0$ and $f(x, 1) = 1$ for all $x \in M$ and such that $f(x, t) = t$ for $x \in \partial M$. Let $\mathcal{E} \subset \mathcal{F}$ be the subset consisting of those functions with no critical points. Let $p : M \times I \rightarrow I$ be the standard projection. The correspondence $G \rightarrow p \circ G$ defines a fibration $\mathcal{P}_{id} \rightarrow \mathcal{E}$ with fibre $\mathcal{H}$ = isotopies of the identity diffeomorphism of $(M, \partial M)$. The path space $\mathcal{H}$ is contractible so $\mathcal{P}_{id} \rightarrow \mathcal{E}$ is a homotopy equivalence. In particular $\pi_0(\mathcal{P}_{id}) = \pi_0(\mathcal{E})$. Moreover $\mathcal{F}$ is contractible so $\pi_0(\mathcal{E}) = \pi_1(\mathcal{F}, \mathcal{E}; p)$. Hence to measure the obstruction to connecting a pseudo-isotopy G to the identity by a path in $\mathcal{P}_{id}$, join p to $p \circ G$ by a path f_t in $\mathcal{F}$ and then try to deform f_t keeping end points fixed to a path lying in $\mathcal{E}$.

For any smooth manifold V^{n+1} the space $\mathcal{F}$ of smooth real valued functions on V^{n+1} has a stratification $\mathcal{F} = \mathcal{F}^0 \cup \mathcal{F}^1 \cup \mathcal{F}^2 \cup \ldots \cup \mathcal{F}^\infty$, where $\mathcal{F}^k$ has codimension k (see [1]). Thus any path f_t in $\mathcal{F}$ can be approximated by a path lying in $\mathcal{F}^0 \cup \mathcal{F}^1$ and any deformation between paths can be approximated by a deformation contained in $\mathcal{F}^0 \cup \mathcal{F}^1 \cup \mathcal{F}^2$. The subset $\mathcal{F}^0$ consists of functions having only non-degenerate critical points (if any at all) and distinct critical values. If q is a critical point of $f \in \mathcal{F}^0$ having index i, there are local coordinates $x_1, \ldots, x_{n+1}$ in a neighbourhood of q with respect to which f has the form $f(x_1, \ldots, x_{n+1}) = f(q) - x_1^2 - \ldots - x_i^2 + x_{i+1}^2 + \ldots + x_{n+1}^2$. The subset $\mathcal{F}^1$ is the union of $\mathcal{F}^1_\alpha$ and $\mathcal{F}^1_\beta$. The set $\mathcal{F}^1_\alpha$ consists of those functions f satisfying

(a) f has distinct critical values, all the critical points of f are non-degenerate except one, say q, and in a neighbourhood of q there are local coordinates $x_1, \ldots, x_{n+1}$ with respect to which f has the form

(2.1) $$f(x_1, \ldots, x_{n+1}) = f(q) - x_1^2 - \ldots - x_i^2 + x_{i+1}^2 + \ldots + x_n^2 + x_{n+1}^3 .$$

By definition we set index $q = i$.

The set $\mathcal{F}^1_\beta$ consists of functions f satisfying

(b) all critical points of f are non-degenerate and all critical values are distinct except that $f(q_1) = f(q_2)$ for exactly one pair of critical points (q_1, q_2).

Any path with end points in $\mathcal{F}^0$ can be approximated (keeping end points fixed) by a "generic" path lying in $\mathcal{F}^0 \cup \mathcal{F}^1$ for which there will be at most finitely many times t where $f_t \in \mathcal{F}^1$ and such that if $f_a \in \mathcal{F}^1_\alpha$ there are local coordinates $x_1, \ldots, x_{n+1}$ around the degenerate critical point q of f_a with respect to which

$$f_t(x_1, \ldots, x_{n+1}) = f_a(q) - x_1^2 - \ldots - x_i^2 + x_{i+1}^2 + \ldots + x_n^2 \pm (t-a)x_{n+1} \pm x_{n+1}^3$$

for t close to a. When considered as a map from I into $\mathcal{F}$ a generic path is transverse to all the strata $\mathcal{F}^k$.

For any generic path f_t, $a \leq t \leq b$, the <u>graphic</u> of f_t is the subset of $[a,b] \times I$ consisting of all pairs (t, u) such that u is a critical value of f_t.

Thus the two basic changes which occur in the critical set as the time t increases are given by the graphics

I (death) I (birth) II

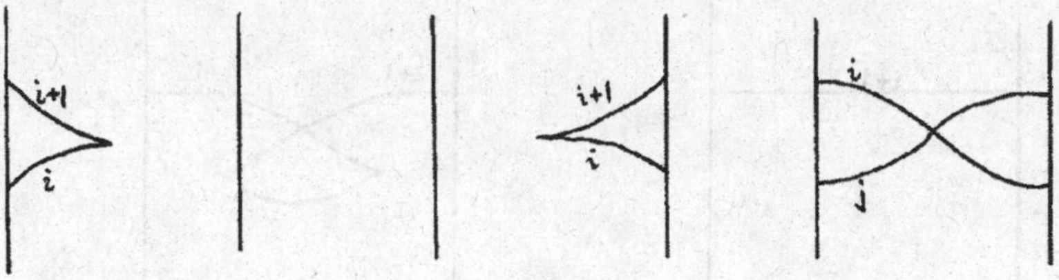

The indices i, i + 1, j etc. indicate that the corresponding lines in the graphic consist of images of critical points of index i, i + 1, j etc. The type I paths are <u>death</u> and <u>birth</u> paths respectively and occur when f_t hits $\mathcal{F}^1_\alpha$. Type II is a <u>crossing</u> path and occurs when the path hits $\mathcal{F}^1_\beta$. A typical graphic for a path between two functions in $\mathcal{E}$ is

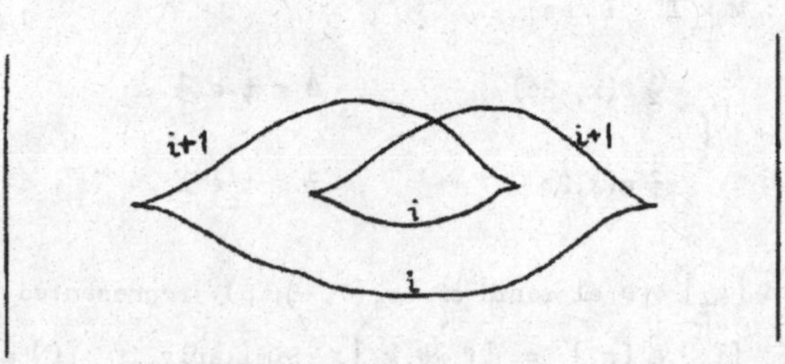

The strategy for deforming a path f_t into $\mathcal{E}$ is to deform f_t a step at a time successively simplifying the graphic until it disappears (cf. §5).

For a complete description of $\mathcal{F}^2$ see [1]. An important example of a function in $\mathcal{F}^2$ is $h_{0,0}$ which occurs as the organising center in the swallow's tail catastrophe given by the two parameter family $h_{t,s} : R^{n+1} \to R$, where

$$h_{t,s}(x_1,\ldots, x_{n+1}) = -x_1^2 - \ldots - x_i^2 + x_{i+1}^2 + \ldots + x_n^2 - (tx_{n+1} + sx_{n+1}^2 + x_{n+1}^4). \tag{2.2}$$

As s goes from positive to negative the transformation in the graphic is

(2.3)

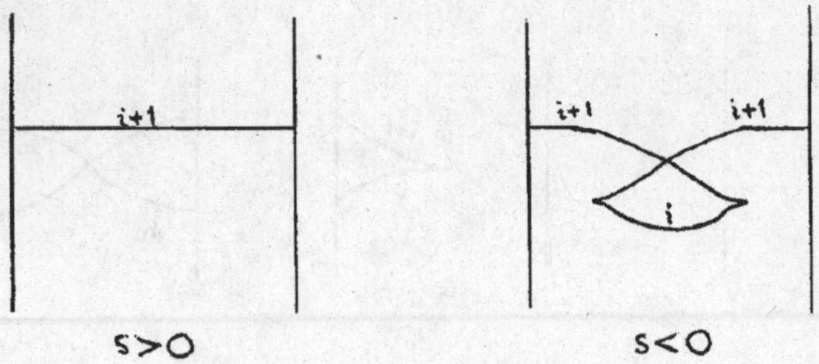

The bijection of $\pi_0(\mathcal{S}_{id})$ with $\pi_0(\mathcal{E})$ and with $\pi_1(\mathcal{F}, \mathcal{E}; p)$ induces a group structure on the latter sets. This can be done directly as follows: Let f and g be in $\mathcal{F}$. Deform f and g slightly so that for a small positive number ϵ they agree with p on $M \times [0, \epsilon]$ and $M \times [1-\epsilon, 1]$. Then define $f \# g : M \times I \to I$ as

$$f \# g\,(x,t) = \begin{cases} \frac{1}{2} f(x, 2t) & , \quad 0 \leq t \leq \frac{1}{2} \\ \frac{1}{2} g(x, 2t-1) + \frac{1}{2} & , \quad \frac{1}{2} \leq t \leq 1 . \end{cases}$$

Now if $[f_t]$ and $[g_t]$ are elements of $\pi_1(\mathcal{F}, \mathcal{E}; p)$ represented by the paths f_t and g_t, define $[f_t] \oplus [g_t] = [f_t \# g_t]$. Similarly if $[f]$ and $[g]$ are in $\pi_0(\mathcal{E})$ let $[f] \oplus [g] = [f \# g]$. This operation $\oplus$ gives the desired group structure on $\pi_1(\mathcal{F}, \mathcal{E}; p)$ and $\pi_0(\mathcal{E})$.

§3. Families of gradient-like vector fields

We shall be interested in studying certain spaces consisting of triples (η, f, μ), where $f : V^{n+1} \to I$ is a C^∞ function on the smooth manifold V and η is a vector field on V that is gradient-like for f with respect to the Riemannian metric μ. The term "gradient-like" means that

a) $df(x)(\eta(x)) > 0$ whenever x is not a critical point of f, and

b) there is a neighbourhood U of each critical point p such that

$\eta(x) = \operatorname{grad} f(x)$ for $x \in U$.

Let $\mathcal{T}$ denote the space of such triples and, when $V = M \times I$, let $\hat{\mathcal{E}} \subset \mathcal{T}$ denote the subspace consisting of those triples (η, f, μ) where f has no critical points. Now fix a Riemannian metric μ_0 on V. Then the map $f \to (\operatorname{grad} f, f, \mu_0)$ induces a homotopy equivalence between the pairs $(\mathcal{F}, \mathcal{E})$ and $(\mathcal{T}, \hat{\mathcal{E}})$. In particular

$$\pi_1(\mathcal{F}, \mathcal{E}; p) \cong \pi_1(\mathcal{T}, \hat{\mathcal{E}}; \hat{p})$$

where $\hat{p} = (\operatorname{grad} p, p, \mu_0)$.

Now let $(\eta, f, \mu) \in \mathcal{T}$ and let $p \in V$ be an isolated (but possibly degenerate) critical point of f. Let ϕ_t be the one-parameter family of diffeomorphisms generated by η. Define the <u>stable trajectory set</u> $W(p)$ of p to be

$$W(p) = \{x \in V \mid \lim_{t \to \infty} \phi_t(x) = p\} .$$

Similarly define the <u>unstable trajectory set</u> $W^*(p)$ of p to be

$$W^*(p) = \{x \in V \mid \lim_{t \to -\infty} \phi_t(x) = p\} .$$

More generally suppose (η_z, f_z, μ_z) is a smooth k-parameter family in $\mathcal{T}$ such that the map $F : I^k \times V \to I^k \times I$ given by $F(z, x) = (z, f_z(x))$ is in general position. The singular set Σ_F of F will be stratified, and if S is a stratum of Σ_F define the <u>stable</u> and <u>unstable</u> <u>trajectory sets</u> of S by $W(S)$ = union of $W(p_z)$ and $W^*(S)$ = union of $W^*(p_z)$ for all $(z, p_z) \in S \subset I^k \times V$. Here $W(p_z) \subset z \times V$ denotes the stable set of p_z for the triple (η_z, f_z, μ_z), and similarly for $W^*(p_z)$. Thus $W(S)$ and $W^*(S)$ are subsets of $I^k \times V$.

<u>Example 1</u>. Each function f_z of the k-parameter family has only isolated and non-degenerate critical points. If p_z is a critical point of f_z with index i, then $W(p_z) \approx R^i$ and $W^*(p_z) \approx R^{n+1-i}$. Each stratum S of Σ_F is the graph of a smooth map of I^k into $I^k \times V$ and $W(S)$ is the union of the smooth family $W(p_z)$. Similarly for $W^*(S)$.

Example 2. Consider the one parameter family (grad f_t, f_t, μ) where $f_t : R^{n+1} \to R$ is defined by

$$f_t(x_1, \ldots, x_{n+1}) = - x_1^2 - \ldots - x_i^2 + x_{i+1}^2 + \ldots + tx_{n+1} + x_{n+1}^3$$

and μ is the standard metric on R^{n+1}. For $t < 0$ let $c_t = \sqrt{-t/3}$. When $t < 0$ f_t has exactly two non-degenerate critical points $a_t = (0, \ldots, 0, c_t)$ and $b_t = (0, \ldots, 0, -c_t)$ of index i and $i + 1$ respectively. For $t > 0$ f_t has no critical points, and f_0 has just $0 \in R^{n+1}$ as a critical point. Thus for $t < 0$

$$W(a_t) = R^i \times 0 \times \{c_t\} \subset R^i \times R^{n-i} \times R$$

$$W^*(a_t) = 0 \times R^{n-i} \times \{-c_t < x_{n+1}\}$$

$$W(b_t) = R^i \times 0 \times \{x_{n+1} < c_t\}$$

$$W^*(b_t) = 0 \times R^{n-i} \times \{-c_t\} .$$

For $t = 0$

$$W(0) = R^i \times 0 \times \{x_{n+1} \leq 0\}$$

$$W^*(0) = 0 \times R^{n-i} \times \{0 \leq x_{n+1}\} .$$

In particular $W(0)$ and $W^*(0)$ are "half-spaces".

Let $\epsilon > 0$ be a small positive number and choose a $d > 0$ so that for all $t \in [-\epsilon, \epsilon]$ the critical values of f_t are contained in $(-d, d)$. The corresponding graphic is

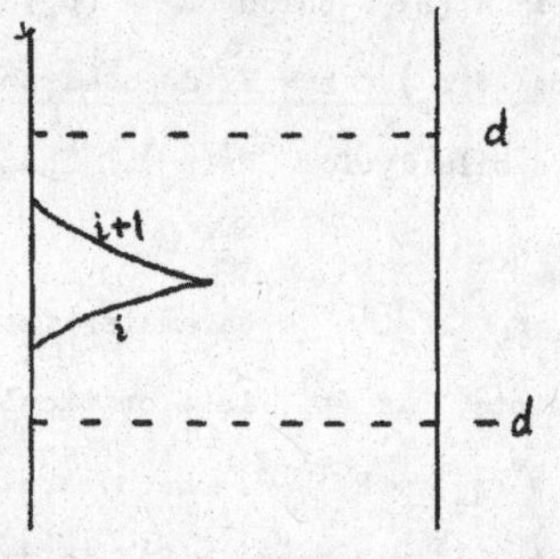

For $t < 0$ let

$$X_t = f_t^{-1}([-d,d]) \cap (W(a_t) \cup W^*(a_t) \cup W(b_t) \cup W^*(b_t)) .$$

Let $X_0 = f_0^{-1}([-d, d]) \cap (W(0) \cup W^*(0))$. For $t \leqslant 0$ let $X_t(\pm d) = f_t^{-1}(\pm d) \cap X_t$.
The following properties hold for $t \leqslant 0$.

a) X_t is contractible to a point.

b) $X_t(d)$ is an $(n - i)$-disc with boundary the $(n - i - 1)$-sphere $W^*(b_t) \cap f_t^{-1}(d)$.

c) $X_t(-d)$ is an i-disc with boundary the $(i - 1)$-sphere $W(a_t) \cap f_t^{-1}(-d)$.

For $t < 0$ we also have

d) the i-sphere $W(b_t) \cap f_t^{-1}(0)$ and the $(n-i)$-sphere $W^*(a_t) \cap f_t^{-1}(0)$ intersect each other transversely in exactly one point within the level surface $f_t^{-1}(0)$.

For $0 \leqslant k \leqslant 2$ any k-parameter family (η_z, f_z, μ_z) can be deformed into "general position" and will then have the following properties :

A) The map $F : I^k \times V \to I^k \times I$ is in general position. The singular set Σ_F is stratified with possibly three types of strata - the non-degenerate singular points of the f_z, the degenerate singular points of codimension one (cf. 2.1), and the degenerate singular points of codimension two (cf. 2.2).

B) For any stratum $S \subset \Sigma_F$, $W(S)$ and $W^*(S)$ are imbedded smooth fiber spaces with fiber either a euclidean space or a half-space. The imbedded fiber $W(p_z)$ or $W^*(p_z)$ over $(z, p_z) \in S$ is contained in $z \times V$.

C) For any two strata S_1 and S_2 of Σ_F, $W^*(S_1)$ and $W(S_2)$ are transverse.

If $C \subset I^k$ is a closed set and U is a neighbourhood of C in I^k such that the family (η_z, f_z, μ_z) restricted to U is in general position, then deformation of (η_z, f_z, μ_z) into general position (as a family over I^k) can be performed keeping things fixed over C.

For example $(\eta, f, \mu) \in \mathfrak{F}$ is in general position provided that all the critical points of f are non-degenerate and, for any two critical points p and q, $W(p)$ is transverse to $W^*(q)$. See [6].

To simplify notation we will omit the reference to the Riemannian metric and denote elements of $\mathcal{F}$ as pairs (η, f).

Let V be an $(n + 1)$ manifold with $\partial V = C^n \cup D^n$. Let $f : (V; C, D) \to (I; 0, 1)$ be a function in $\mathcal{F}^0 \cup \mathcal{F}^1$. We say that f is <u>ordered</u> provided index $p >$ index q implies $f(p) > f(q)$ for critical points p and q of f. Now suppose $f \in \mathcal{F}^0 \cup \mathcal{F}^1_\beta$ is ordered and let η be a gradient-like vector field for f. Choose paths from the critical points of f to a base point in V and choose orientations for the stable manifolds of the critical points. This data determines as in [5] a based chain complex $(C_*, \partial(\eta, f))$ where each chain group C_i is free over $Z[\pi_1 V]$ with one basis element for each critical point of index i. The homology of this complex is $H_*(\tilde{V}, \tilde{C})$ where $\theta : \tilde{V} \to V$ is the universal cover and $\tilde{C} = \theta^{-1}(C)$. If η is deformed to another gradient-like vector field for f the boundary matrices $\partial_i(\eta, f) : C_i \to C_{i-1}$ will in general change by multiplication by elementary matrices.

§4. <u>Geometry of the Steinberg group</u>

Let V be an $(n + 1)$-manifold with two boundary components C and D. Suppose that there is a function $g : (V; C, D) \to (I; 0, 1)$ satisfying

(*) g has exactly r critical points each of which is non-degenerate of index i.

The critical values of g need not be distinct. Let $\mathcal{A}$ be the space of pairs (η, f) where f satisfies (*) and η is gradient-like for f. Let $\mathcal{B} \subset \mathcal{A}$ be the subspace of pairs (η, f) such that η is in general position; that is, for any pair of critical points p and q of f we have

$$[W(p) \cup W^*(p)] \cap [W(q) \cup W^*(q)] = \phi .$$

<u>Theorem 4.1</u> <u>Let</u> $(\eta_0, f_0) \in \mathcal{B}$. <u>If</u> $3 \leq i \leq n - 3$, <u>there is a bijection</u>

$$\Delta : St(r, Z[\pi_1 V]) \to \pi_1(\mathcal{A}, \mathcal{B}; (\eta_0, f_0)).$$

The formula for Δ will be given in (4.6).

Consider any $(\xi, h) \in \mathcal{B}$. Let $p_1, \ldots, p_r$ be an ordering of the critical points of h. Assume that the p_j are connected to a base point by paths γ_j, and choose an orientation for each stable manifold $W(p_j)$. This data determines a basis $\epsilon_1, \ldots, \epsilon_r$ for $H_i(\tilde{V}, \tilde{C})$. Let p_α and p_β be two critical points of h with $h(p_\alpha) > h(p_\beta)$. Let $x^\lambda_{\alpha\beta}$ be a Steinberg generator with $\lambda \in Z[\pi_1 V]$.

<u>Lemma 4.2.</u> <u>Under the above conditions there is a well defined path</u> $x^\lambda_{\alpha\beta} \cdot (\xi, h) = (\xi_t, h_t)$ $(0 \leq t \leq 1)$ <u>such that</u> $h_t = h$ <u>for all</u> t, $\xi_0 = \xi$, $(\xi_1, h_1) \in \mathcal{B}$, <u>and such that the stable manifolds of</u> ξ_1 <u>determine the basis</u> $\epsilon_1, \ldots, \epsilon_\alpha + \lambda \cdot \epsilon_\beta, \ldots, \epsilon_\beta, \ldots \epsilon_r$ <u>of</u> $H_i(\tilde{V}, \tilde{C})$.

<u>Remark</u> The critical points of h_1 will be ordered and based and the stable manifolds oriented according to the following convention. Let (ξ_t, h_t) be any path in $\mathcal{A}$ with starting point (ξ_0, h_0). The critical set Σ_F of $F : I \times V \to I \times R$ where $F(t, v) = (t, h_t(v))$ consists of an ordered collection of arcs $a_1(t), \ldots, a_r(t)$ where $a_1(0) = p_1, \ldots, a_r(0) = p_r$. The jth critical point q_j of h_1 is defined by $q_j = a_j(1)$. The base path for q_j is the base path γ_j followed by the path $a_j(t)$. The orientation for $W(q_j)$ is the one which transports along $a_j(t)$ to the given orientation of $W(p_j)$. Also, if p is a critical point of h_t let $W_t(p)$ and $W^*_t(p)$ denote the stable and unstable manifolds for p determined by ξ_t.

<u>Proof of 4.2</u> This follows the handle addition theorem in the theory of s-cobordisms (cf. [1], [3]). For convenience we recall the argument when $\lambda = \pm g$ for $g \in \pi_1 V$. Choose a value c between $h(p_\alpha)$ and $h(p_\beta)$. Let $S^{i-1} = W(p_\alpha) \cap h^{-1}(c)$ and $S^{n-i} = W^*(p_\beta) \cap h^{-1}(c)$. Orient S^{i-1} as the boundary of $W(p_\alpha) \cap \{x | h^{-1}(x) \geq c\}$. The orientation of $W(p_\beta)$ determines an orientation of the normal bundle of S^{n-i} in $h^{-1}(c)$. The spheres S^{i-1} and S^{n-i} can be considered as based by γ_α and γ_β. Choose an arc γ in $h^{-1}(c)$ between a point of S^{i-1} and a point of S^{n-i} such that $g = \gamma_\alpha * \gamma * \gamma_\beta^{-1}$ and such that γ misses $(W(p_j) \cup W^*(p_j)) \cap h^{-1}(c)$ for $j \neq \alpha$ or β. There is an imbedding $\theta : S^{i-1} \times I \to h^{-1}(c)$, with image contained in a neighbourhood N of $S^{i-1} \cup \gamma$, and such that $\theta(S^{i-1} \times 0) = S^{i-1}$, $\theta(S^{i-1} \times I) \cap S^{n-i}$ is exactly one transverse point $\theta(x, a)$, $0 < a < 1$, and the product orientation of $\theta(S^{i-1} \times I)$ either agrees or disagrees with the orientation of the normal bundle of S^{n-i} depending on the choice of sign for $\pm g$. The imbedding ϕ can be realized by an isotopy of

$h^{-1}(c)$ which has support in N. This isotopy then determines the desired deformation (ξ_t, h_t) of (ξ_0, h_0) as in [3].

The path (ξ_t, h_t) is in general position. For $t \neq a$ we have $(\xi_t, h_t) \in \mathcal{B}$, but $W_a^*(p_\beta) \cap W_a(p_\alpha) \neq \phi$ while $W_a^*(p) \cap W_a(q) = \phi$ for any other pair of critical points. This situation is shown on the graphic by a vertical arrow as follows :

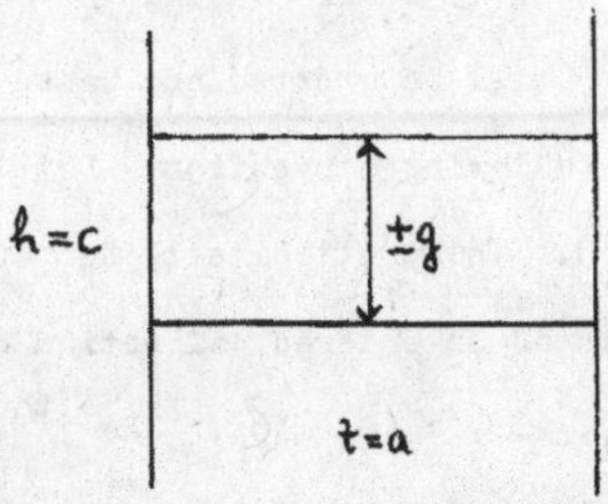

Such a path will be called an <u>elementary gradient crossing path</u> or <u>Type III path</u>.

The construction of $x_{\alpha\beta}^{\lambda} \cdot (\xi, h)$ for a general $\lambda \in Z[\pi_1 V]$ consists of putting a sequence of elementary gradient crossing paths together end to end. The Whitney process for elimination of intersection points together with (4.5) below is used to show that the class of $x_{\alpha\beta}^{\lambda} \cdot (\xi, h')$ in $\pi_1(\mathcal{A}, \mathcal{B}; (\xi_0, h_0))$ is well defined.

<u>Remark 4.3</u> It is necessary to define $x_{\alpha\beta}^{\lambda} \cdot (\xi, h)$ as in (4.2) without the assumption that $h(p_\alpha) > h(p_\beta)$. To do this, first use (4.4) below to choose a path (ξ_t', h_t') in $\mathcal{B}$ from (ξ, h) to (ξ', h') where $h'(p_\alpha) > h'(p_\beta)$. Then apply the process of (4.2) to (ξ', h').

<u>Lemma 4.4</u> <u>Let</u> $(\xi, h) \in \mathcal{B}$ <u>and let</u> $p_1, \ldots, p_r$ <u>be the critical points of</u> h. <u>Let</u> $x_1, \ldots, x_r \in (0, 1)$ <u>be arbitrary. Then there is a path</u> (ξ_t, h_t) <u>in</u> $\mathcal{B}$ <u>starting at</u> (ξ, h) <u>such that</u> $p_1, \ldots, p_r$ <u>are the critical points of</u> h_t <u>for each</u> t <u>and</u> $h_1(p_j) = x_j$.

The proof is essentially that of the Rearrangement Theorem in [3]. A one-parameter version of that argument shows

<u>Lemma 4.5</u> <u>Let</u> (ξ_t, h_t) $(0 \leq t \leq 1)$ <u>be a path in</u> $\mathcal{B}$. <u>Any deformation of the graphic of</u> h_t <u>in</u> $I \times I$ <u>which keeps end points fixed and respects projection onto</u> $I \times 0$ <u>can be covered by a deformation of</u> (ξ_t, h_t) <u>lying in</u> $\mathcal{B}$ <u>and keeping end points fixed.</u>

For example, the deformation indicated by

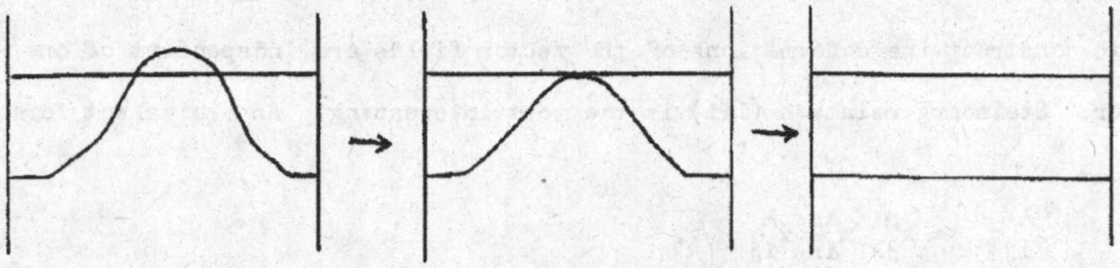

can be realized by a deformation in $\mathcal{B}$. Lemma 4.5 is an example of the more general "principle of non-intersecting trajectory sets". Roughly this says that if the trajectory sets corresponding to two parts of a graphic don't intersect in $I \times V$, then "any" deformation of these two parts of the graphic relative to one another can be covered by a deformation in $\mathcal{F}$.

<u>Construction of</u> Δ.

Let $x \in St(r, \pi_1 V)$ be represented as a word $\prod_{\alpha=1}^{m} x_\alpha$ where each x_α is a generator x_{ij}^{λ} with $\lambda \in Z[\pi_1 V]$. Then in the notation of (4.2) and (4.3) define

(4.6) $$\Delta(x) = x_1 \cdot (\ldots x_m \cdot (\eta_0, f_0) \ldots)$$

For example, the path (ξ_t, h_t) corresponding to the word $x_{12}^{1}\, x_{21}^{-1}\, x_{12}^{1}$ in $St(3, Z)$ has the following graphic :

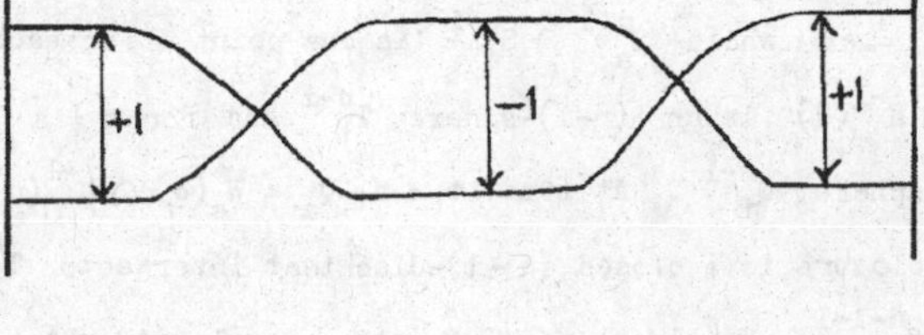

We shall briefly outline what is involved in showing that (4.6) gives a well-defined map which is a bijection as desired in (4.1).

Δ is well-defined

The main point here is to show that the Steinberg relations give rise to paths which can be deformed into $\mathcal{B}$. This is true of the Steinberg relations (i) and (ii) essentially because in these two cases the isotopies in the proof of (4.2) which are used to construct the deformations of the vector fields are independent of one another. Steinberg relation (iii) is the most interesting. An equivalent form of (iii) is

(iv) $$x_{ij}^{\lambda} x_{jk}^{\mu} = x_{jk}^{\mu} x_{ik}^{\lambda\mu} x_{ij}^{\lambda} \quad .$$

Suppose $\lambda = \pm g$ and $\mu = \pm h$. Then (iv) can be stated in terms of deformations of graphics keeping end points fixed as

(4.7)

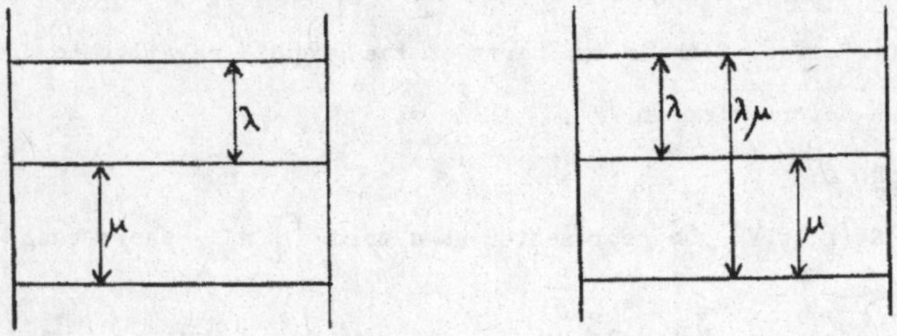

Consider an elementary gradient crossing path (ξ_t, h_t) starting at $(\xi, h) \in \mathcal{B}$ which passes the i-handle for the critical point p over the i-handle for the critical point q at time $t = a$. Assume for simplicity that $h_t = h$ for all t. Choose values c and d satisfying $h(q) < c < h(p) < d < 1$. For each $t \in [0,1]$ $W_t(p) \cap h^{-1}(c) \simeq S_t^{i-1}$ and $W_t^*(q) \cap h^{-1}(c) \simeq S_t^{n-i}$. For $t \neq a$, $S_t^{i-1} \cap S_t^{n-i} = \phi$ while $S_a^{i-1} \cap S_a^{n-i}$ is one point. For each $t \in [0,1]$ $W_t^*(p) \cap h^{-1}(d)$ is an $(n-i)$-sphere T_t^{n-i} and for $t \neq a$ $W_t^*(q) \cap h^{-1}(d)$ is also an $(n-i)$-sphere Q_t^{n-i}. At time $t = a$, $Q_a = W_a^*(q) \cap h^{-1}(d)$ is an open $(n-i)$-disc whose closure is a closed $(n-i)$-disc that intersects T_t^{n-i} in an $(n-i-1)$-sphere S^{n-i-1}. The local model for the singularity which occurs in Q_t as t passes through time a is described by the following diagram :

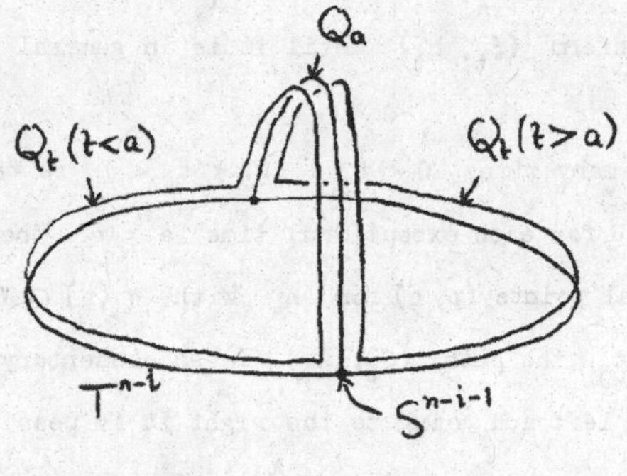

To illustrate the deformation (4.7) when $n = 1$ and $i = 1$ consider a Morse function h on a 2-manifold having three critical points p_1, p_2, and p_3 with $h(p_1) < h(p_2) < d < h(p_3)$. The following diagram describes the evolution with time t of $X_t = W_t^*(p_1) \cap h^{-1}(d)$, $Y_t = W_t^*(p_2) \cap h^{-1}(d)$, and $Z_t = W_t(p_3) \cap h^{-1}(d)$ corresponding to the left hand side of (4.7). The dotted line shows how to choose the level preserving isotopy of $I \times h^{-1}(d)$ which will give rise to the deformation (4.7) :

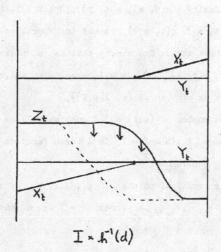

$I \times h^{-1}(d)$

Bijectivity of Δ

To show surjectivity of Δ consider a path (ξ_t, h_t) representing an element of $\pi_1(\mathcal{A}, \mathcal{B}; (\eta_0, f_0))$. Deform (ξ_t, h_t) until it is in general position. This can be done so that

(4.8) There are finitely many times $0 < t_0 < \ldots < t_m < 1$ so that for $t \neq t_j$ $(\xi_t, h_t) \in \mathcal{B}$ and for each exceptional time $a = t_j$ there is exactly one pair of critical points (p, q) for h_a with $W_a(p) \cap W_a^*(q) \neq \phi$.

At each exceptional time t_j the path (ξ_t, h_t) is an elementary gradient crossing path. Hence starting from left and going to the right it is possible to read off a word in the free group F_Λ on the generators x_{ij}^λ which maps to (ξ_t, h_t) under Δ.

For injectivity consider two words x and y representing elements of $St(r, Z[\pi_1 V])$ such that $\Delta(x) = \Delta(y)$. Let $\Delta(x)$ and $\Delta(y)$ be represented by paths (ξ_t, h_t) and (η_t, f_t) respectively which are in general position and satisfy (4.8). Since $\Delta(x) = \Delta(y)$ there is a two parameter family $(\alpha(t, s), g(t, s))$ such that $(\xi_t, h_t) = (\alpha(t, 0), g(t, 0))$ and $(\eta_t, f_t) = (\alpha(t, 1), g(t, 1))$ for $0 \leq t \leq 1$. Put the two parameter family into general position. This can be done so that there will be finitely many times $0 < s_0 < \ldots < s_m < 1$ such that for $s \neq s_j$ the one-parameter family $(\alpha(t, s), g(t, s))$, $0 \leq t \leq 1$, is in general position. Each family $(\alpha(t, s_j), g(t, s_j))$ is not in general position, but only fails to be so in one of three ways. One checks that as s passes by s_j the word corresponding to $(\alpha(t, s), g(t, s))$, $0 \leq t \leq 1$, changes by one of the three Steinberg relations. Hence $x = y$ in $St(r, Z[\pi_1 V])$.

There is an obvious extension of (4.1) which goes as follows : Let $\mathcal{A}'$ be the space of pairs (η, f) where $f : M \times I \to I$ is a Morse function and η is gradient-like for f satisfying the conditions

(A) f has exactly r critical points $p_1, \ldots, p_r$ of index i and r critical points $q_1, \ldots, q_r$ of index $i - 1$ such that $f(q_j) < \frac{1}{2} < f(p_i)$; and

(B) $W(q_\alpha) \cap W^*(q_\beta) = 0$ for $q_\alpha \neq q_\beta$.

Let $\mathcal{B}' \subset \mathcal{A}'$ be the subspace of pairs (η, f) such that the stable and unstable manifolds of the critical points p_α also satisfy (B).

Theorem 4.9 Let $3 \leq i \leq n - 3$ and $(\eta_0, f_0) \in \mathcal{B}'$. There is a bijection

$$\Delta' : St(r, Z[\pi_1 M]) \to \pi_1(\mathcal{A}', \mathcal{B}' ; (\eta_0, f_0)).$$

Addendum Let $x \in St(r, Z[\pi_1 M])$ be represented by the word $\prod_\alpha x_{i_\alpha j_\alpha}(\lambda_a)$ and let $(\eta_t, f_t) = \Delta'(x)$. Then

$$\partial(\eta_1, f_1) = (\prod_\alpha e_{i_\alpha j_\alpha}(\lambda_\alpha)) \cdot \partial(\eta_0, f_0) .$$

This follows from Lemma 4.2. To define $\partial(\eta_1, f_1)$ the critical points of f_1 must be ordered and based and the stable manifolds of η_1 must be oriented. This is done by the convention used in (4.2).

§5. Simplification of the graphic

This section gives some of the lemmas used to successively simplify the graphic of a path f_t or a path (η_t, f_t). In (5.1) through (5.3) the claim is that the indicated change in the graphic can be realized by a deformation (keeping end points fixed) of the path f_t or of (η_t, f_t).

Lemma (5.1) (Uniqueness of birth; cf. [1])

If $1 \leq i \leq n - 1$, then

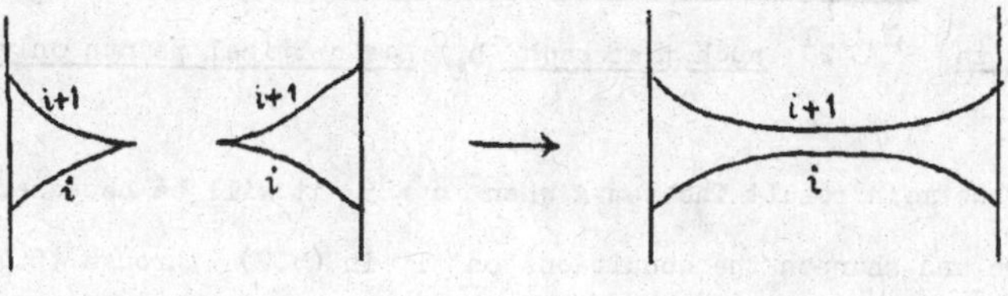

Lemma 5.2 (Swallow's tail) If $2i + 2 \leq n$ or $4 \leq i \leq n - 3$, then

It is easy to introduce a "swallows tail" (cf.(2.1)) but relatively difficult to make one disappear. Lemma 5.2 is valid no matter what gradient crossings may occur between the two $(i + 1)$-handles. Hence the vertical arrows are omitted from the diagram. See [1] for a proof of (5.2) when $\pi_1 M = 0$.

Lemma 5.3 (Exchange). Suppose $4 \leq i \leq n - 4$ and $\partial(\eta_0, f_0) = \text{id}$. Then

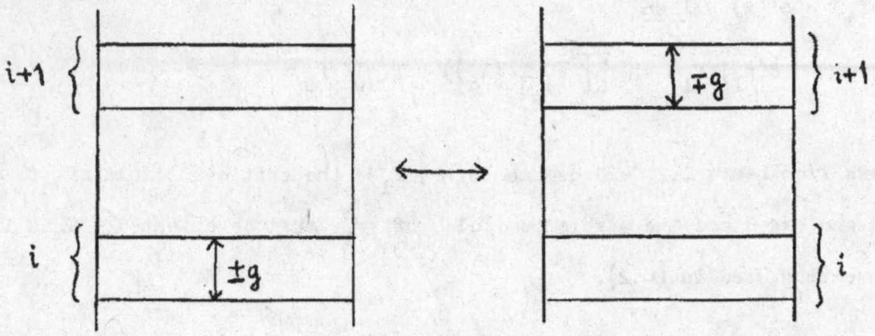

A sharper form of (5.3) says that if $0 \leq i \leq n - 4$ and if $\partial(\eta_0, f_0)$ is geometrically (and not just algebraically) the identity, then (η_t, f_t) can be deformed so as to satisfy $W^*(q_t) \cap W(p_t) = \phi$ for all t, where p_t and q_t are the two critical points of f_t of index i such that $f_t(p_t) > f_t(q_t)$.

These lemmas (plus a few others) are used to prove

Proposition 5.4 Let $n \geq 8$ and $4 \leq i \leq n - 2$. Any path $f_t : M \times I \to I$ with f_0 and f_1 in $\mathcal{E}$ can be deformed with end points fixed to a path h_t of ordered functions lying in $\mathcal{F}^0 \cup \mathcal{F}^1$ such that each h_t has critical points only of index i or $i - 1$.

To obtain the main result Theorem A when $n \geq 5$ it will be necessary to take a little more care and sharpen the conditions on i in (5.2) through (5.4). A.Chenciner has independently obtained (5.2) and (5.4) and has made some improvements in the constraints on i (cf. [2]). Also see [1] for a result similar to (5.4) when $\pi_1 M = 0$. An interesting use of (5.2) is

(5.5)

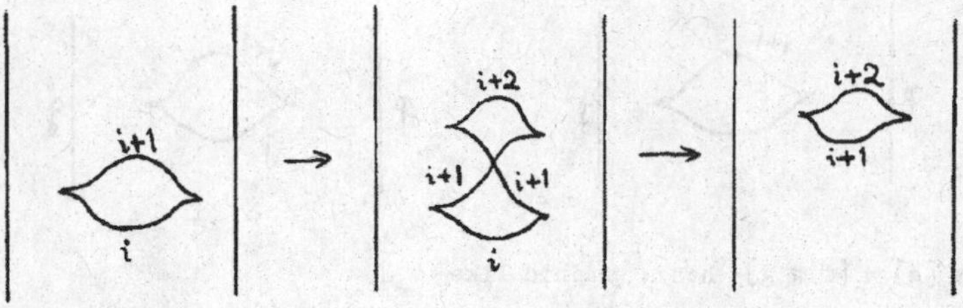

Another example is

(5.6)

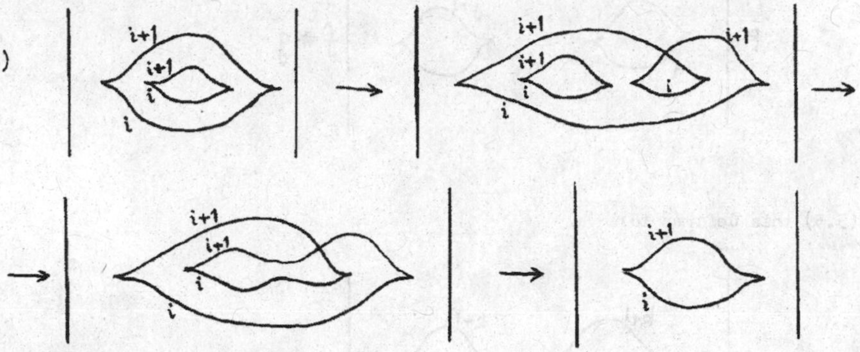

The unicity of death subgroup $\mathcal{D}$.

Let $\mathcal{D} \subset \pi_0(\mathcal{E})$ be the set of [f] such that for some $0 \leq i \leq n$ the function $f \in \mathcal{E}$ lies in the same component of $\mathcal{E}$ as the endpoint of a path f_t starting at p and having a graphic like

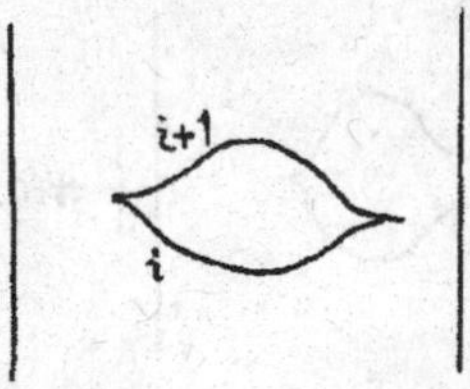

Lemma 5.7 $\mathcal{D}$ is a subgroup.

Proof We show why $\mathcal{D}$ is closed under addition. Let [f] and [g] in $\mathcal{D}$ have graphics like

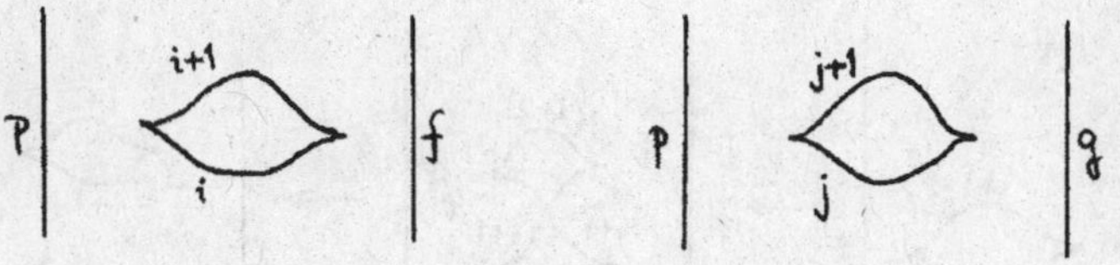

Then $[f] \oplus [g] = [f \# g]$ has a graphic like

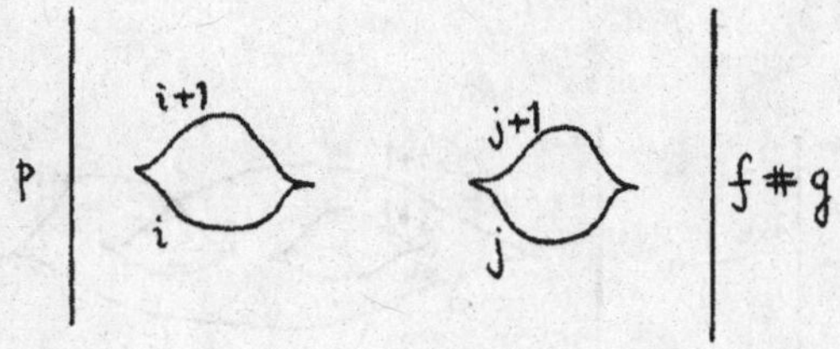

By (5.5) this deforms to

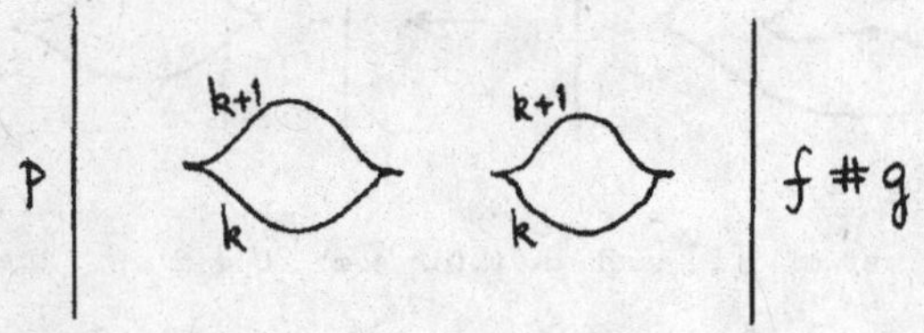

By (5.1) this last graphic deforms to

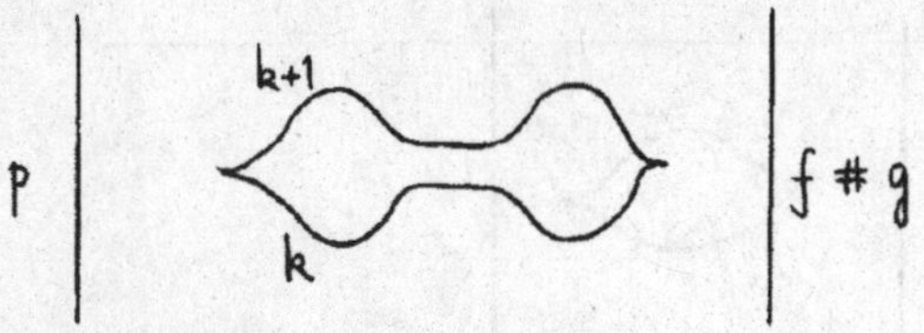

Hence $[f] \oplus [g] \in \mathfrak{D}$.

The term "uniqueness of death" is used because $\mathfrak{D}$ measures the extent to which the Whitney - Smale process for cancelling two critical points with algebraic intersection number one gives rise to functions lying in different components of $\mathcal{E}$.

§6. Construction of ρ

Throughout this section Λ will denote $Z[\pi_1 M]$ and G will denote $\pi_1 M$. Let $E_G \subset E\ell(\Lambda)$ denote the subgroup of those elements of $E\ell(\Lambda)$ which can be written in the form $P \cdot D$ where P is a finite permutation matrix and D is a diagonal matrix with diagonal entries of the form $\pm g$ for $g \in G$. Let $W(\Lambda) \subset St(\Lambda)$ be the subgroup of those $x \in St(\Lambda)$ such that $\pi(x) \in E_G$.

Theorem A. Let $n \geq 8$.

(I) There is a homomorphism

$$\rho : W(\Lambda) \rightarrow \pi_0(\mathcal{S}_{id}) \bmod \mathcal{D}$$

such that $W(\pm G) \subset \ker \rho$.

(II) The natural homomorphism

$$K_2(\Lambda) \rightarrow W(\Lambda) \bmod W(\pm G)$$

is onto.

In this section we define ρ (see 6.1) and indicate the proof of Theorem A.

Let $x \in W(\Lambda)$. Then $x \in St(r, \Lambda)$ for some $r < \infty$. Let $\eta_0 = \text{grad } p$ with respect to a fixed choice of Riemannian metric μ_0 on $M \times I$. Let $4 \leq i \leq n - 3$. Use the standard model for birth paths (recall §3, e.g. 2) to construct a deformation (η_t, f_t), $0 \leq t \leq 1$, of (η_0, p) with the following properties :

(a) The deformation has support contained in r disjoint balls in $M \times I$ and the graphic has no gradient crossings and looks like

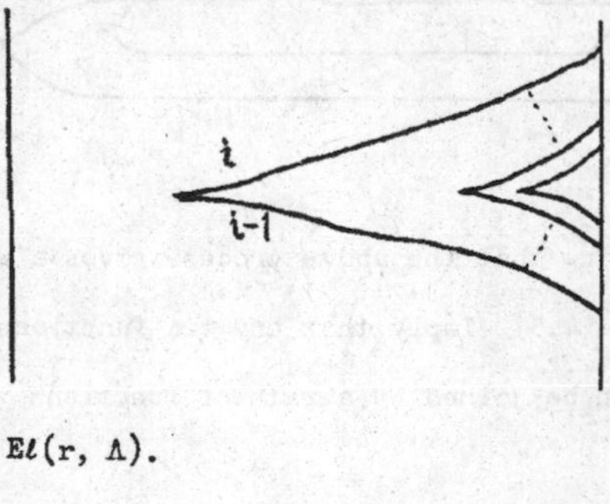

(b) $\partial(\eta_1, f_1) = id \in E\ell(r, \Lambda)$.

Now apply (4.9) to obtain a deformation $\Delta^1(x) = (\xi_t, h_t)$ for $1 \le t \le 2$ of (η_1, f_1). Since $x \in W(\Lambda)$ we know that $\partial(\xi_2, h_2) \in E\ell(r, \Lambda)$ is a permutation of an $r \times r$ diagonal matrix with diagonal entries $\pm g_1, \ldots, \pm g_r$. Hence the Whitney process for eliminating intersection points and the Smale cancellation lemma imply that the critical points of h_2 can be cancelled in pairs to produce a function $g_3 : M \times I \to I$ with no critical points. More precisely, there is a deformation (α_t, g_t) for $2 \le t \le 3$ of (ξ_2, h_2) such that $(\alpha_t, g_t) \in \mathcal{B}'$ for $2 \le t \le 2.5$ and the graphic of (α_t, g_t) for $2.5 \le t \le 3$ has no gradient crossings and looks like

(c)

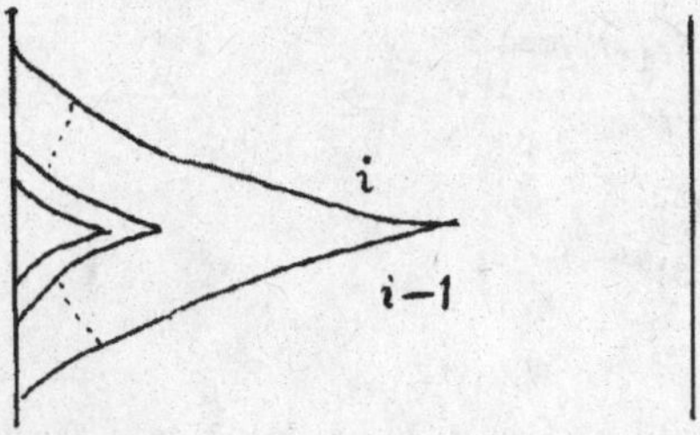

We define

(6.1) $\rho(x) = g_3$.

For example, if $x = w_{12}(g) = x_{12}(g) \cdot x_{21}(-g^{-1}) \cdot x_{12}(g) \in St(3, \Lambda)$ this process yields the following graphic :

(6.2)

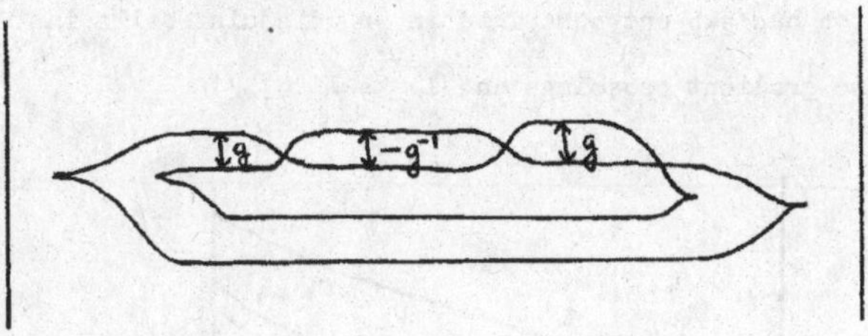

The main point in showing that the above process gives a well defined map ρ is this : (5.1), (4.9), and (4.5) imply that any two functions g_0 and g_1 obtained from $x \in W(\Lambda)$ can be joined by a path of functions g_t having a graphic like

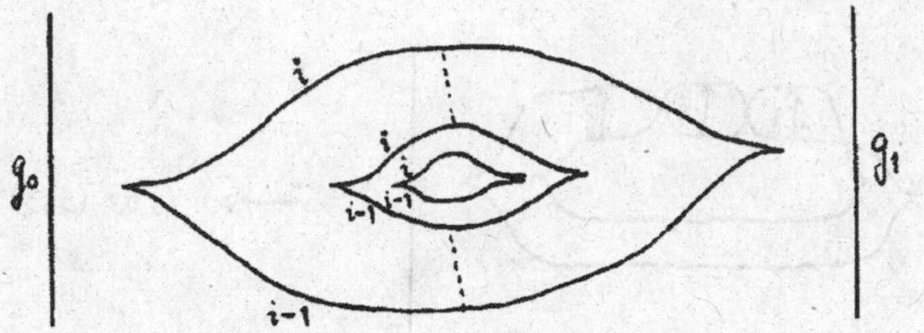

Using (5.6) several times this reduces to

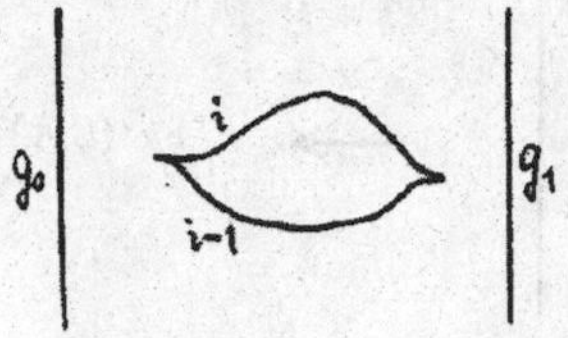

Hence $g_0 \equiv g_1 \bmod \mathfrak{D}$.

To see that $W(\pm G) \subset \ker \rho$ it suffices to show how to kill modulo $\mathfrak{D}$ a generator $w_{ij}(g)$ having (6.2) as a graphic :

(1)

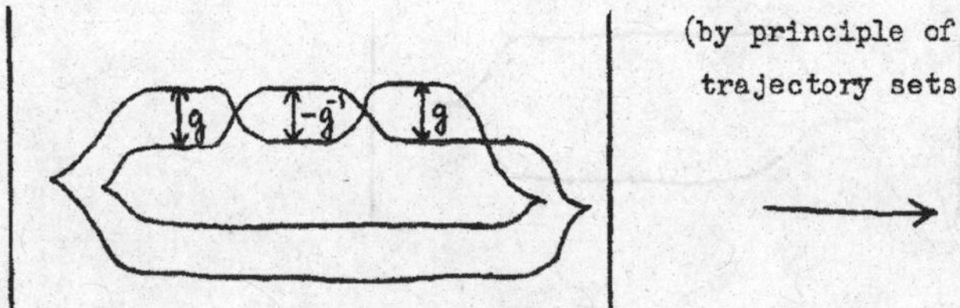

(by principle of missing trajectory sets)

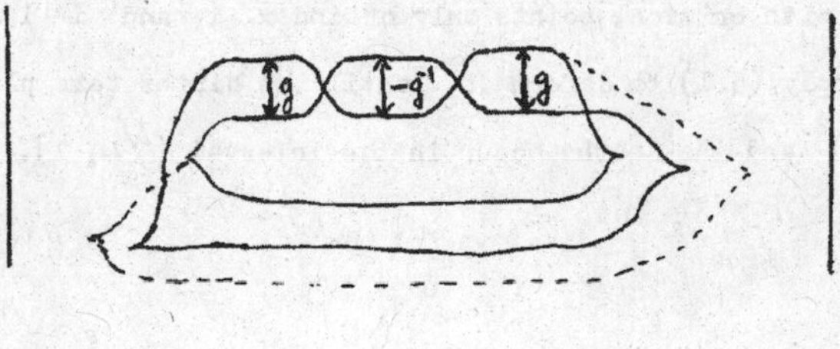

=

(2)

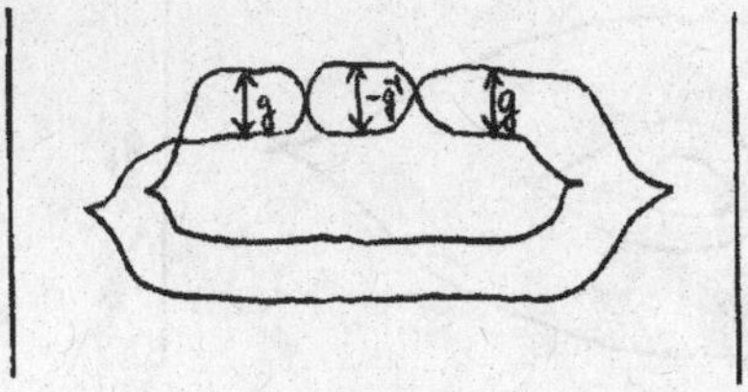

$\longrightarrow$ by (5.3)

(3)

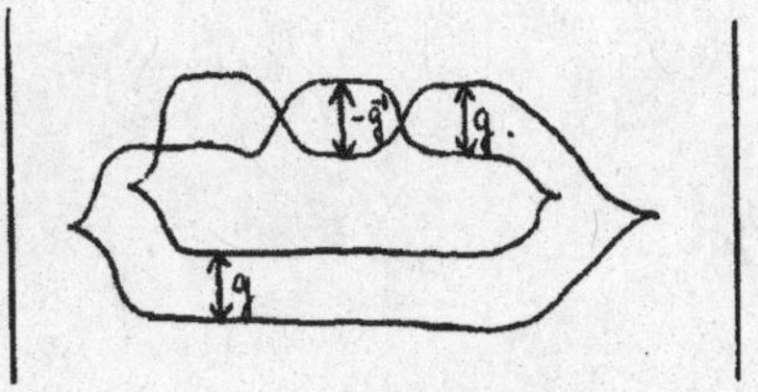

$\longrightarrow$ by (4.5)

(4)

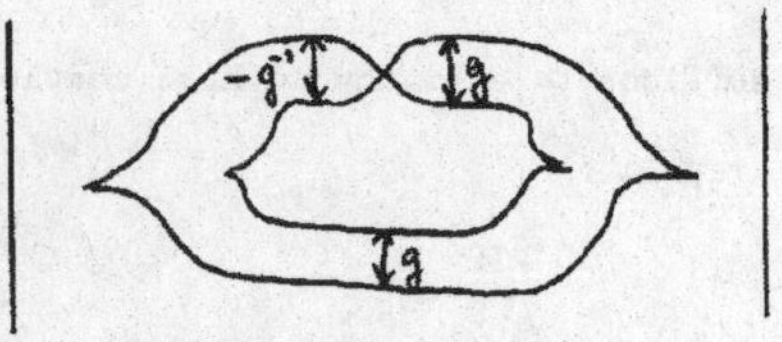

$\longrightarrow$ by (5.2)

(5)

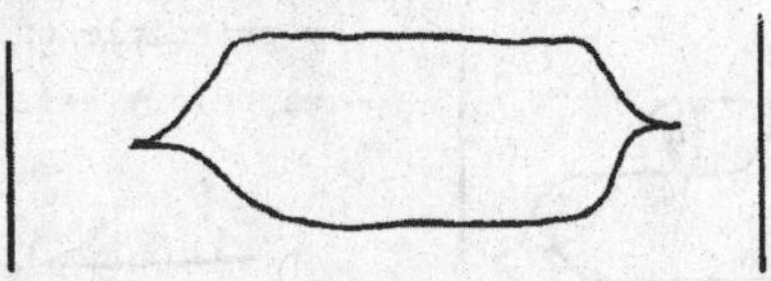

To see why $\rho : W(\Lambda) \rightarrow \pi_0(\mathcal{S})$ mod $\mathcal{D}$ is onto, take $g \in \mathcal{E}$ and connect g to p by a path f_t in $\mathcal{F}$. By (5.4) the path f_t can be taken to be a path of ordered functions with critical points only of index i and $i-1$ where $4 \leq i \leq n-3$. Apply (5.1) to deform f_t until the births take place in the interval $[0, {}^1/_4)$ and the deaths occur in the interval $({}^3/_4, 1]$. The graphic will be

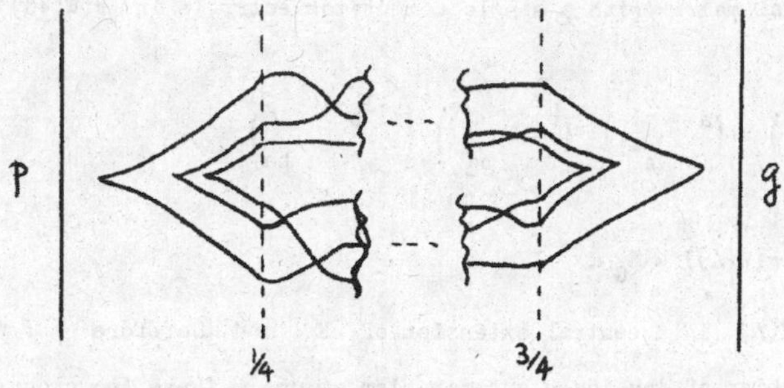

Now choose a family η_t of gradient-like vector fields for f_t such that the family (η_t, f_t) is in general position. This can be done so that there are no gradient crossings in the intervals $[0, 1/4)$ and $(3/4, 1]$. At this stage there will be gradient crossings possibly among both the i-handles and the (i-1)-handles. Use (5.3) to eliminate the latter. Now read off from the successive gradient crossings of the i-handles the word in $W(\Lambda)$ that maps to under ρ to g modulo $\mathfrak{D}$.

To prove (II) of Theorem A it suffices to show that

(6,3) $\quad \pi(W)(\pm G)) = E_G$, and

(6.4) $\quad W(\pm G)$ is a normal subgroup of $W(\Lambda)$.

<u>Proof of 6.3</u> : Let $x \in W(\Lambda) \cap St(r, \Lambda)$ and let $A = \pi(x)$. Then $A = P.D$ where P is a permutation times a diagonal matrix with ± 1 as diagonal entries and D is a diagonal matrix with entries $g_1, \ldots, g_r$ $(g_i \in G)$. Let $\epsilon : Z[G] \to Z$ be the augmentation such that $\epsilon(g) = 1$ for all $g \in G$. Since A is a product of elementary matrices so is $\epsilon(A) = P \in GL(r, Z)$. Hence $\det P = 1$ and $P \in \pi(W(\pm 1)) \subset \pi(W(\pm G))$. Now consider D. Multiplying D on the right by the 2×2 matrix

$$\begin{pmatrix} g_1^{-1} & \\ & g_1 \end{pmatrix} = \pi(w_{12}(g_1^{-1}) . w_{12}(-1))$$

produces a diagonal matrix with entries $1, g_2 g_1, \ldots, g_r$. Continuing the process we reduce the matrix D to a diagonal matrix D' with only one entry $g = g_r \ldots g_2 g_1$ which is possibly not 1. However g is a product of commutators because D' has determinant 1 over the ring $Z[H]$ where $H = G/[G,G]$. Finally any diagonal matrix with a single commutator entry is in $\pi(W(\pm G))$ because of the identity

$$\begin{pmatrix} aba^{-1}b^{-1} & \\ & 1 \end{pmatrix} = \begin{pmatrix} a & \\ & a^{-1} \end{pmatrix} \begin{pmatrix} b & \\ & b^{-1} \end{pmatrix} \begin{pmatrix} a^{-1}b^{-1} & \\ & ba \end{pmatrix} .$$

Hence $\pi(W(\pm G)) = \pi(W(\Lambda)) = E_G$.

Proof of 6.4 $W(\Lambda)$ is a central extension of E_G and therefore $\pi : W(\Lambda) \to E_G$ induces an isomorphism of the inner automorphism groups of these two groups. Now let $x \in W(\Lambda)$ and $w \in W(\pm G)$. By (6.3) choose a $u \in W(\pm G)$ with $\pi(u) = \pi(x)$. Then $x \cdot w \cdot x^{-1} = u \cdot w \cdot u^{-1} \in W(\pm G)$. Hence $W(\pm G)$ is normal in $W(\Lambda)$.

REFERENCES

[1] Cerf, J. — La stratification naturelle des espaces de fonctions différentiables réelles et le théorème de la pseudo-isotopie, to appear in Annales Scientifiques de l'I.H.E.S.

[2] Chenciner, A. and Laudenbach, F. — Théorie de Smale à un parametre dans le cas non simplement connexe, C.R. Acad. Sci. Paris, 270, série A, p.176 and p.307.

[3] Milnor, J. — Lectures on the h-cobordism theorem. Princeton Math. Notes, 1965.

[4] Milnor, J. — Notes on Algebraic K-Theory, M.I.T. 1969.

[5] Milnor, J. — Whitehead Torsion, Bull. Amer. Math. Soc. 72, 358-426 (1966)

[6] Smale, S. — Differential Dynamical Systems. Bull. Amer. Math. Soc. 73, 747-817 (1967).

[7] Siebenmann, L. — Notices A.M.S., p.852 and p.942, (1967)

[8] Wagoner, J.B. — On K_2 of the Laurent Polynomial Ring, to appear in Amer. Jour. of Math.

REFLECTIONS ON GRADIENT VECTOR FIELDS

C.T.C. Wall

The equation $X = \text{grad } f$ expresses a relation between a vector field X, a metric s, and a function f all defined on a manifold M. If s is given, then f determines X; conversely a vector field X determines a 1-form ω and $X = \text{grad } f \Leftrightarrow \omega = df$ which is soluble locally if and only if $d\omega = 0$, and then determines f up to a constant (for global solution, the class of ω in $H^1(M; \mathbb{R})$ - or its periods - gives the remaining obstruction). More interesting is the case when s is not given.

Given a vector field X, when can we solve for f and s? If M is compact, a necessary condition is that the α- (or ω-) limit set of any orbit is a zero of X. This may not be sufficient, but I know of no counterexample. The only known result seems to be one due to Smale [S]: if we assume

(i) the zeros of X are transverse

(ii) each limit set is a zero

(iii) the stable and unstable manifolds of the zeros meet transversely,

then X is a gradient vector field. (The equivalence of (i) with the condition actually given by Smale follows from the discussion below.) This covers the generic case, but one would like at least an extension to small codimensions.

There remains the problem, given X and f when is there a metric s with $X = \text{grad } f$? The question is purely local, since piecing together locally defined solutions using a partition of unity gives a solution to the global problem (though if M is non-compact, there appears to be no way to get a complete metric). It thus suffices to consider a neighbourhood of the origin in $\mathbb{R}^n$.

For any metric,

$$\text{grad } f(f)_p = \text{grad } f_p(df_p) = \langle df_p, df_p \rangle \geq 0$$

so a first necessary condition is that $Xf \geq 0$ everywhere. Moreover, if $Xf = 0$

at P, then $X_p = df_p = 0$. At a point where $Xf > 0$, no further condition is needed. For the bundle of metrics is an open convex cone in a vector bundle on M, the condition that df_p corresponds to X_p gives an affine subbundle, and the condition $Xf(P) > 0$ is equivalent to the affine subspace at P meeting the open cone. Thus for ordinary points, all we need do is find a section of a bundle with contractible fibre.

In general, we must look deeper. Let $\mathcal{E}_n$ be the ring of germs at 0 of C^∞-functions on $\mathbb{R}^n$. For $f \in \mathcal{E}_n$ define $I(f)$ to be the set of functions Xf, where X is a germ at 0 of smooth vector field on $\mathbb{R}^n$. Since

$$X(f) = \sum_{i=1}^{n} X(x_i) \frac{\partial f}{\partial x_i},$$

$I(f)$ is the ideal in $\mathcal{E}_n$ generated by the $\partial f/\partial x_i$. Similarly if X is a vector field near 0 the functions $X(f)$, $f \in \mathcal{E}_n$ generate (but need not fill) an ideal in $\mathcal{E}_n$. The same equation shows that this ideal, $I(X)$, is generated by the $X(x_i)$. Now if $X = \operatorname{grad} f$ then, with the usual notation,

$$\frac{\partial f}{\partial x_i} = \sum_j g_{ij}(x)\, X(x_j) .$$

Since the matrix (g_{ij}) over $\mathcal{E}_n$ is invertible, it follows that $I(f) = I(X)$. In particular, if we write $\mu(f)$, $\mu(X)$ for the codimensions in $\mathcal{E}_n$ of these ideals, $\mu(f) = \mu(X)$.

I had originally conjectured that the inequality $Xf \geq 0$, together with $\mu(f) = \mu(X)$, implied $I(f) = I(X)$. This is trivial for $\mu(X) \leq 1$, but can be false with $\mu(X) = 3$: an example is

$$f = x^3 + xy^2$$
$$X = (x^2 + y^2)\, \partial/\partial x + xy\, \partial/\partial y .$$

It is however true when $\mu(X) = 2$, but this result is hardly worth the effort needed to prove it.

Now let us suppose $I(f) = I(X)$. Then the equations

$$\frac{\partial f}{\partial x_i} = \sum_j g_{ij}(x)\, X(x_j)$$

can be solved for $g_{ij} \in \mathcal{E}_n$, but do not in general give a symmetric matrix. Let

$\{k_{ij}\}$ be a solution. If now $(k_{ji} - k_{ij}) \in I(X)$ for all i, j, we can write

$$k_{ji} - k_{ij} = \sum_k \lambda_{ijk} X(x_k)$$

with λ_{ijk} skew-symmetric in i and j. Then, as is easily verified, writing

$$g_{ij} = k_{ij} + \tfrac{1}{2}\sum_k (\lambda_{ijk} + \lambda_{jki} - \lambda_{kij}) X(x_k)$$

gives a symmetric solution for the above equation. The most interesting case here is where $\mu(X) < \infty$: then the conditions $k_{ji} - k_{ij} \in I(X)$ amount to a finite set of linear equations over $\mathbb{R}$. In this case, too, we can prove a converse. This depends on the

<u>Lemma</u> (proof later) <u>If $f_1, \ldots, f_n \in \mathcal{E}_n$ generate an ideal of finite codimension, and $\sum_{i=1}^n f_i g_i = 0$, then we can write $g_i = \sum_{j=1}^n \lambda_{ij} f_j$, with $\lambda_{ij} \in \mathcal{E}_n$ skew-symmetric in i and j.</u>

In the case above, the $X(x_j)$ generate an ideal of finite codimension. Hence if

$$\frac{\partial f}{\partial x_i} = \sum_j g_{ij}(x) X(x_j) = \sum_j k_{ij}(x) X(x_j),$$

we can write

$$g_{ij}(x) - k_{ij}(x) = \sum_k \lambda_{ijk}(x) X(x_k)$$

(with λ_{ijk} skew-symmetric in j and k). If now g_{ij} is symmetric, it follows that $(k_{ij} - k_{ji}) \in I(X)$.

Note also that if $I(X)$ is proper, this also shows that the values $k_{ij}(0)$ are uniquely determined by the equations. Even if we assume these form a symmetric matrix, it need not be positive definite. Nor does the positivity condition $Xf \geq 0$ imply this : a counterexample is

$$f = x^3 + xy^2$$

$$X = (3x^2 + y^2)\frac{\partial}{\partial x} - 2xy\frac{\partial}{\partial y}.$$

Again, this trouble cannot occur if $\mu(X) = 1$ (easy), nor if $\mu(X) = 2$ (less easy).

This concludes our discussion of the problem; it remains to prove the lemma. Suppose $f_i \in \mathcal{E}_n$ generate the ideal I :

$$I = \langle f_1, \ldots, f_n \rangle \supset \mathfrak{m}^N,$$

where $\mathfrak{m}$ is the maximal ideal. Choose polynomials $f'_i \equiv f_i \pmod{\mathfrak{m}^{N+1}}$. Then

$$< f'_1, \ldots, f'_n > + \mathfrak{m}^{N+1} \supset I \supset \mathfrak{m}^N,$$

so by Nakayama's lemma, the ideal of the f' contains $\mathfrak{m}^N$, hence equals f. We can change from f to f' by multiplying a matrix congruent to 1 mod $\mathfrak{m}$: since the result is independent of such changes, it suffices to prove it when f_i is a polynomial.

Let Λ be the Koszul complex given by the exterior algebra on symbols $e_1, \ldots, e_n$ with de_i defined to be f_i. Our lemma is equivalent to exactness of Λ at λ^1. Let θ_n denote the ring of analytic functions in $x_1, \ldots, x_n$. Since $\mathcal{E}_n$ is flat over θ_n [M], it is enough to prove the result over θ_n. Since $\mathbb{C}$ is faithfully flat over $\mathbb{R}$, it suffices to work over $\theta_n \otimes \mathbb{C}$. The lemma will thus follow from the following, essentially known,

<u>Theorem</u> <u>Let</u> $f_1, \ldots, f_n$ <u>be germs at</u> 0 <u>of holomorphic functions on</u> $\mathbb{C}^n$. <u>The following conditions are equivalent</u> :

i) <u>The complex</u> Λ <u>is exact at</u> λ^1.

ii) <u>It is exact at all</u> λ^i, $i > 0$.

iii) <u>For each</u> $i \leq n$, f_i <u>is not a zero-divisor modulo</u> $f_1, \ldots, f_{i-1}$.

iv) <u>For each</u> $i \leq n$, <u>the germ of analytic set defined by</u> $f_1 = \ldots = f_i = 0$ <u>is unmixed of dimension</u> $n - i$.

v) <u>0 is an isolated point of</u> $f_1 = \ldots = f_n = 0$.

vi) <u>The ideal generated by</u> $f_1, \ldots, f_n$ <u>has finite codimension in the ring</u> $\theta_n^{\mathbb{C}}$ <u>of germs at</u> 0 <u>of holomorphic functions</u>.

<u>Proof</u> The equivalence of (i) - (iii) is contained in [N, p.374, Theorem 8].

<u>(iii) <=> (iv)</u> We show by induction on j that (iii) holds for all $i \leq j$ if and only if (iv) does : this is clear for $j = 0$, assume it for $j - 1$. We can regard $\theta_n^{\mathbb{C}}$ modulo $<f_1, \ldots, f_{j-1}>$ as the algebra of germs at 0 of analytic functions on the analytic set given by

$$f_1 = \ldots = f_{j-1} = 0$$

which is unmixed, of dimension $n - j + 1$. Then f_j is a non-zero divisor iff it is not identically zero on any component of this set iff the subset defined by $f_j = 0$ has dimension $n - j$; when this holds, the subset is clearly unmixed.

(iv) <=> (v). Clearly (iv) => (v) (take $i = n$); for the converse, if the set $f_1 = \ldots = f_i = 0$ had dimension $> n - i$, then $f_1 = \ldots = f_n = 0$ would have dimension > 0. Since the dimensions are exactly right, the sets are (as before) of unmixed dimension.

(v) <=> (vi) Clearly (vi) => (v), for it implies that the ideal contains a power of each coordinate function. The converse follows from Hilbert's zero theorem.

[N] D.G. Northcott, Lectures on Rings, Modules and Multiplicities. Cambridge University Press, 1968.

[S] S. Smale, On gradient dynamical systems. Ann. of Math. 74 (1961) 199-206.

[M] B. Malgrange, Ideals of differentiable functions. Oxford University Press, 1966.

SINGULARITIES IN SPACES

D. Sullivan

Introduction

In this note we hope to outline a few results and intuitions about singularities among various classes of spaces.

We will work in the context of topological spaces with some extra geometrical structure. We may start with a piecewise linear structure, an analytic structure, or some stratification of the space into equisingular manifolds.

We will be concerned with various geometrical and algebraic problems associated to the singularities. For example we consider the singularities of geometric cycles and their possible stratified structure. One theorem gives a canonical form for the singularities after resolving. These are join-like singularities based on an a-priori sequence of almost-complex manifolds.

Another theorem gives an implicit description of the singularities in a generic embedded cycle.

Finally, there is a geometric obstruction theory for reducing the dimension of singularities in a given situation - the most powerful application being to homology manifolds.

These last two topics are very elementary and mostly interesting because of their geometric appeal. The discussion of canonical forms is at the same time geometrical and algebraic. One is led simultaneously to a geometric approach to generalised homology theory and to certain difficult questions about the algebraic significance of certain singularities.

Resolving Singularities . Canonical forms

First consider the general problem of resolving singularities. We assume that our space with singularities V is a geometric cycle, that is for some triangulation V is the union of its top-dimensional simplices and that these can be oriented so that their sum is a cycle.

By a <u>blowing up</u> of V we mean an onto stratifiable map of geometric cycles

$$W \xrightarrow{f} V$$

such that

i) f (singularity W) $\subseteq$ singularity V

ii) f induces an isomorphism

$$f^{-1}(V - \text{singularity } V) \rightarrow V - \text{singularity } V .$$

(We can also usually assume that the left hand side of ii) is dense in W.)

Since f has degree one we know from the classic paper of Thom on cobordism that for some V there is no non-singular blow-up. Thom shows that it is not true that every homology class in a manifold say contains a non-singular cycle.

The investigations below stemmed from our curiosity about these <u>innately singular homology classes</u> discovered by Thom. What do they look like geometrically?

The first "innately singular" example occurs in dimension seven, for example the torsion product

$$x_1 * x_5 \in H_7(K(Z/3 \times Z/3, 1)) \quad \text{(Thom)} .$$

The theory below implies that this class contains a geometric cycle V whose singularity Σ_V is a two-dimensional equisingular submanifold of V. In fact, a neighbourhood of Σ_V in V is isomorphic to

$$\Sigma_V \times \text{cone } \mathbb{C}P^2 ,$$

$\mathbb{C}P^2$ the complex projective plane.

Any such V in an innately singular homology class cannot be completely resolved. However, any seven dimensional geometric cycle can be blown up so that the result has only this untwisted $\mathbb{C}P^2$ - singularity.

In order to give the theorem we need to discuss join-like singularities. First of all, let me say that these singularities are not canonical for this problem - other geometric ideas might work. However they provide a fairly elegant means of cancelling the Thom phenomenon. Also their simplicity (and success) is based on a deep theorem from algebraic topology - the structure of the complex cobordism ring <u>over Z.</u>

The product structures and a-priori description of the links of strata for these singularities allow these cycles to be treated as simply as manifolds in some geometric contexts.

Now we look at these join-like singularities. Consider any sequence of distinct closed manifolds

$$\Sigma_1 , \Sigma_2 , \ldots , \Sigma_i , \ldots \quad .$$

If $I = (i_1, \ldots, i_r)$ is any finite set of distinct indices, consider the join

$$\Sigma_I = \Sigma_{i_1} * \Sigma_{i_2} * \ldots * \Sigma_{i_{r+1}} .$$

Recall that the points of Σ_I are the points of all possible r - simplices whose vertices lie (respectively) in the disjoint union $\Sigma_{i_1} \cup \Sigma_{i_2} \cup \ldots \cup \Sigma_{i_{r+1}}$.
Suppose that one of the manifolds in the sequence (say the last) is a positive dimensional sphere.

Then Σ_I is singular at those boundary points of the simplex not in the open face opposite the vertex in $\Sigma_{i_{r+1}}$.

We can stratify Σ_I according to the natural stratification scheme of a closed quadrant in Euclidean space of r-dimensions.

The stratification is achieved by removing the closure of that open face from each simplex and identifying the result with the quadrant.

There are certain points to be made about this type of stratification -

i) each point p in Σ_I has a neighbourhood isomorphic to

(euclidean space) $\times$ (cone Σ_J)

with $J \subset (i_1, \ldots, i_r)$. J is the set of indices for which the natural barycentric coordinates of p (excluding i_{r+1}) vanish.

ii) the natural (cone Σ_J) bundle giving the neighbourhood of the stratum of p has a given product structure. This bundle and its product structe extend to the closure of the stratum - giving a neighbourhood of the closure.

iii) along a stratum in the adherence of larger strata the various product structures are related by the embeddings

$\mathbb{C}_J \subseteq \mathbb{C}_L \ , \quad J \subseteq L \ .$

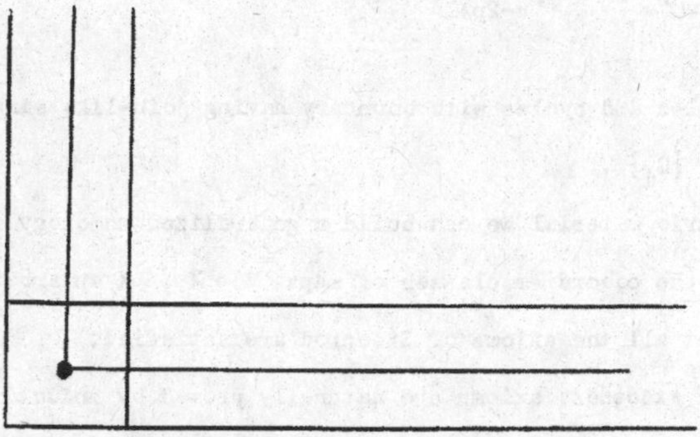

Actually, these consideration also apply to the inclusion of p and its stratum into any larger stratum.

If W can be stratified so that these properties hold then we say that W has *join-like singularities with respect to the sequence* $\{\mathbb{C}_i\}$.

Now we can state the resolution theorems. First the more precise geometric version.

Theorem A *Let* $\{\mathbb{C}_i\}$ *be an irredundant sequence of almost-complex manifolds generating the complex cobordism ring. Suppose our space with singularities* V *has an almost complex structure on its non-singular points. Then there is blow up*

$$W \overset{f}{\to} V$$

so that W *has only join-like singularities with respect to the sequence* $\{\mathbb{C}_i\}$.

The same proof gives the following representing result for homology.

Theorem B *In a manifold any homology class of less than half the dimension contains an embedded geometric cycle having only join-like singularities for the sequence* $\{\mathbb{C}_i\}$.

We note in passing that the representation of Theorem B is unique up to a cobordism with join-like singularities. Thus if $V \subset M$ represents $x \in H_v(M)$ the closures of strata of V define lower dimensional homology classes in M canonically associated to x. Some of these classes are determined by dual cohomology operations. For example we can assume for p a prime $\mathbb{C}_{p-1} = \mathbb{C}P^{p-1}$, and the closure of the stratum of V with a normal (cone $\mathbb{C}P^{p-1}$) singularity represents

$$\beta_* \mathcal{P}^1_* x \in H_{v-2p+1}(M; Z/p) .$$

Sketch of Proof

Consider the cycles and cycles with bouhdary having join-like singularities based on the sequence $\{\mathbb{C}_i\}$.

From this geometric material we can build a generalized homology theory by forming groups out of the cobordism classes of maps $V \to X$, X an arbitrary space.

One can show that all the axioms of Steenrod are satisfied.

The excision and exactness axioms are naturally proved by inductive transversality arguments.

The dimension axiom is more delicate. It follows by deriving an exact sequence relating this theory to that with one manifold left out of the sequence. We can then peel off the singularities one by one and get back to the vacuous sequence, cobordism theory, and the beautiful complex Thom cobordism ring.

Thus we have integral homology theory represented by quasi-complex manifolds with join-like singularities based on $\{\mathbb{C}_i\}$.

This and general position proves theorem B.

The first theorem is proved by looking at a nice neighbourhood N of the singularity of V . The inclusion $\partial N \subset N$ is homologous to zero so may be replaced by a join-like homology in N. This homology is glued to the exterior of N to obtain W. There is a natural collapsing map $W \to V$ which may be shifted slightly to obtain the precise properties required of a blow-up.

Geometric Homology Theories

The join-like singularities construction may be considered from another point of view.

Any such construction for any sequence of manifolds leads to a homology theory satisfying all the Steenrod axioms but that of dimension.

For example, consider any irredundant sequence of almost complex manifolds generating the ideal of manifolds having zero Todd genus. In this case we obtain a theory (V_*, V^*) which is a version of connective complex K-theory,

$$V^0(X) \simeq \text{complex } K(X)$$
$$V^n(X) \simeq \{0\}, \quad n > \dim X .$$

Similar remarks apply to oriented manifolds, the signature, and real K-theory (ignoring the prime 2).

In the general case it is possible to compute the groups for a point in case $\{\mathbb{C}_i\}$ is a regular sequence.

<u>Theorem C</u> Suppose $\{\mathbb{C}_i\}$ satisfies

$$\mathbb{C}_{k+1} \text{ is not a zero divisor of } \Omega/(\mathbb{C}_1,\ldots,\mathbb{C}_k).$$

Then for the point the homology theory based on $\{\mathbb{C}_i\}$ join-like singularities has the value

$$\Omega/(\mathbb{C}_1, \mathbb{C}_2, \ldots) .$$

(Ω the complex cobordism ring.)

We can generalize all this and contemplate constructing hordes of homology theories from the geometric material of cycles and homologies with specified singularities.

In fact there is a functor

$$\left\{\begin{array}{c}\text{category of}\\ \text{singularity}\\ \text{schema}\end{array}\right\} \xrightarrow{\ \pi\ } \left\{\begin{array}{c}\text{generalized}\\ \text{homology}\\ \text{theories}\end{array}\right\} .$$

We specify stratified sets which are the cycles and homologies of the theory by saying what schemes of stratifications are allowed, what the normal cones to strata are, and what the normal cone bundles can be. The specification can be completely explicit as in the join-like singularities above. The description can be rather implicit - for example "the cycles have the local homology properties of manifolds for some coefficients". The point is that the specification be essentially geometric in character.

This geometric approach to generalized homology theory is certainly distinct from the homotopy theoretical one begun in the basic paper of G.W. Whitehead. Geometric considerations on the cycle level seem more precise than constructions with spectra - the objects of stable homotopy theory. However one is led to a Pandora's box of unknown and difficult questions relating the local and semi-local geometry of a space and its global algebraic properties.

For example one might try to analyse the Zeeman spectral sequence showing how Poincaré duality is affected by the singularities. One might then be able to solve the problem of which singularities do not disturb the homology invariance of the signature of a cycle (K-theory is the limit of all geometric theories based on these singularities).

Another example is the functor π itself - what are the global properties of a singularity schema as manifested in the corresponding homology theory?

The Generic Cycle in a Homology class

We can try to find an embedded cycle in a homology class of high dimension. We cannot dictate the singularities as in Theorem B because the algebraic obstructions to a non-singular cycle are not understood at all - there is only the homotopy theoretical criterion of Thom in terms of a pair of finite Grassmannians. However, there is a kind of generic representation theorem. The result and method is completely naive but they lead to a certain pictorial intuition for singularities - even the idea of a stratified space.

Suppose $x \in H_v(M)$ corresponds to a handle in some handlebody decomposition of the manifold M.

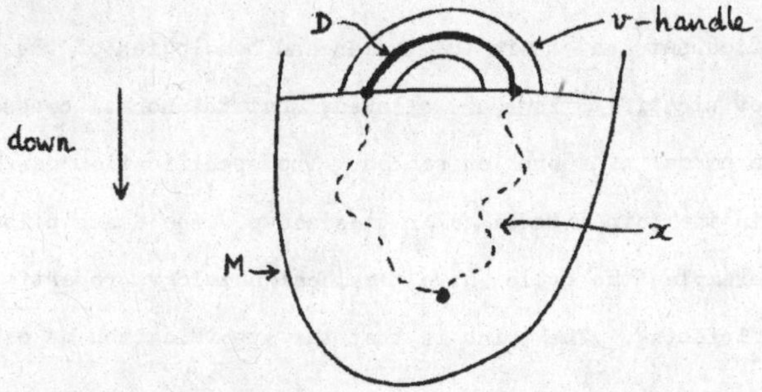

If D represents the v-disc core of the handle, we try to construct a cycle representing x by dragging (homologically) ∂D down through M to a point. The inductive obstruction to proceeding past a lower $(m-\lambda)$ - handle turns out to be the transversal intersection of $(\partial D)'$ (the inductively constructed edge of the deformation) with the boundary of the transverse disc to the handle.

Any homology in D^{λ} of this intersection gives the core of a deformation of $(\partial D)'$ allowing it to slip past the handle. If we use a cone on the intersection (which creates the singularities) we obtain Theorem B' below.

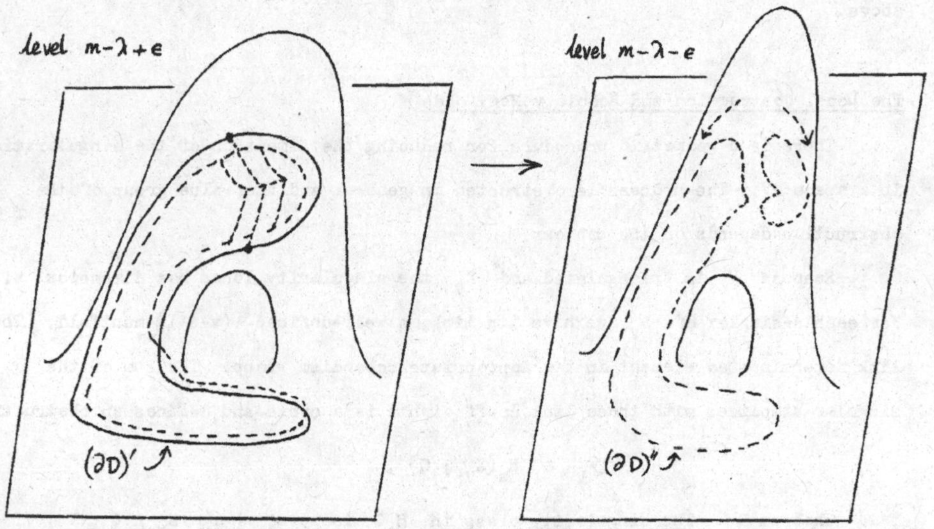

If we define a generalized handle to be -

"attach cone V × disc to Q along an embedding of V × disc in ∂Q"

then we have

Theorem B' Any homology class of M contains an embedded geometric cycle constructed inductively from a disc by attaching generalized handles.

The singularities in such a representation are almost general singularities. We may assume that the singularities have codimension at least three and that each transverse cone to a stratum has boundary irreducible geometric cycle of the appropriate dimension. The theorem gives an inductively explicit description of what singularities "look like".

From a more algebraic point of view the "dragging down" process allows a geometric interpretation of the differentials, indeterminacy, filtration, etc. in the bordism spectral sequence

$$H_*(X, \Omega_*) \Rightarrow \Omega_* X .$$

It might be interesting to compare this cycle with that obtained by forming the closure of the union of the descending trajectories through the critical point corresponding to D for a nice Morse function giving the handle decomposition used above.

The Local Obstruction and Homology Manifolds

There is a geometric procedure for reducing the dimension of the singularities in a space V. The process is obstructed in general and the value group of the obstruction depends on the context.

Suppose V is triangulated and Σ_V the singularity locus has dimension s. For each s-simplex of Σ_V we have its link, a well-defined $(v-s-1)$-manifold. The link determines an element in the appropriate cobordism group. The sum of the singular simplices with these link coefficients is a cycle and defines an obstruction

$$\vartheta_V \in H_s(\Sigma_V ; \Omega) .$$

__Theorem D__ __The singularity class in__ $H_* Q$ __is zero, where__ $\Sigma_V \subseteq Q \subseteq V$, __if and only if there is a blow-up of__ V __(in the context)__

$$W \xrightarrow{f} V$$

__so that__ f __is an isomorphism outside__ Q __and the singularity of__ W __has dimension smaller that__ s.

So, given this W we can look at its $(s-1)$ dimensional singularity obstruction and so on.

__Example i)__ (General singularities - oriented case)

If V is a geometric cycle, the natural obstructions lie in

$$H_s(V; \Omega_r) \qquad r + s + 1 = \dim V$$

where Ω_r is the oriented cobordism group.

If V is a complex variety, the first obstruction vanishes because the chain vanishes identically. The links are quasi-complex manifolds of odd dimension and therefore cobordant to zero.

Example ii) (Homology manifolds)

Let R be a subring of Q with unit and suppose V has the local homology properties of a manifold (coefficients in R).

Then we can consider blow-ups

$$W \xrightarrow{f} V$$

where W is also a "homology manifold" and $f^{-1}(p)$ is "R-acyclic" for each point of V.

The local obstructions lie in

$$H_s(V, \vartheta_r) \qquad r + s + 1 = \dim V$$

where ϑ_r is the group of H-cobordism classes of r-dimensional homology spheres (namely r-manifolds having the R-homology groups of S^r). H-cobordism means oriented cobordism where the cobordism is R-homologically like $S^r \times$ unit interval.

a) In the case R is the ring of integers, we have ordinary homology manifolds. Then the coefficient groups are all zero (by surgery arguments) except when r is three. The group ϑ_3 is unknown except for a famous surjection

$$\vartheta_3 \xrightarrow{r} Z/2 .$$ [1]

If V happens to be a topological manifold the dual cohomology class

$$(r_* \vartheta) \in H^4(V, Z/2)$$

is the obstruction to a combinatorial triangulation discussed by Kirby and Siebenmann.[2]

b) In case R = Q we have rational homology manifolds. The obstruction groups in this case are not even finitely generated in all dimensions of the form 4k - 1. This is seen by using the determinant invariant in $Q^*/(Q^*)^2$ and the Brieskorn singularities.

---

[1] This Kervaire-Milnor-Rochlin invariant is one-eighth of the signature of W reduced mod 2, where ∂W is the homology sphere in question and W is assumed to be parallizable.

[2] Historically the classes ϑ and Bockstein $(r_* \vartheta)$ were studied in 1967 by the author in work on the Hauptvermutung. In fact the desire to understand the elusive Bockstein obstruction led to the discussion of this section.

Sketch of Proof of Theorem D

To see that the singularity chain is a cycle we look at the link of each $(s-1)$ - simplex. The non-singular part of this link provides just the cobordism needed to deduce that the coefficient of the boundary for this simplex is zero.[1]

The resolution $W \xrightarrow{f} V$ is constructed by replacing the normal cone to various s-simplices by a cobordism of its link to zero. (This assumes the chain is zero. If it is only homologous to zero there is some initial preparatory replacement along various $(s+1)$ - simplices.) The rest of W is constructed by coning. f is constructed by a natural collapse.

[1] This cycle argument has natural extensions down through the singularities. It seems that there is a host of a-priori obstructions with complicated coefficients somehow related to the higher order obstructions encountered in this process.

PONTRJAGIN CLASSES OF RATIONAL HOMOLOGY MANIFOLDS AND THE SIGNATURE OF SOME AFFINE HYPERSURFACES

F. Hirzebruch

Let X be a rational homology manifold in the sense of Thom (Symp. Intern. Top. Alg. 1956, p. 54-67, Universidad de México 1958). Thom defines Pontrjagin classes

$$p_i(X) \in H^{4i}(X, \mathbb{Q}).$$

Instead of defining

$$p(X) = 1 + p_1(X) + p_2(X) + \dots \in H^*(X, \mathbb{Q})$$

one can define

$$\mathcal{L}(X) = 1 + L_1(X) + L_2(X) + \dots \in H^*(X, \mathbb{Q})$$

(see Hirzebruch, Topological methods in algebraic geometry, third edition, Springer 1966, §1). The classes $p(X)$ and $\mathcal{L}(X)$ determine each other.

Let G_m be the group of mth roots of unity and

$$G_b = G_{b_0} \times \dots \times G_{b_n} \quad \text{where} \quad b = (b_0, \dots, b_n),\ b_k \geq 1.$$

Let G_b act on the complex projective space $P_n(C)$ (homogeneous coordinates $t_0, \dots, t_n$) as follows:

$$\alpha(t_0, \dots, t_n) = (\alpha_0 t_0, \dots, \alpha_n t_n),$$

$$\text{where } \alpha = (\alpha_0, \dots, \alpha_n) \in G_b.$$

The orbit space $P_n(C)/G_b$ is a rational homology manifold. The map $p : P_n(C) \to P_n(C)/G_b$ induces an isomorphism

$$p^* : H^*\big(P_n(C)/G_b, \mathbb{Q}\big) \to H^*\big(P_n(C), \mathbb{Q}\big).$$

$H^*\big(P_n(C), \mathbb{Q}\big)$ is the truncated polynomial ring $\mathbb{Q}[x]/(x^{n+1})$ where $x \in H^2\big(P_n(C), \mathbb{Z}\big)$ is the Poincaré dual of the hyperplane. Bott (not yet published) has calculated the

Pontrjagin classes of $P_n(C)/G_b$.

(1) $$p^*\mathcal{L}\left(P_n(C)/G_b\right) = \frac{1}{[b_0,\dots,b_n]} \sum_{0\leq\xi<\pi} \prod_{k=0}^{n} \frac{b_k x}{\tanh b_k(x+i\xi)}$$

where $[b_0, \dots, b_n]$ is the greatest common divisor of $b_0, \dots, b_n$. The sum is over all real numbers ξ with $0 \leq \xi < \pi$. Observe, however, that for any natural number $a \geq 1$ the term $\left(\tanh a(x + i\xi)\right)^{-1}$ is a power series in x if $a\xi \not\equiv 0 \mod \mathbb{Z}\pi$. Therefore, $\prod_{k=0}^{n} \frac{b_k x}{\tanh b_k(x+i\xi)}$ is divisible by x^{n+1}, and thus vanishes in the truncated polynomial ring, if for all k we have $b_k\xi \not\equiv 0 \mod \mathbb{Z}\pi$. Therefore, the above sum (1) is actually only over the finitely many ξ for which

$$b_k\xi \equiv 0 \mod \mathbb{Z}\pi$$

for at least one k with $0 \leq k \leq n$.

Let N be a common multiple of $b_0, \dots, b_n$ and consider hypersurface

$$X^N : t_0^N + \dots + t_n^N = 0$$

in $P_n(C)$. Then $\alpha(X^N) = X^N$ for $\alpha \in G_b$ and $X^N/G_b \subset P_n(C)/G_b$. This "submanifold" X^N/G_b of $P_n(C)/G_b$ has a normal bundle ν in the sense of Thom (loc. cit.). It is a U(1)-bundle whose lift to X^N is the normal bundle of X^N in $P_n(C)$. Observe that for each $\alpha \in G_b$ the set $P_n(C)^\alpha$ of fixed points is transversal to X^N. We obtain for the signature of X^N/G_b

$$\operatorname{sign} X^N/G_b = \text{coefficient of } x^n \text{ in}$$

$$\tanh (Nx). \sum_{0\leq\xi<\pi} \prod_{k=0}^{n} \frac{x}{\tanh b_k(x+i\xi)} .$$

Here we used the fact that the map p has degree $b_0 \dots b_n/[b_0, \dots, b_n]$. Since N is a multiple of all the b_k, easy shifting of coordinates yields

(2) $$\operatorname{sign} X^N/G_b =$$

$$\sum_{0\leq\xi<\pi} \operatorname{res}_{i\xi} \left[\tanh Nz \cdot \prod_{k=0}^{n} \frac{1}{\tanh b_k z}\right]$$

where the sum is over those ξ with $0\leq\xi<\pi$ such that $\left(\prod_{k=0}^{n} \tanh b_k z\right)^{-1}$ has a pole in $i\xi$.

Let us denote the expression between brackets in (2) by $g(z)$. We integrate $g(z)$ along the following path γ_M in the z-plane.

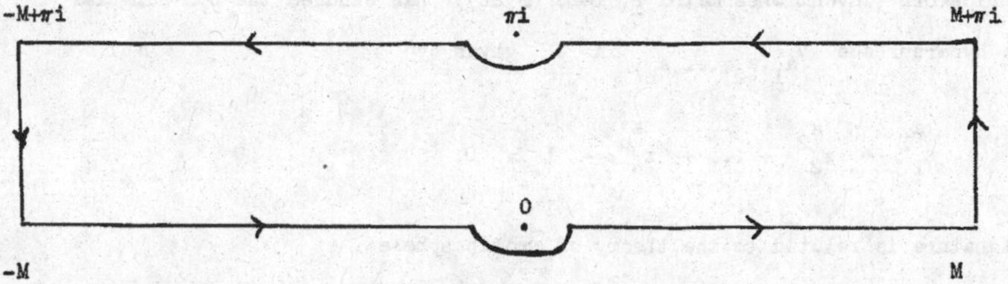

The integrals along the "horizontal" parts of γ_M cancel each other since $g(z + \pi i) = g(z)$. The sum of the integrals along the vertical parts of γ_M converges for $M \to \infty$ to

$$\pi i\left(1 - (-1)^n\right),$$

because $\tanh(M + iy)$ converges uniformly to 1 or -1 respectively if M converges to $+\infty$ or $-\infty$ respectively. The real dimension of X^N/G_b is $2(n - 1)$. Therefore, we suppose n odd from now on. Otherwise the signature vanishes by definition. We obtain from (2)

$$\text{(3)} \qquad \operatorname{sign} X^N/G_b = 1 - \sum_{0 \leq \eta < \pi} \operatorname{res}_{i\eta} g(z)$$

where the sum is now over those η with $0 \leq \eta < \pi$ such that $\tanh Nz$ has a pole in $i\eta$, which means

$$\eta = \frac{\pi j}{2N}, \quad j \text{ odd}, \ 1 \leq j < 2N.$$

The function $\tanh Nz$ has poles of order 1 at the $i\eta$. The residue in all these poles is $\frac{1}{N}$. By (3) we get (with $a_k = N/b_k$)

$$\text{(4)} \qquad \operatorname{sign} X^N/G_b - 1 =$$

$$\frac{(-1)^{(n-1)/2}}{N} \sum_{\substack{j \text{ odd} \\ 1 \le j < 2N}} \cot \frac{\pi j}{2a_0} \cot \frac{\pi j}{2a_1} \cdots \cot \frac{\pi j}{2a_n} .$$

Brieskorn (Inventiones math. 2, 1-14 (1966)) has studied the non-singular affine hypersurface $V_{a_1,a_2,\ldots,a_n}$ in C^n given by

$$(5) \qquad z_1^{a_1} + z_2^{a_2} + \ldots + z_n^{a_n} + 1 = 0;$$

its signature is related to the theory of exotic spheres.

Let E be the hyperplane $t_0 = 0$ in $P_n(C)$. Let N be any common multiple of $a_0, a_1, \ldots, a_n$ and

$$(6) \qquad b_k = \frac{N}{a_k} \quad (\text{for } k = 1, \ldots, n), \quad b_0 = 1, \; a_0 = N.$$

$Y^N = X^N - X^N \cap E$ is given in $P_n(C) - E$ by

$$t_1^N + t_2^N + \ldots + t_n^N + 1 = 0 \quad (\text{put } t_0 = 1).$$

By the map $z_k = t_k^{b_k}$ from $P_n(C) - E$ to C^n $(k = 1, \ldots, n)$ we have

$$(7) \qquad V_{a_1,\ldots,a_n} = Y^N/G_b$$

where $G_b = G_{b_0} \times G_{b_1} \times \ldots \times G_{b_n}$ and $b = (1, b_1, \ldots, b_n)$ as in (6).

The Lefschetz theorem on hyperplane sections implies the following fact. $X^N \cap E$ has a tubular neighbourhood T in X^N invariant under G_b. The middle dimensional homology group of T is infinite cyclic with a generator (invariant under G_b) of self-intersection number + 1. By the Novikov additivity of the signature we get from (4) and (7)

<u>Theorem</u> <u>Let</u> n <u>be odd and</u> N <u>any common multiple of</u> $a_1, \ldots, a_n$. <u>Then the signature of the Brieskorn variety</u> $V_{a_1,\ldots,a_n}$ <u>is given by the formula of Zagier</u>

(8) $$\operatorname{sign} V_{a_1,\ldots,a_n} = \frac{(-1)^{(n-1)/2}}{N} \sum_{\substack{j \text{ odd} \\ 1\le j<2N}} \cot\frac{\pi j}{2N} \cot\frac{\pi j}{2a_1} \ldots \cot\frac{\pi j}{2a_n} .$$

Brieskorn (loc. cit.) gives the following formula

$$\operatorname{sign} V_{a_1,\ldots,a_n} =$$

(9) $$\# \left\{ 0 < x_k < a_k \;\middle|\; 0 < \sum_{k=1}^{n} \frac{x_k}{a_k} < 1 \mod 2 \right\}$$

$$- \# \left\{ 0 < x_k < a_k \;\middle|\; 1 < \sum_{k=1}^{n} \frac{x_k}{a_k} < 2 \mod 2 \right\} ,$$

[$(x_1, \ldots, x_n)$ are n-tuples of integers].

Zagier has proved by Fourier series and by other methods that the two expressions in (8) and (9) equal each other. The interesting formula (8) is due to him. In virtue of Zagier's result, we have given a new method (involving Pontrjagin classes) to calculate sign $V_{a_1,\ldots,a_n}$ and to prove (9).

We can identify (8) and (9) in the following way (which is essentially Zagier's method). Put

$$((x)) = x - [x] - \tfrac{1}{2} , \quad \text{if } x \text{ is not an integer ;}$$

$$((x)) = 0 , \quad \text{if } x \text{ is an integer.}$$

Then the expression in (9) is

(10) $$\operatorname{sign} V_{a_1,\ldots,a_n} =$$

$$2 \sum_{0<x_k<a_k} \left(\left(\frac{x_1}{2a_1} + \ldots + \frac{x_n}{2a_n} + \tfrac{1}{2} \right)\right) - \left(\left(\frac{x_1}{2a_1} + \ldots + \frac{x_n}{2a_n} \right)\right) .$$

If $r = \frac{p}{q}$ is any positive rational number (where p, q are natural numbers, not necessarily coprime), then

$$((r)) = \frac{i}{2q} \sum_{j=1}^{q-1} \cot \frac{\pi j}{q} \, e^{2\pi i . jr} .$$

This is a formula of Eisenstein (see Rademacher, Lectures on Analytic Number Theory, Notes, Tata Institute, Bombay 1954-55, p.276). Feeding it into (10) gives (8).

Remark: The Dedekind sums studied by Rademacher are in close relation to formula (8) and the Atiyah-Bott-Singer fixed point theorem applied to the "signature operator" as will be explained elsewhere.

H. A. Hamm (Dissertation Bonn-Göttingen : see also the following paper) has studied the following affine varieties (given by r equations in C^n)

$$c_{j1} z_1^{a_1} + \dots + c_{jn} z_n^{a_n} + c_{j,n+1} = 0,$$

$j = 1, \dots, r$ and $r \leq n$.

If all $s \times s$ subdeterminants of the $r \times (n+1)$-matrix (c_{jk}) are different from 0 for $1 \leq s \leq r$, then the affine variety is a non-singular complete intersection of hypersurfaces. (Our conditions are stronger than those of Hamm.) We denote such a variety by $V^r_{a_1,\dots,a_n}$. Its complex dimension is $n - r$. If we assume $n - r$ to be even, then the same method as above yields

(11) $$\operatorname{sign} V^r_{a_1,\dots,a_n} =$$

$$- \sum_{\substack{1 \leq j < 2N \\ j \text{ odd}}} \operatorname{res}_{\pi i j/2N} \left((\tanh Nz)^r \coth z \prod_{k=1}^{n} \coth \frac{Nz}{a_k} \right)$$

where N is any common multiple of $a_1, \dots, a_n$. But it seems harder to get a formula similar to (8) or (9) because we have poles of order > 1. Bott's proof of (1) involves also residue calculations and there is in fact a short cut to (8) or (9) from a point on the way to (1). But it seemed amusing to adopt the view of somebody knowing (1) and not its proof and to begin to calculate.

TOPOLOGY OF ISOLATED SINGULARITIES OF COMPLEX SPACES

H. A. Hamm

In this talk I should like to present some results which are mainly contained in my doctoral dissertation [5] and will be published in [6], [7] and [8]. These results concern the local topology of complex spaces. As some theorems will be formulated not for a single complex space, but for a pair of them, we shall assume that the following data are given:

Let $g_1,\dots, g_s, f_1,\dots f_r$ $(r, s \geq 0)$ be holomorphic functions which are defined on a neighbourhood of the origin in $\mathbb{C}^m$, and let ϵ be a positive real number which we shall assume to be "sufficiently small". Define $D = \{z\in \mathbb{C}^m \mid \|z\|\leq\epsilon\}$, $\overline{X}^* = \{z\in D \mid g_1(z) = \dots = g_s(z) = 0\}$, $\overline{X} = \{z\in\overline{X}^* \mid f_1(z) = \dots = f_r(z) = 0\}$, $\Sigma^* = \overline{X}^* \cap \partial D$, $\Sigma = \overline{X} \cap \partial D$, $X^* = \overline{X}^* - \Sigma^*$, $X = \overline{X} - \Sigma$. It will always be assumed that $X^* - X$ is regular, i.e. a complex manifold, and that $0 \in X$.*

If 0 is the only singular point of X, we can ask the following question:

Can X be a topological (but not complex) manifold?

Milnor has shown (cf. [10] 2.9 and 2.10) that Σ is a compact orientable C^∞-manifold and $\overline{X}$ is homeomorphic to the cone $\{t\cdot z \mid t \in [0,1], z \in \Sigma\}$ over Σ. Therefore X is a topological manifold if and only if Σ is homeomorphic to a sphere (but Σ may have an exotic differentiable structure). The first examples have been found in the case where X is a complex hypersurface, i.e. where $s = 0$, $r = 1$ (see Brieskorn [1], [3]).

In my doctoral dissertation [5] I have proved the following theorem which shows that there exist also examples where X is not a complex hypersurface (cf. also [6], [7]). It is a natural generalization of a theorem of Brieskorn ([1] Satz 1). First let us introduce the following notations. Let $a_1,\dots,a_m$ be natural numbers ≥ 2, $G_a = G_{a_1,\dots,a_m}$ the graph (i.e. one-dimensional simplicial complex) which contains m vertices $x_1,\dots,x_m$ (corresponding to $a_1,\dots,a_m$) and in which two different vertices x_i and x_j are joined by a line segment if and only if the greatest common divisor (a_i, a_j) is ≥ 2. Let (B) be the following condition for connectedness components K of G_a: K consists of an odd number of points, and for all corresponding a_i, a_j $(i \neq j)$ one has $(a_i, a_j) = 2$. (Of course, there can be only one K of this kind having more than one point.)

Theorem 1: If $(\alpha_{\mu\nu})$ is a real $(r \times m)$ - matrix whose $(r \times r)$-subdeterminants do not vanish, and if $f_\mu(z) = \sum_{\nu=1}^{m} \alpha_{\mu\nu} z_\nu^{a_\nu}$, $\mu = 1,\dots, r$; $s = 0$; $0 < r \leq m - 3$, then

(i) If there are at least $r + 1$ components of G_a which fulfil condition (B), then Σ is a topological sphere.

(ii) If $2 \leq r \leq m + 2$, the converse of (i) is also true.

The proof is performed by generalizing considerations due to Milnor [10] in the case of hypersurfaces. Because of the generalized Poincaré conjecture it is - in the case $\dim \Sigma \geq 5$ - sufficient to check the fundamental group and the homology groups of Σ with integer coefficients. Therefore the following theorem is useful, which holds without the assumption that $X - \{0\}$ is regular.

Theorem 2 (cf. [5] 2.6, [8] 1.3 and Milnor [10] 5.2):
If X is a complete intersection of dimension n, then Σ is $(n - 2)$-connected.

X is a complete intersection, if all irreducibility components of X have dimension $m - s - r$. Theorem 2 is a corollary of the following theorem. Let $\dim^{-} X^*$ be the smallest dimension of the irreducibility components of X^*.

Theorem 3 (cf. [8] 2.9): (Σ^*, Σ) is $(\dim^{-} X^* - r - 1)$-connected.

This theorem is proved by applying Morse theory to $\|f\|^2 : \Sigma^* - \Sigma \to \mathbb{R}$. For the proof of theorem 2, we can assume $s = 0$ (without loss of generality). But then it can be derived from theorem 3 by application of the exact homotopy sequence for (Σ^*, Σ).

Now let us return to the case where X is a complete intersection of dimension $n \geq 3$, $X - \{0\}$ regular. In order to decide whether Σ is a topological sphere or not, it is only necessary to check the group $H_{n-1}(\Sigma; \mathbb{Z})$. In the case $s = 0$, $r = 1$ the problem of computing this group has been treated by Milnor (cf. [10] §8). In the general case I have tried two procedures:

a) Reduction to the case $s = 0, r = 1$ by application of the Mayer-Vietoris sequence (cf. [5] §4 and [6]): Assume e.g. $s = 0, r = 2$. Then $\Sigma = \Sigma_1 \cap \Sigma_2$, where $\Sigma_i = \{z \in \partial D | f_i(z) = 0\}$, $i = 1, 2$, and $\Sigma_1 \cup \Sigma_2 = \{z \in \partial D | f_1(z) \cdot f_2(z) = 0\}$. As for Σ_1, Σ_2 and $\Sigma_1 \cup \Sigma_2$, we have the case of complex hypersurfaces; but $\Sigma_1 \cup \Sigma_2$ has a non-isolated singularity. Nevertheless we can essentially apply the methods described by Milnor; and if Σ_1 and Σ_2 are topological spheres, then we can get information on $H_{n-1}(\Sigma; \mathbb{Z})$ by looking at the Mayer-Vietoris sequence for (Σ_1, Σ_2). For greater r, there exists a similar procedure.

b) Suppose we have a pair (Σ^*, Σ), where X^* is also a complete intersection, $X^* \neq X$, $r = 1$, $X^* - \{0\}$ regular. If we can prove that $H_n(\Sigma^*, \Sigma; \mathbb{Z}) = 0$, then we can conclude by theorem 2 and Poincaré duality that Σ is a topological sphere (and Σ^* too).

For procedure b it is important to have a theorem which shows how to compute $H_n(\Sigma^*, \Sigma; \mathbb{Z})$. In the following theorem, we make no assumption that $X^* - \{0\}$ and $X - \{0\}$ are regular or that X^* or X are complete intersections; but we do assume that X^* and X are pure-dimensional (i.e. that each irreducibility component of X^* resp. X has the same dimension; this restriction is not important), and that $\dim X^* = \dim X + r$. We denote the dimension of X by n.

Theorem 4 (cf. [5] 3.1, 3.6, 4.1; [8] 1.6, 1.7, 1.8 and Milnor [10] 4.8, 5.1 and 8.4):

If $r = 1$, then $\frac{f_1}{|f_1|} : \Sigma^* - \Sigma \to S^1$ defines a C^∞-fibre bundle, whose fibres F have the homotopy type of a finite n-dimensional cell complex. The Wang sequence leads to the following exact sequence (with integer coefficients):

$$\dots \to H_{n+2}(\Sigma^*, \Sigma) \to H^{n-1}(F) \to H^{n-1}(F) \to H_{n+1}(\Sigma^*, \Sigma)$$
$$\to H^n(F) \xrightarrow{h^* - id} H^n(F) \to H_n(\Sigma^*, \Sigma) \to 0,$$

where h^* is the so-called monodromy, defined by the action of the fundamental group of S^1 on the cohomology of the fibre. Therefore $H_n(\Sigma^*, \Sigma; \mathbb{Z}) = 0$ if and only if $h^* - id$ is an epimorphism.

It is essential in theorem 4 that we assume $r = 1$: the mapping

$\frac{f}{\|f\|} : \Sigma^* - \Sigma \to S^{2r-1} \subset \mathbb{C}^r$ - where $f = (f_1,\ldots,f_r)$ - does not define a C^∞-fibre bundle for $r > 1$ (cf. [4] 1.2.4). Here we have the very reason why it is inconvenient to put $X^* = \mathbb{C}^m$, if X is not a hypersurface.

In the special example considered in theorem 1, I have followed procedure b and used the generalized method of Pham [11] for the computation of the monodromy (cf. [5] §5 and [7]). In general, this method of calculating the monodromy does not exist. However, Brieskorn [4] has developed another method, using differential forms: by this method I could prove part (i) of theorem 1 (cf. [5] §7, [6]). In this case, I have followed procedure a; but I hope that the method can also be applied to procedure b, perhaps generalizing Brieskorn [2].

(Added September 1970) It is also possible to get a general statement concerning the connectivity properties of $\Sigma^* - \Sigma$ (cf. [8] 1.5 and 2.19), which might be interesting if X consists of the singular points of X^*.

At the end let us say a few words about the differentiable structure of Σ if Σ is a topological sphere. Assume that $X - \{0\}$ and $X^* - \{0\}$ are regular and $r = 1$. Note that Σ bounds a parallelizable (n-1)-connected manifold, namely the closure $\overline{F}$ of the fibre F in X^* (cf. [8] 1.7), provided that X and X^* are complete intersections of dimension n resp. $n+1$. If n is odd and Σ^* is a topological sphere too, then the monodromy determines the differentiable structure of Σ (up to diffeomorphism; cf. [8] 1.11), and for the example of theorem 1 I can give an explicit condition when Σ is a Kervaire sphere (cf. [7]). If n is even, you must look at the signature of $\overline{F}$; for the special example of theorem 1 the reader is here referred to Hirzebruch's talk [9].

References

[1] Brieskorn, E. Beispiele zur Differentialtopologie von Singularitäten. Inventiones Math. 2, 1-14 (1966)

[2] Brieskorn, E. Die Monodromie der isolierten Singularitäten von Hyperflächen. Manuscripta Math. 2, 103-161 (1970)

[3] Brieskorn, E. Examples of Singular Normal Complex Spaces which are Topological Manifolds. Proc. Nat. Acad. Sci. U.S.A., 55, 1395-1397 (1966)

[4] Brieskorn, E. Singularitäten von Hyperflächen. Manuscript.

[5] Hamm, H. A. Die Topologie isolierter Singularitäten von vollständigen Durchschnitten komplexer Hyperflächen. Doctoral dissertation, Bonn 1969.

[6] Hamm, H. A. Ein Beispiel zur Berechnung der Picard - Lefschetz - Monodromie für nichtisolierte Hyperflächensingularitäten. To appear.

[7] Hamm, H. A. Exotische Sphären als Umgebungsränder in speziellen komplexen Räumen. To appear.

[8] Hamm, H. A. Lokale topologische Eigenschaften komplexer Räume. To appear.

[9] Hirzebruch, F. Pontrjagin classes of rational homology manifolds and the signature of some affine hypersurfaces. This volume, pp. 207 - 212.

[10] Milnor, J. Singular Points of Complex Hypersurfaces. Ann. of Math. Studies 61, Princeton University Press (1968)

[11] Pham, F. Formules de Picard - Lefschetz généralisées et ramification des intégrales. Bull. Soc. Math. de France 93, 333-367 (1965)

DEFORMATIONS EQUISINGULIÈRES DES IDÉAUX JACOBIENS DE COURBES PLANES

F. Pham

Introduction

Pour un topologue, un des problèmes les plus naturels à se poser en Géométrie algébrique est celui de l'équisingularité: étant donné une famille algébrique de germes d'ensembles algébriques, définir un ouvert dense de Zariski de valeurs du paramètre (l'ouvert d' "équisingularité") pour lesquelles la famille sera triviale du point de vue topologique [4].

Dans le cas des familles de courpes planes on dispose de nombreuses définitions équivalentes de l'équisingularité, dues à Zariski [7]. Remarquant que l'une de ces définitions (celle qui fait intervenir la résolution des singularités) se généralise immédiatement aux familles d'idéaux dans le plan, nous montrons que même pour les idéaux elle est susceptible d'une interprétation topologique (§2, théorème de trivialité). Nous montrons ensuite (§3) qu'une famille de courbes dont les idéaux jacobiens sont équisinguliers est une famille équisinguliere de courbes, maix que la reciproque est fausse (§3, exemple 3) - ce dernier résultat a peut-être de quoi surprendre le topologue naif: car comment interpréter les valeurs "spéciales" du paramètre pour lesquelles le jacobien d'une famille "equisingulière" de courbes sera "plus singulier" qu'ailleurs?

Nos démonstrations seront transcendantes, et utiliseront une technique de construction de champs de vecteurs (étudiée en Appendice) que j'espère pouvoir appliquer à l'étude de l'équisingularité en dimension quelconque.

Precautions de langage

Nous utiliserons le langrage de la Géométrie analytique, plus adapté aux démonstrations transcendantes (mais pas plus général pour ce qui nous intéresse, puisque toute singularité analytique de courbe plane est isomorphe a une singularité

Communication envoyée au : Premier Congrès de Mathématiques du Viêt Nam (Hanoi, Automne 1970).

algébrique). Bien que le problème qui nous intéresse soit local il sera commode d'utiliser le langage de la Géométrie analytique globale, mais pour éviter d'avoir à prendre des précautions "à l'infini", nous supposerons, chaque fois que nous parlerons d'un "Idéal sur un espace analytique", qu'il s'agit en fait de la restriction d'un Idéal défini sur un autre espace analytique dans lequel l'espace donné se plonge comme ouvert relativement compact.

1 - Résolutions d'idéaux plans

Soit $\mathcal{I}$ un "Idéal" non trivial (faisceau cohérent d'idéaux non nuls) sur une variété analytique complexe (lisse) Z. On dit que l'Idéal $\mathcal{I}$ est, au point $z \in Z$, un "diviseur à croisements normaux" (ou plus brièvement, que l'Idéal $\mathcal{I}$ est "résolu" au point z) si la variété Z admet un système, centré en z, de coordonnées analytiques locales $(z_1, z_2, \ldots, z_n)$ telles que $\mathcal{I}_z = z_1^{a_1} z_2^{a_2} \ldots z_n^{a_n} \mathcal{O}_{Z,z}$ (où $\mathcal{O}_{Z,z}$ désigne l'anneau local de Z en z, et les a_i sont des entiers non négatifs). En particulier, on dira que $\mathcal{I}$ est, au point z, un "diviseur lisse" (ou plus brièvement, que l'Idéal $\mathcal{I}$ est "lisse" au point z) s'il peut s'écrire $\mathcal{I}_z = z_1^{a_1} \mathcal{O}_{Z,z}$.

Un morphisme analytique bi-méromorphe

$$\epsilon : Z' \to Z$$

est appelé "résolution" de l'Idéal $\mathcal{I}$ si Z' est lisse et si l'Idéal $\mathcal{I} \mathcal{O}_{Z'}$ est résolu en tout point de Z' ($\mathcal{I} \mathcal{O}_{Z'}$ est une abreviation pour $\epsilon^*(\mathcal{I}). \mathcal{O}_{Z'}$).

Fixons quelques notations :

$|\mathcal{I}|$ = "support" de $\mathcal{I} = \{z \in Z \mid \mathcal{I}_z \neq \mathcal{O}_{Z,z}\}$;

$[\mathcal{I}]$ = "support singulier" de $\mathcal{I} = \{z \in Z \mid \mathcal{I}_z$ non lisse$\}$.

Ces deux ensembles $|\mathcal{I}|$ et $[\mathcal{I}]$ sont des sous-espaces analytiques (réduits) de Z. De plus la codimension de $[\mathcal{I}]$ est toujours ≥ 2.

Nous nous intéresserons ici au cas où la variété Z est à deux dimensions; nous dirons dans ce cas que $\mathcal{I}$ est un "Idéal plan" (comme on dit une "courbe plane"), par référence au modèle local où Z est un ouvert du plan $\mathbb{C}^2$

(d'ailleurs notre vrai problème est local, et le langage des variétés n'est introduit ici que pour permettre des énoncés plus concis).

Lorsque $\dim Z = 2$, le support singulier $[\mathcal{I}]$ est un ensemble analytique de dimension zéro, ce qui permet d'associer canoniquement à $\mathcal{I}$ une suite de "transformations monoïdales" (éclatements à centres ponctuels)

$$\cdots \xrightarrow{\epsilon_3} Z_2 \xrightarrow{\epsilon_2} Z_1 \xrightarrow{\epsilon_1} Z$$

ou ϵ_1 est l'éclatement de centre $[\mathcal{I}]$,

ϵ_2 est l'éclatement de centre $[\mathcal{I}\, \mathcal{O}_{Z_1}]$,

ϵ_3 est l'éclatement de centre $[\mathcal{I}\, \mathcal{O}_{Z_2}]$,

etc... .

On démontre qu'au bout d'un nombre fini de ces opérations l'idéal est résolu. Le morphisme composé $\epsilon = \epsilon_1 \circ \epsilon_2 \circ \ldots \circ \epsilon_r$ est alors appelé <u>résolution canonique de l'Idéal</u> $\mathcal{I}$.

2 - <u>Familles équisingulières d'idéaux plans</u>

<u>Définition 0</u> : <u>Une famille</u> $(\mathcal{I},\pi)$ <u>à un paramètre d'Idéaux plans</u> est un couple $(\mathcal{I},\pi)$, où $\mathcal{I}$ est un Idéal sur une variété analytique Z à trois dimensions, tandis que $\pi : Z \to T$ est un morphisme analytique <u>lisse</u> de Z sur une variété analytique T de dimension 1; on exigera de plus que pour tout $t \in T$, $\mathcal{I}_t = \mathcal{I}|\pi^{-1}(t)$ soit un Idéal non trivial de $Z_t = \pi^{-1}(t)$.

<u>Définition 0'</u>: La famille $(\mathcal{I},\pi)$ est dite <u>propre</u> si le morphisme π est propre.

<u>Définition 1</u> : La famille $(\mathcal{I},\pi)$ est dite <u>équisingulière</u> s'il existe un entier r tel que la suite de conditions 0), 1), 2), ..., r) suivantes soit satisfaite:

0) $\pi|[\mathcal{I}]$ est unmorphisme lisse (c'est-à-dire, $[\mathcal{I}]$ est une courbe lisse se projetant avec rang maximum sur T) ;

1) en désignant par $\epsilon_1 : Z_1 \to Z$ l'éclatement de centre $[\mathcal{I}]$, le morphisme $\pi \circ \epsilon_1 |[\mathcal{I}\, \mathcal{O}_{Z_1}]$ est lisse ;

2) en désignant par $\epsilon_2 : Z_2 \to Z_1$ l'éclatement de centre $[\mathscr{I}\,\mathcal{O}_{Z_1}]$, le morphisme $\pi \circ \epsilon_1 \circ \epsilon_2 \,|[\mathscr{I}\,\mathcal{O}_{Z_2}]$ est lisse ;

... etc ..., jusqu'à :

r) en désignant par $\epsilon_r : Z_r \to Z_{r-1}$ l'éclatement de centre $[\mathscr{I}\,\mathcal{O}_{Z_{r-1}}]$, le morphisme $\pi \circ \epsilon_1 \circ \epsilon_2 \circ \ldots \circ \epsilon_r \,|[\mathscr{I}\,\mathcal{O}_{Z_r}]$ est lisse, et l'Idéal $\mathscr{I}\,\mathcal{O}_{Z_r}$ est résolu.

Le morphisme composé $\epsilon = \epsilon_1 \circ \epsilon_2 \circ \ldots \circ \epsilon_r$ est alors appelé <u>résolution équisingulière canonique de la famille</u> $(\mathscr{I},\pi)$.

<u>Définition 1'</u>: La famille $(\mathscr{I},\pi)$ est dite <u>équisingulière au point</u> $z \in Z$ si ce point admet un voisinage $U \subset Z$ tel que $(\mathscr{I}|U,\ \pi|U)$ soit une famille équisingulière.

<u>Remarque 1</u> : <u>Pour toute famille</u> $(\mathscr{I},\pi)$, <u>l'ensemble des points de</u> Z <u>où cette famille n'est pas équisingulière a pour image dans</u> T <u>un ensemble de mesure nulle, et même, dans le cas d'une famille propre, un ensemble localement fini.</u>

En effet, tout Idéal $\mathscr{I}$ dans une variété Z de dimension 3 peut être résolu par une succession finie d'éclatements dont les centres sont des courbes lisses ou des points (courbes composant le support singulier, ou points "spéciaux" de ces courbes, ou points isolés du support singulier : c'est - généralisé aux Idéaux - le procédé de résolution de Beppo Levi [3]). A chaque étape de cette résolution, l'ensemble des points où le morphisme π restreint au support singulier n'est pas lisse est évidemment un ensemble analytique de dimension zéro.

<u>Exemple</u>: Dans le cas particulier où $(\mathscr{I},\pi)$ est une famille d'idéaux principaux, représentant une famille de courbes planes, la Définition 1 n'est autre que l'une des définitions équivalentes de l'équisingularité des courbes proposées par Zariski dans [7].

<u>Théorème de Trivialité</u> : <u>Si</u> $(\mathscr{I},\pi)$ <u>est une famille propre et équisingulière d'Idéaux plans, l'éclatement de</u> $\mathscr{I}$ <u>est fibré topologiquement par</u> π.

De facon précise, désignons par $\hat{\epsilon} : \hat{Z} \to Z$ le morphisme d'éclatement de l'Idéal $\mathscr{I}$ dans Z. Par l'expression "l'éclatement est fibré topologiquement par π" nous entendons ceci :

- tout $t \in T$ admet un voisinage V tel qu'on ait un diagramme commutatif

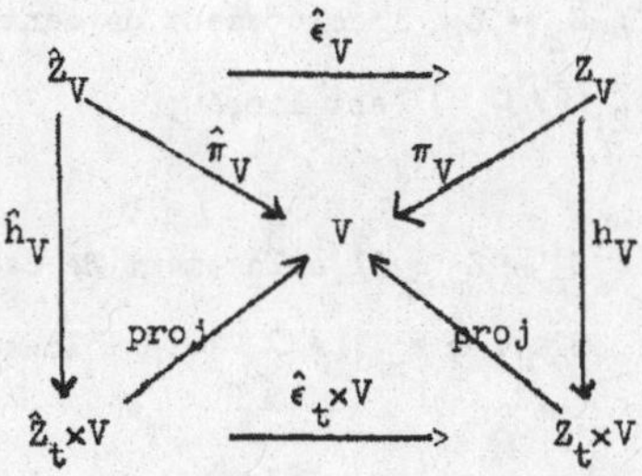

où h_V et $\hat{h}_V$ sont des homéomorphismes, et où h_V transforme $|\mathcal{I}_V|$ en $|\mathcal{I}_t| \times V$, de sorte que $\hat{h}_V$ transforme $|\mathcal{I}_V \mathcal{O}_{\hat{Z}_V}| = \hat{\epsilon}_V^{-1}(|\mathcal{I}_V|)$ en $\hat{\epsilon}_t^{-1}(|\mathcal{I}_t|) \times V$.
(L'indice V (resp. t) au-dessous d'une lettre signifie la restriction, au-dessus de l'ouvert V (resp. au-dessus du point t), de l'objet désigné par cette lettre).

Exemple : Cas d'une famille de courbes planes : c'est le cas où $\mathcal{I}$ est un Idéal localement principal dans Z; l'éclatement de $\mathcal{I}$ est alors l'application identique de Z. Le Théorème de Trivialité signifie alors simplement qu'une famille équisingulière de courbes planes est topologiquement triviale (mais pas toujours analytiquement, cf. le contre-exemple de Whitney [5], exemple 13.1) : c'était un résultat de Zariski.

Conjecture : Réciproquement, si $(\mathcal{I}, \pi)$ est une famille d'Idéaux plans telle que l'éclatement de $\mathcal{I}$ soit fibré topologiquement par π, cette famille est-elle équisingulière au sens de la Définition 1 ?

Cette conjecture est vraie dans le cas des courbes.

Preuve du Théorème :

Lemme 1 - Soit $\epsilon : Z' \to Z$ la résolution équisingulière canonique d'une famille équisingulière $(\mathcal{I}, \pi)$.
Alors la projection $\pi \circ \epsilon : Z' \to T$ est, au voisinage de tout point $z' \in Z'$, analytiquement équivalente à un modèle local

$$\mathbb{C}^3 \to \mathbb{C}$$
$$(x,y,t) \mapsto t$$

avec $\mathcal{I}\mathcal{O}_{Z'} = x^a y^b \mathcal{O}_{Z'}$ (a,b entiers non négatifs).

En effet, notons $H = \epsilon^{-1}([\mathcal{I}])$ le "diviseur exceptionnel" (réduit). Le Lemme

1 est une conséquence du théorème des fonctions implicites, en choisissant les fonctions x et y de la facon suivante :

1er cas : $z' \notin H$; alors ϵ est, au voisinage de z', un isomorphisme, et il suffit de prendre pour x,y n'importe quel couple de coordonnées "générales" sur Z' (alors $a=b=0$) ;

2e cas : $z' \in H$ qui est, au voisinage de z', une surface lisse ; cette surface est alors isomorphe au diviseur exceptionnel d'un éclatement (à centre lisse) ϵ_i, à savoir le dernier éclatement de la suite $\epsilon_1,\epsilon_2,\ldots,\epsilon_r$ qui n'est pas un isomorphisme local en z'; alors $\epsilon_i|H$ est une application de rang 1, dont l'image (le centre de l'éclatement ϵ_i) s'envoie isomorphiquement sur T (en vertu de la i-ème condition de la Définition 1), de sorte que rang $(\pi \circ \epsilon|H) = 1$; on en déduit le Lemme 1 (avec $a>0$, $b=0$), en prenant pour x n'importe quelle équation locale de H et pour y une coordonnée "générale" quelconque ;

3e cas : $z' \in H_1 \cap H_2$, intersection de deux surfaces lisses, se coupant transversalement, dont l'union est le diviseur exceptionnel ; la Définition 1 demande à $\pi \circ \epsilon|H_1 \cap H_2$ d'être une application de rang 1, ce qui implique le Lemme 1 (avec $a>0$, $b>0$) en prenant pour x (resp. y) n'importe quelles équations locales de H_1 (resp. H_2).

Lemme 2 - Soit $\epsilon : Z' \to Z$ la résolution équisingulière canonique d'une famille équisingulière $(\mathscr{I},\pi)$. Alors tout champ de vecteurs D_T holomorphe sur T se relève dans Z en un champ de vecteurs ϵ-extensible (Appendice, Définition 3'), respectant l'Idéal $\mathscr{I}$ (Appendice, Définition 4).

En effet, recouvrons Z' par des ouverts U_i' dans lesquels un modèle comme celui du Lemme 1 s'applique. On peut évidemment relever D_T, dans chaque ouvert U_i', en un champ de vecteurs holomorphe

$$D_i' : \mathcal{O}_{U_i'} \to \mathcal{O}_{U_i'}$$

tel que

$$D_i'\mathscr{I}\mathcal{O}_{U_i'} \subset \mathscr{I}\mathcal{O}_{U_i'} ;$$

il suffit de prendre pour D_i' le champ de vecteurs égal à $g(t)\,\frac{\partial}{\partial t}$ dans les coordonnées (x,y,t) du Lemme 1, où $g(t)\frac{\partial}{\partial t}$ est l'expression de D_T dans la coordonnée t de T.

Ceci fait, posons $D' = \sum_i \phi_i D_i'$, où (ϕ_i) est une partition différentiable de l'unité subordonnée au recouvrement (U_i'). Evidemment D' est une dérivation de $\mathcal{O}_{Z'}$ dans $\mathcal{E}_{Z'}$, et $D'\mathcal{I}\mathcal{O}_{Z'} \subset \mathcal{I}\mathcal{E}_{Z'}$.

Il ne reste plus qu'à montrer que D' est compatible avec la projection ϵ, de facon à définir un champ de vecteurs (non nécessairement différentiable) sur Z. C'est évident dans l'ouvert dense où ϵ est un isomorphisme, c'est-à-dire en-dehors du "diviseur exceptionnel" $H = \epsilon^{-1}([\mathcal{I}])$. Reste à le montrer sur H. Or ϵ projette H sur la courbe $[\mathcal{I}]$, et π restreint à cette courbe est un isomorphisme analytique local (théorème des fonctions implicites); donc le champ de vecteurs D_T se relève de facon unique dans $[\mathcal{I}]$ et la commutativité du diagramme

$$\begin{array}{ccc} H & \xrightarrow{\epsilon|H} & [\mathcal{I}] \\ & {\scriptstyle \pi\circ\epsilon|H}\searrow & \downarrow {\scriptstyle \pi|[\mathcal{I}]} \\ & & T \end{array}$$

montre que les champs de vecteurs D_i' restreints à H ont tous la même projection dans $[\mathcal{I}]$, de sorte que leur superposition D' est elle aussi compatible avec la projection.

Le Lemme 2 est entierement démontré. Nous pouvons maintenant démontrer le théorème.

Puisque $\mathcal{I}\mathcal{O}_{Z'}$ est un Idéal <u>inversible</u> (c'est-à-dire localement principal), on a une factorisation unique (propriété universelle de l'éclatement d'un Idéal):

$$\begin{array}{ccc} Z' & \xrightarrow{\epsilon'} & \hat{Z} \\ & {\scriptstyle \epsilon}\searrow & \downarrow {\scriptstyle \hat{\epsilon}} \\ & & Z \end{array}$$

D'après le Lemme 2, tout champ de vecteurs D_T holomorphe dans T se relève dans Z en un champ de vecteurs ϵ-extensible D respectant l'Idéal $\mathcal{I}$. Grâce au Théorème 2 de l'Appendice, on en déduit dans $\hat{Z}$ un champ de vecteurs ϵ'-extensible $\hat{D}$ qui relève D, et qui respecte l'Idéal $\mathcal{I}\mathcal{O}_{\hat{Z}}$. D'après le Théorème 1 de l'Appendice, $\hat{D}$ est localement intégrable dans $\hat{Z}$, donc intégrable sur l'image réciproque de tout ouvert $V \subset T$ assez petit (puisque π est propre). Son intégration réalise la trivialité topologique de $\hat{Z}$.

3 - Application aux idéaux jacobiens des courbes planes

Soit $f(x,y)$ l'équation réduite d'une courbe plane dans $\mathbb{C}^2$. L'idéal jacobien de la courbe plane est l'idéal J engendré par $\frac{\partial f}{\partial x}$, $\frac{\partial f}{\partial y}$ et f. Le support de cet idéal est l'ensemble des points singuliers de la courbe. Vu les liens nombreux qui semblent exister entre l'idéal jacobien et le "type de singularité" de la courbe, il est naturel de se demander s'il y a un rapport entre la résolution d'une courbe et la résolution de son idéal jacobien. Les deux exemples qui suivent semblent répondre : aucun rapport !

Exemple 1 : où la résolution d'une courbe ne résoud pas son idéal jacobien

$$f = x^5+y^2$$
$$J = (x^4,y,x^5 ty^2) = (x^4,y)$$

Si l'on commence la résolution canonique de f, on s'aperçoit que dès le troisième éclatement le support singulier de J contient des points qui ne sont pas singuliers pour f.

Schéma du calcul : effet de trois éclatements, calculé dans une certaine carte de la variété éclatée :

$$\begin{pmatrix} x_3 = x_2 \\ \\ y_3 = \frac{y_2}{x_2} \end{pmatrix} \xmapsto{\epsilon_3} \begin{pmatrix} x_2 = x_1 \\ \\ y_2 = \frac{y_1}{x_1} \end{pmatrix} \xmapsto{\epsilon_2} \begin{pmatrix} x_1 = x \\ \\ y_1 = \frac{y}{x} \end{pmatrix} \xmapsto{\epsilon_1} \begin{pmatrix} x \\ \\ y \end{pmatrix}$$

c'est-à-dire $\epsilon_1 \circ \epsilon_2 \circ \epsilon_3 : (x_3,y_3) \longmapsto (x = x_3,\ y = x_3^3 y_3)$.

$f = x_3^5(1 + x_3y_3^2) \sim x_3^5$ au voisinage du point $(x_3 = y_3 = 0)$, donc l'idéal (f) est lisse en ce point. Par contre $J = (x_3^4,\ x_3^3y_3)$ n'est pas résolu en ce point.

Exemple 2 : où la résolution de l'idéal jacobien ne résoud pas la courbe

$$f = x^7+y^3$$
$$J = (x^6,y^2,x^7+y^3) = (x^6,y^2).$$

Si l'on commence la résolution canonique de J, on s'aperçoit que dès le troisième éclatement le support singulier de f contient des points qui ne sont pas singuliers pour J.

Schéma du calcul : effet de trois éclatements, calculé dans une autre carte que celle de l'exemple 1 :

$$\begin{pmatrix} x_3' = \dfrac{x_2}{y_2} \\ \\ y_3' = y_2 \end{pmatrix} \xrightarrow{\epsilon_3} \begin{pmatrix} x_2 = x_1 \\ \\ y_2 = \dfrac{y_1}{x_1} \end{pmatrix} \xrightarrow{\epsilon_2} \begin{pmatrix} x_1 = x \\ \\ y_1 = \dfrac{y}{x} \end{pmatrix} \xrightarrow{\epsilon_1} \begin{pmatrix} x \\ \\ y \end{pmatrix}$$

c'est-à-dire $\epsilon_1 \circ \epsilon_2 \circ \epsilon_3 : (x_3', y_3') \mapsto (x = x_3' y_3' \, , \, y = x_3'^2 y_3'^3)$.

$J = (x_3'^6 y_3'^6 , \, x_3'^4 y_3'^6)$ est résolu au point $(x_3' = y_3' = 0)$.

Par contre $f = x_3'^7 y_3'^7 + x_3'^6 y_3'^9 = x_3'^6 y_3'^7 (x_3' + y'^2)$ n'est évidemment pas résolu.

Et pourtant, on a le

<u>Théorème d'équisingularité jacobienne :</u>

<u>Si les idéaux jacobiens d'une famille de courbes planes forment une famille équisingulière d'idéaux (§2, Définition 1), la famille de courbes est équisingulière.</u>

Il serait sûrement très instructif de chercher une démonstration algébrique de ce théorème. En voici une démonstration transcendante.

Notons $(\mathcal{J}, \pi)$ la famille des Idéaux jacobiens, et $(\mathfrak{m}, \pi)$ la famille des Idéaux maximaux suivant les singularités de la courbe (dans un modèle local où la famille de courbes est donnée par $f(x,y,t) = 0$, et où la singularité pour tout t est à l'origine du plan, $\mathcal{J}$ est l'Idéal engendré par $\dfrac{\partial f}{\partial x}$, $\dfrac{\partial f}{\partial y}$, f tandis que $\mathfrak{m}$ est l'Idéal engendré par x et y).

Supposant que $(\mathcal{J}, \pi)$ est une famille équisingulière, notons $\epsilon : Z' \to Z$ sa résolution canonique.

Evidemment $|\mathfrak{m}| = |\mathcal{J}|$, de sorte que dès le premier éclatement $\mathfrak{m}$ est résolu. ϵ est donc aussi la résolution équisingulière canonique de $\mathfrak{m} . \mathcal{J}$, de sorte que d'après le théorème de trivialité (§2), l'éclatement $\hat{\epsilon} : \hat{Z} \to Z$ de l'Idéal $\mathfrak{m} . \mathcal{J}$ est fibré topologiquement par π, c'est-à-dire que le morphisme $\pi \circ \hat{\epsilon} : (\hat{Z}, \hat{H}) \to T$ (où $\hat{H}$ désigne le diviseur exceptionnel de $\hat{\epsilon}$) est une fibration topologique de paires. En particulier $(\pi \circ \hat{\epsilon}) | \hat{H}$, étant une fibration topologique, est à fibres équidimensionnelles, ce qui implique la trivialité topologique de la famille de courbes (résultat non publié d'Hironaka, d'après lequel une propriété d'équidimensionnalité analogue pour toute famille de singularités <u>isolées</u> en dimension quelconque implique l'incidence régulière de Whitney [5] et la trivialité topologique ; voir aussi [1] [2]).

<u>Exemple 3</u> : <u>où l'on voit que la réciproque du théorème d'équisingularité jacobienne est fausse.</u>

$$f = y^3 + tx^8 y + x^{11}$$

$$J = (11x^{10} + 8tx^7 y\ ,\ 3y^2 + tx^8\ ,\ f).$$

f est l'équation d'une famille de courbes planes dépendant du paramètre t. Cette famille est équisingulière : plutôt que de le vérifier directement en appliquant la Définition 1 (ce qui serait assez pénible), je préfère calculer le discriminant de f comme polynôme (du 3e degré) en y :

$$\mathrm{Disc}_y f = 4(tx^8)^3 + 27(x^{11})^2 = x^{22}(27 + 4t^3x^2) \sim x^{22}$$

et appliquer un critère de Zariski [7] qui affirme que si le discriminant est équivalent à une puissance de x la famille de courbes est équisingulière.

Essayons maintenant de résoudre l'idéal J.

Au cours des trois premiers éclatements, tout se passe bien. Au quatrième, on obtient un modèle local

$$\epsilon_1 \circ \epsilon_2 \circ \epsilon_3 \circ \epsilon_4 : (x_4, y_4, t) \mapsto (x = x_4,\ y = x_4^4 y_4,\ t = t)$$

$$J = (11x_4^{10} + 8tx_4^{11}y_4,\ 3x_4^8 y_4^2 + tx_4^8,\ x_4^{12}y_4^3 + tx_4^{12}y_4 + x_4^{11})$$

$$= x_4^8\ J'\ ,$$

où

$$J' = (x_4^2(11 + 8tx_4 y_4),\ 3y_4^2 + t,\ x_4^4 y_4^3 + tx_4^4 y_4 + x_4^3$$

$$= (x_4^2,\ 3y_4^2 + t)\ .$$

Le support singulier de l'idéal J est donc la parabole $(x_4 = 0,\ 3y_4^2 + t = 0)$, dont la projection sur l'axe des t est critique à l'origine. La condition 4) de la Définition 1 est donc violée, et la famille des idéaux jacobiens n'est pas équisingulière.

APPENDICE

UNE CLASSE DE CHAMPS DE VECTEURS INTEGRABLES SUR LES ESPACES ANALYTIQUES COMPLEXES

<u>Notations</u> pour quelques faisceaux définis fonctoriellement sur un espace analytique complexe réduit X :

$$\begin{array}{ccccccccc}
\mathbb{R}_X & \supset & \mathcal{E}^{\mathbb{R}}_X & \supset & \mathcal{A}_X & & \supset & & \mathbb{R} \\
\downarrow & & \downarrow & & \downarrow & & & & \downarrow {\scriptstyle \otimes \mathbb{C}} \\
\mathbb{C}_X & \supset & \mathcal{E}_X & \supset & \mathcal{A}_X \otimes \mathbb{C} & \supset & \mathcal{O}_X & \supset & \mathbb{C}
\end{array}$$

$\mathbb{R}$, $\mathbb{C}$: faisceaux constants (réel resp. complexe);

$\mathbb{R}_X$, $\mathbb{C}_X$: faisceaux des germes de fonctions numériques (continues ou non) à valeurs réelles resp. complexes;

$\mathcal{E}^{\mathbb{R}}_X$, $\mathcal{E}_X$: faisceaux des germes de fonctions indéfiniment différentiables (au sens de Whitney [6]) a valeurs réelles resp. complexes :

$\mathcal{A}_X$: faisceau des germes de fonctions analytiques réelles;

$\mathcal{O}_X$: faisceau des germes de fonctions holomorphes.

<u>Définition 1</u>: <u>Germe de champ de vecteurs sur</u> X = dérivation $D_x : \mathcal{O}_{X,x} \to \mathbb{C}_{X,x}$

Une telle derivation s'étend de façon unique, par conjugaison complexe et linéarité, en une dérivation $D_x : \mathcal{A}_{X,x} \otimes \mathbb{C} \to \mathbb{C}_{X,x}$ invariante par conjugaison complexe, d'où l'on déduit par restriction aux réels une dérivation $D_x : \mathcal{A}_{X,x} \to \mathbb{R}_{X,x}$, et inversement. Il y a donc identité entre les champs de vecteurs "complexes" de la Définition 1 et les champs de vecteurs "réels" sur l'espace analytique réel sous-jacent à X.

<u>Définition 1'</u> : Un germe de champ de vecteurs $D_x : \mathcal{O}_{X,x} \to \mathbb{C}_{X,x}$ est dit "<u>différentiable</u>" resp. "<u>analytique réel</u>" resp. "<u>holomorphe</u>" si D_x envoie $\mathcal{O}_{X,x}$ dans $\mathcal{E}_{X,x}$ resp. $\mathcal{A}_{X,x} \otimes \mathbb{C}$ resp. $\mathcal{O}_{X,x}$.

Les champs de vecteurs de la Définition 1', bien que localement intégrables, sont malheureusement trop "rigides" pour pouvoir être utilisés dans le problème de l'équisingularité (cf. le contre-exemple de Whitney [5], exemple 13-1). Le but de ce qui suit est d'introduire des classes de champs de vecteurs plus "souples".

Soit $\epsilon : Y \to X$ un morphisme propre et surjectif d'espaces analytiques complexes réduits.

<u>Définition 2</u> : <u>Germe de fonction ϵ-différentiable sur</u> X = germe $f_x \in \mathcal{C}_{X,x}$ tel que $\epsilon_y^* f_x \in \mathcal{E}_{Y,y}$ pour tout $y \in \epsilon^{-1}(x)$.

<u>Exemple</u> : Si ϵ est l'éclatement d'un point dans X, toute fonction différentiable sur Y et constante sur le diviseur exceptionnel définit sur X une fonction ϵ-différentiable; il est facile de fabriquer des exemples de telles fonctions qui ne sont pas différentiables sur X au point éclaté.

<u>Remarque 1</u> : <u>Toute fonction ϵ-différentiable est continue.</u>

En effet, puisque ϵ est surjectif, toute suite de points de X tendant vers une limite peut se relever en une suite de points de Y, qui admet au moins un point d'accumulation puisque ϵ est propre; il n'y a plus qu'à remarquer qu'une fonction différentiable est continue !

<u>Définition 3</u> : <u>Germe de champ de vecteurs ϵ-différentiable sur</u> X = dérivation $D_x : \mathcal{O}_{X,x} \to \mathcal{C}_{X,x}$ telle que pour tout $y \in \epsilon^{-1}(x)$, $\epsilon_y^* D_x$ envoie $\epsilon_y^* \mathcal{O}_{X,x}$ dans $\mathcal{E}_{Y,y}$.

En d'autres termes, il s'agit des champs de vecteurs sur X dont les "composantes" sont des fonctions ϵ-différentiables.

<u>Définition 3'</u> : <u>Germe de champ de vecteurs ϵ-extensible sur</u> X = germe de champ de vecteurs ϵ-différentiable tel que pour tout $y \in \epsilon^{-1}(x)$, $\epsilon_y^* D_x$ s'étend (au moins d'une façon) en une dérivation $D_y : \mathcal{O}_{Y,y} \to \mathcal{E}_{Y,y}$.

En d'autres termes, il s'agit des champs de vecteurs sur X que l'on peut remonter en des champs de vecteurs différentiables sur Y.

<u>Exemple</u> : Si ϵ est l'éclatement d'un point dans X, les champs de vecteurs ϵ-extensibles sont les projections des champs de vecteurs différentiables sur Y tangents au diviseur exceptionnel; un champ de vecteurs ϵ-extensible doit donc

s'annuler au point éclaté, ce qui montre que si tout champ de vecteurs différentiable sur X est (évidemment) ϵ-différentiable, il n'est par contre pas nécessairement ϵ-extensible.

<u>Remarque 2</u> : Quand le morphisme ϵ est <u>biméromorphe</u> (c'est-à-dire un isomorphisme dans un ouvert dense), la dérivation D_y de la Définition 3', si elle existe, est unique.

<u>Théorème 1</u> : <u>Tout champ de vecteurs ϵ-extensible est localement intégrable sur</u> X.

<u>Preuve</u> : En multipliant le champ de vecteurs par une fonction différentiable à support compact égale à 1 au voisinage du point qui nous intéresse, nous pouvons nous ramener au cas d'un champ de vecteurs ϵ-extensible à support compact dont il s'agit de prouver l'intégrabilité (globale). Or le champ de vecteurs qui le remonte dans Y est différentiable, à support compact (puisque ϵ est propre), donc integrable dans Y d'après un théorème classique d'analyse. En l'intégrant on obtient une famille de difféomorphismes de Y qui transforment les fibres de ϵ les unes dans les autres, donc une famille d'automorphismes ϵ-différentiables de l'ensemble X: d'après la remarque 1 ce sont donc des homéomorphismes, ce qui signifie l'intégrabilité du champ.

<u>Définition 4</u> : On dira que le germe de champ de vecteurs ϵ-différentiable D_x "<u>respecte un Idéal $\mathcal{J}$ de X</u>" si pour tout $y \in \epsilon^{-1}(x)$ la dérivation $\epsilon^*_y D_x$ envoie $\epsilon^*_y \mathcal{J}_x$ dans $\epsilon^*_y \mathcal{J}_x . \mathcal{E}_{Y,y}$.

<u>Exemple</u> : $\epsilon = 1_X$, X lisse, $\mathcal{J}_x = z\,\mathcal{O}_{X,x}$ où z est une fonction coordonnée, (Idéal d'une hypersurface lisse sur une variété). Cet exemple montre que les champs de vecteurs différentiables respectant l'Idéal $\mathcal{J}$ (dérivations qui envoient $z\mathcal{O}_{X,x}$ dans $z\mathcal{E}_{X,x}$) forment une classe plus restreinte que les champs de vecteurs différentiables "tangents a l'hypersurface" (dérivations qui envoient $z\,\mathcal{O}_{X,x}$ dans $(z,\bar{z})\ \mathcal{E}_{X,x}$).

<u>Théorème 2</u> : <u>Soit</u>

$$\begin{array}{ccc} Y & \xrightarrow{\epsilon'} & X' \\ & {\scriptstyle \epsilon}\searrow & \downarrow {\scriptstyle \eta} \\ & & X \end{array}$$

<u>un triangle commutatif de morphismes biméromorphes (d'espaces réduits), où</u> η <u>est l'éclatement d'un Idéal</u> $\mathcal{J}$ <u>de</u> X. <u>Alors tout champ de vecteurs ϵ-extensible sur</u> X, <u>respectant</u> $\mathcal{J}$, <u>se remonte de façon unique en un champ de vecteurs ϵ'-extensible sur</u> X', <u>respectant</u> $\eta^*\mathcal{J}.\mathcal{O}_{X'}$.

Preuve : Considérons les dérivations définies par restriction de D_y (où l'on pose $x = \epsilon(y)$, $x' = \epsilon'(y)$):

$$\begin{array}{lccccl}
D_y : & \mathcal{O}_{Y,y} & \to & \mathcal{E}_{Y,y} & & \\
 & \cup & & \| & & \\
\overline{D}_y : & \epsilon'^*_y \mathcal{O}_{X',x'} & \to & \mathcal{E}_{Y,y} & (\text{à valeurs dans } \epsilon'^*_y \mathbb{C}_{X',x'} \ ???) & \\
 & \cup & & \| & \cup & \\
\overline{\overline{D}}_y : & \epsilon^*_y \mathcal{O}_{X,x} & \to & \mathcal{E}_{Y,y} & (\text{à valeurs dans } \epsilon^*_y \mathbb{C}_{X,x}) . &
\end{array}$$

Tout le problème est le suivant : on sait que $\overline{\overline{D}}_y = \epsilon^*_y D_x$ prend ses valeurs dans $\epsilon^*_y \mathbb{C}_{X,x}$; peut-on en déduire que $\overline{D}_y$ prend ses valeurs dans $\epsilon'^*_y \mathbb{C}_{X',x'}$?
Or en considérant $\overline{\overline{D}}_y$ comme une dérivation de $\epsilon^*_y \mathcal{O}_{X,x}$ dans $\mathcal{F}_{Y,y} = \mathcal{E}_{Y,y} \cap \epsilon'^*_y \mathbb{C}_{X',x'}$ et en appliquant le Lemme 1 ci-dessous, on voit que $\overline{\overline{D}}_y$ s'étend (de façon unique) en une dérivation

$$D'_y : \epsilon'^*_y \mathcal{O}_{X',x'} \to \mathcal{F}_{Y,y} .$$

Alors $\overline{D}_y$ et D'_y sont deux dérivations de $\epsilon'^*_y \mathcal{O}_{X',x'}$ dans $\mathcal{E}_{Y,y}$, ces deux dérivations coïncident évidemment dans l'ouvert dense de Y où ϵ et ϵ' sont des isomorphismes, et par conséquent elles coïncident partout puisque $\mathcal{E}_Y$ est un faisceau de germes de fonctions continues.

Donc $\overline{D}_y = D'_y$ qui prend ses valeurs dans $\epsilon'^*_y \mathbb{C}_{X',x'}$, ce qu'il fallait démontrer.

La démonstration a utilisé le

Lemme 1 - Soit $\begin{array}{ccc} Y & \xrightarrow{\epsilon'} & X' \\ & \searrow^{\epsilon} & \downarrow \eta \\ & & X \end{array}$ un triangle commutatif de morphismes analytiques, où η est l'éclatement d'un Idéal $\mathcal{I}$ dans X. Soit $\mathcal{F}_Y$ un faisceau d'anneaux locaux sur Y, contenant $\epsilon'^* \mathcal{O}_{X'}$. Alors toute dérivation $D_y : \epsilon^*_y \mathcal{O}_{X,x} \to \mathcal{F}_{Y,y}$ qui "respecte l'Idéal $\mathcal{I}$" (c'est-à-dire qui envoie $(\epsilon^* \mathcal{I})_y$ dans $(\epsilon^* \mathcal{I})_y . \mathcal{F}_y$) s'étend de façon unique en une dérivation $D'_y : \epsilon'^*_y \mathcal{O}_{X',x'} \to \mathcal{F}_{Y,y}$ qui respecte l'Idéal $(\eta^* \mathcal{I})_{x'} \mathcal{O}_{X',x'}$.

<u>Preuve</u> : Notons pour abréger

$$A = \epsilon_y^* \mathcal{O}_{X,x}$$
$$\cap$$
$$A' = \epsilon_y'^* \mathcal{O}_{X',x'}$$
$$\cap$$
$$B = \mathcal{J}_{Y,y}$$
$$\text{et} \quad I = (\epsilon^* \mathcal{I})_y \subset A .$$

A' peut s'écrire comme la localisation à l'origine d'une extension de A par des fonctions méromorphes :

$$A' = A\left[\frac{f_1}{f_o}, \frac{f_2}{f_o}, \ldots, \frac{f_k}{f_o}\right]_o$$

où $(f_o, f_1, \ldots, f_k)$ est un système de générateurs de l'idéal I dans A. Alors $IA' = f_o A'$.

Le lemme est immédiat si l'on remarque que toute dérivation $D : A \to B$ telle que $DI \subset IB$ admet une extension unique $D' : A' \to B$ déterminée par les formules

$$D'\left(\frac{f_i}{f_o}\right) = \frac{Df_i}{f_o} - \frac{f_i}{f_o}\,\frac{Df_o}{f_o} \qquad i = 1, 2, \ldots, k ,$$

formules qui définissent bien des éléments de B puisque f_i, Df_i et Df_o appartiennent à $IB = f_o B$.

REFERENCES

[1] H. Hironaka — Equivalences and deformations of isolated singularities, Woods hole seminar in algebraic geometry (1964)

[2] H. Hironaka — Normal cones in analytic Whitney stratifications, Publications Mathematiques de l'I.H.E.S. Volume dédié a O. Zariski (1969)

[3] Beppo Levi — Sulla riduzione delle singolarità puntuali delle superficie algebriche dello spazio ordinario per transformationi quadratiche. Ann. Mat. pura appl. II-s- Vol. 26 (1897)

Risoluzione delle singolarità puntuali delle superficie algebriche. Atti Accad. Sci. Torino Vol. 33 (1897)

[4] R. Thom — Sur la stabilité topologique des applications polynomiales, L'Enseignement mathématique, tome VIII, fasc. 1,2 (1962)

[5] H. Whitney — Local properties of analytic varieties, in Differential and Combinatorial Topology, Morse Symposium (Princeton Math. Series n°27, (1965))

[6] H. Whitney — Differentiable functions defined in closed sets Trans. Amer. Math. Soc., Vol. 36, (1934)
Voir aussi :
B. Malgrange - Ideals of differentiable functions, Tata Institute, Oxford Univ. Press (1966)

[7] O. Zariski — Studies in Equisingularity I, Amer. J. Math. Vol. 87 (1965)

NON-SINGULAR DEFORMATIONS OF SINGULAR SPACES WITH NORMAL CROSSINGS

J. Morrow

§1. Introductory Remarks

In this paper we shall state some results which will be proved in more generality in a later paper [3]. Here, we shall sketch a construction in some special cases which solves the following problem. To state the problem we need some preliminary remarks. We wish to study one-parameter families. By a one-parameter family we mean a triple (M, π, Δ) where M, Δ are connected complex manifolds and $\pi : M \to \Delta$ is a proper, surjective, holomorphic map. We assume $\dim_{\mathbb{C}} \Delta = 1$ and that $M_t = \pi^{-1}(t)$ is connected. Given $t \in \Delta$, let $\mathfrak{m}_t$ be the ideal sheaf of germs of holomorphic functions on Δ which vanish at t. Then let $\mathcal{I}_t = \mathcal{O}\pi^* \mathfrak{m}_t$, which is the ideal sheaf on M generated by the pull-backs of elements of $\mathfrak{m}_t$ by π, where $\mathcal{O}$ is the sheaf of germs of holomorphic functions on M. Then $\mathcal{O}/\mathcal{I}_t$ is a sheaf vanishing outside of M_t and we let $\mathcal{R}_t = \mathcal{O}/\mathcal{I}_t$ restricted to M_t. Then $(M_t, \mathcal{R}_t)$ is a complex space (which may have nilpotent elements in its structure sheaf). We also speak of $(M_t, \mathcal{R}_t)_{t\in\Delta}$ as a one parameter family.

Let $T = \{t \in \Delta \mid M_t$ is singular or $\mathcal{R}_t$ is not reduced$\}$. There is the following elementary theorem.

Theorem 1. (Bertini's theorem) T has no accumulation point (in Δ).

Proof: It follows easily from the fact that π is proper and that an analytic set has finitely many components through any given point. We leave it as an exercise.

This theorem means that given $t \in T$ one can find a neighbourhood N centred at t such that $T \cap (N-\{t\}) = \emptyset$. Given $t \in T$ we want to see what influence $(M_t, \mathcal{R}_t)$ has on M_s for $s \notin T$. In fact we have the following theorem. Let $(M, \mathcal{R})$ be a given complex space. Suppose $(M, \mathcal{R})$ can occur as a fibre in some one parameter family $(N_t, \mathcal{S}_t)_{t\in D}$, say $(M, \mathcal{R}) = (N_{t_0}, \mathcal{S}_{t_0})$.

<u>Theorem 2. One can construct a topological manifold</u> $\nu(M, \mathcal{R})$ <u>such that for any one-parameter family</u> (N, π, Δ) <u>in which</u> $(M, \mathcal{R})$ <u>occurs as a fibre,</u> $\nu(M, \mathcal{R})$ <u>is homeomorphic to</u> N_t <u>for any</u> $t \notin T$.

One can quickly reduce this problem to the case in which M has only normal crossing singularities by using Hironaka's resolution of singularities [1], which is now known to be valid for arbitrary complex spaces. We recall that a complex space $(M, \mathcal{R})$ has <u>only normal crossing singularities</u> if the following conditions are satisfied.

1) $M = \bigcup_{i=1}^{r}$, where each M_i is a (non-singular) complex manifold

2) The M_i intersect normally. This means that given $p \in M$ one can find a neighbourhood of p in M which is locally isomorphic to a set in $\mathbb{C}^{n+1}$ described by an equation of the form $\prod_{j\in J} z_j = 0$, where J is some subset of $\{1, \ldots, n\}$ and p corresponds to the origin. For simplicity suppose $p \in \bigcap_{1\le i\le a} M_i$, and $p \notin M_i$ $i > a$. Then we assume that near p, the sheaf $\mathcal{R}$ is isomorphic to $\mathcal{O}/(z_1^{e_1} \ldots z_a^{e_a})$, where $\mathcal{O}$ is the sheaf of germs of holomorphic functions on $\mathbb{C}^{n+1}$, $(z_1^{e_1} \ldots z_a^{e_a})$ is the ideal sheaf generated by $z_1^{e_1} \ldots z_a^{e_a}$, and the e_j are positive integers.

We refer to the integer e_j as the <u>multiplicity</u> of M_j.

§2. Construction of $\nu(M, \mathfrak{R})$ in Cases I and II.

In this section we shall describe the construction of $\nu(M, \mathfrak{R})$ in the following two special cases.

Case I. $M = M_1$ is a non-singular complex manifold and $e = e_1 > 1$.

Case II. $M = M_1 \cup M_2$ with multiplicities e_1 and e_2.

Case I goes as follows. We are assuming that $(M, \mathfrak{R})$ occurs as a fibre in some one-parameter family (N, ρ, D). Let $\{U_j\}$ be an open covering of M so that $(U_j, \mathfrak{R}|_{U_j})$ is isomorphic to $(V_1, \mathcal{O}/(z_1^e))$ where $V_1 = \{z = (z_1, \ldots, z_{n+1}) \in \mathbb{C}^{n+1} | z_1 = 0, |z| < \epsilon\}$. Thus we may use $(z_2, \ldots, z_{n+1}) = (z_{j2}, \ldots, z_{j,n+1})$ as coordinates on U_j. On $U_j \cap U_k$ we have

$$z_{j1} = f_{jk} \cdot z_{k1} \tag{1}$$

where f_{jk} is a non-vanishing holomorphic function of $(0, z_{j2}, \ldots, z_{j,n+1})$. We may suppose that $z_j = (z_{j1}, \ldots, z_{j,n+1})$ also defines a local coordinate on N in which $\rho = z_j^e$ Then from (1) we easily see that $f_{jk}^e = 1$. Thus $\{f_{jk}\}$ is a one-cocycle on the nerve of the covering $\{U_j\}$ with values in the group of eth roots of unity. This one-cocycle defines an e-sheeted covering of M which is the non-singular model $\nu(M, \mathfrak{R})$. In this case $\nu(M, \mathfrak{R})$ has a natural differentiable structure as a covering of M. In §3 we shall show that N_t is <u>diffeomorphic</u> to $\nu(M, \mathfrak{R})$ if $t \notin T$.

We now describe $\nu(M, \mathfrak{R})$ in case II. First we construct some (complex) line bundles on M_1 and M_2. Let $R = M_1 \cap M_2$ and let $E_{12} = [R]_1$, $E_{21} = [R]_2$ where $[R]_i$ is the bundle of the divisor R in M_i. Using the local description of the structure sheaf $\mathfrak{R}$ we construct a one-cocycle $\{f_{jk}\}$ on M_1 where f_{jk} satisfies (1) on a covering of M_1 and M_1 is defined locally in $\mathbb{C}^{n+1}$ by $z_{j1} = 0$. We may suppose that in the appropriate local coordinates $\rho = z_{j1}^{e_1}$ or $\rho = z_{j1}^{e_1} z_{j2}^{e_2}$ where $z_{j2} = 0$ defines R locally in M_1. The one-cocycle $\{f_{jk}\}$ defines a line bundle E_{11} on M_1. We see easily that $E_{11}^{e_1} E_{12}^{e_2} = 1$ where $F^\ell = F \otimes \ldots \otimes F$ (F tensored ℓ times) and 1 denotes the trivial bundle. This follows from the fact that

$$z_{j1}^{e_1} z_{j2}^{e_2} = z_{k1}^{e_1} z_{k2}^{e_2}, \text{ etc.}$$

Now let $F_1 = E_{11}^{-1}$. Then we have

(2) $$F_1^{e_1} = E_{12}^{e_2}.$$

Let ξ_j be a local fibre coordinate for F_1 and $\varphi_j = 0$ be the local defining equation for R on the open set $U_j(\varphi_j \equiv 1$ if $U_j \cap R = \emptyset)$. Then $\xi_j^{e_1} = \varphi_j^{e_2}$ defines a subvariety V_j of $b^{-1}(U_j)$ where $b : F_1 \to M_1$ is the fibre projection. One easily checks that the V_j fit together to give a subvariety V_1 of F_1. The map $b : V_1 \to M_1$ makes V_1 into an e_1-sheeted covering of M_1, ramified over R (This is a generalized kind of branched covering where for example $z_1^8 = z_2^6$ is an 8-sheeted branched covering of $\mathbb{C} = \{z_1 = 0\} \subseteq \mathbb{C}^2$, branched over 0.). We do the same for M_2 and get an e_2-sheeted covering V_2, ramified over R. Consider the normal disk bundle of R in M_1. Pull it back via b to V_1. This is the normal bundle of R in V_1. One does the same to get the normal bundle of R in V_2. The boundaries of these normal bundles are bundles over R with fibre a union of $d = (e_1, e_2)$ circles, where (x,y) is the greatest common divisor of x and y. If we cut the interior of the normal bundle of R out of V_1 we get a manifold with boundary $\tilde{V}_1$. Similarly we get $\tilde{V}_2$ Then $\partial\tilde{V}_2$ is homeomorphic to $\partial\tilde{V}_1$ via an orientation reversing homeomorphism ($\tilde{V}_1$ and $\tilde{V}_2$ are naturally oriented since V_1 and V_2 are complex varieties). Patch $\tilde{V}_1$ and $\tilde{V}_2$ together along their boundaries to get $\nu(M, \mathfrak{R}) = \tilde{V}_1 + \tilde{V}_2$. Note that if $e_i = 1$, $V_i = M_i$.

§3. Proof of the Deformation Theorem for Case I

We give a sketch of one of the two proofs we know for the following theorem.

Theorem 3. Suppose $(M, \mathcal{H})$ is a given Case I space which occurs as a fibre in some one-parameter family. Let (N, π, Δ) be any one-parameter family which has $(M, \mathcal{H})$ as a fibre. Then if $t \notin T$ (see §1), N_t is diffeomorphic to $\nu(M, \mathcal{H})$. (This is a special case of Theorem 2)

Proof: By Bertini's theorem we may assume that $\Delta = \{z \in \mathbb{C} \mid |z| < 1\}$ and $T = \{0\}$. Let $\tilde{\Delta}$ be another copy of Δ and $\sigma : \tilde{\Delta} \to \Delta$ the map which sends z to z^e. We define the pull back $\tilde{N}$ of N via σ as follows. In $\tilde{\Delta} \times N$ we have a subvariety $Q = \{(z,x) \mid \sigma(z) = \pi(x)\}$. This subvariety is not a manifold; it looks like e sheets crossing along $\{0\} \times \pi^{-1}(0)$. If we separate these sheets we get a non-singular manifold $\tilde{N}$ which has a natural projection $\tilde{\pi} : \tilde{N} \to \tilde{\Delta}$. This map $\tilde{\pi}$ has rank1; hence $\tilde{\pi}^{-1}(s)$ is diffeomorphic to $\tilde{\pi}^{-1}(t)$ for $t, s \in \tilde{\Delta}$. However $\tilde{\pi}^{-1}(0) = \nu(M, \mathcal{H})$ and $\tilde{\pi}^{-1}(t) = N_t$ if $t \neq 0$. Thus $\nu(M, \mathcal{H})$ is diffeomorphic to N_t for $t \neq 0$. Q.E.D.

§4. Examples

We will not indicate how to prove Theorem 2 in Case II. However we will give some interesting examples of the construction of $\nu(M, \mathcal{H})$ and check that Theorem 2 is valid in some special cases.

Example 1. Equation (2) of §2 gives us a necessary condition for a complex space in Case II to be a fibre in some one-parameter family. We use this criterion to verify that $M = \{(\zeta_0, \zeta_1, \zeta_2, \zeta_3) \in \mathbb{P}^3 \mid \zeta_0 = 0 \text{ or } \zeta_1 = 0\}$ with the reduced structure sheaf $\mathcal{H}$ cannot occur as a fibre in a one-parameter family. $M = W_1 \cup W_2$ where each W_i is a copy of $\mathbb{P}^2$ and $W_1 \cap W_2 = \mathbb{P}^1$. Let N_i be the bundle of the divisor $\mathbb{P}^1$ in W_i restricted to $\mathbb{P}^1$. Equation (2) implies that $N_1N_2 = 1$. However $N_1N_2 = [2p]$ where p is a point of $\mathbb{P}^1$ and $[2p]$ is the bundle of the divisor $2p$. This is certainly a non-trivial bundle on $\mathbb{P}^1$, hence $(M, \mathcal{H})$ cannot occur as a fibre in a one-parameter family.

<u>Example 2</u>. Let us recall some definitions. By an <u>elliptic</u> <u>surface</u> we mean a triple (V, Φ, R) where V is a connected, complex, compact manifold of complex dimension 2, R is a non-singular algebraic curve (compact Riemann surface), Φ is a proper, surjective, holomorphic map and the general fibre $\Phi^{-1}(u)$ is a non-singular elliptic curve. Thus an elliptic surface is a one-parameter family of complex spaces of dimension one. Assuming that all of the fibres are free from rational curves with self intersection -1, Kodaira has given a list of every kind of singular fibre which can occur in an elliptic surface (see Kodaira [2], Theorem 6.2). We shall take a selection of these singular spaces, construct $\nu(M, \mathfrak{R})$ for each of them, and verify that $\nu(M, \mathfrak{R})$ is a torus in each case thus checking Theorem 2 for these examples.

(i) The first type of singular fibre listed by Kodaira is a non-singular elliptic curve Θ with multiplicity $m > 1$. An m-sheeted unramified covering of Θ is again a torus so Theorem 2 is verified for this case.

(ii) Next we consider a rational curve Θ with an ordinary double point with multiplicity $m > 0$. Θ is not yet a space with normal crossings. However if we blow up the double point we get a space $m\Theta_1 + 2m\Theta_2$ where Θ_2 is a rational curve, the integers preceding the curves represent their multiplicities, and Θ_1, Θ_2 intersect normally in two points. Θ_1 is now a non-singular rational curve, the proper transform of Θ. The fact that $g.c.d.(m, 2m) = m$ implies that the boundaries of the varieties $\tilde{V}_1$ and $\tilde{V}_2$ (described in §2) are bundles with fibre a union of m circles. The varieties V_1 and V_2 from which $\tilde{V}_1$ and $\tilde{V}_2$ are constructed can be described as follows. V_1 is just a union of m copies of Θ_1 pinched together at two points (notice this is not a branched covering in the usual sense). V_2 is a union of m copies of a 2-sheeted covering of Θ_2 branched over 2 points with branching order 2 at each point and then these m varieties are pinched together at two points, each point lying over one of the points of the branch locus in Θ_2. Thus $\tilde{V}_1$ is a disjoint union of m manifolds with boundary, each manifold being a copy of Θ_1 with two disks cut out. Using the Riemann-Hurwitz formula we see that $\tilde{V}_2$ is also a union of m copies of a sphere with two disks cut out. The manifolds are pasted together along their boundaries according to the following scheme.

Let $\tilde{V}_1 = \bigcup_{i=1}^{m} \tilde{V}_{1i}$, $\tilde{V}_2 = \bigcup_{i=1}^{m} \tilde{V}_{2i}$, where each $\tilde{V}_{ij}$ is a sphere with two disks cut out. Then $\tilde{V}_{1i}$ is pasted to $\tilde{V}_{2i}$ along one of the boundary circles and $\tilde{V}_{2i}$ is pasted to $\tilde{V}_{1,i+1}$ along the other boundary circle where we consider $i + 1 \bmod m$ so $m + 1 \equiv 1$. The resulting manifold is clearly a 2-torus.

(iii) Let us now consider the case of a rational curve Θ with a cusp. The curve Θ can only have multiplicity 1 as a fibre (i.e. the structure sheaf $\mathcal{R}$ is the reduced structure sheaf). In a neighbourhood of the cusp the curve Θ is isomorphic to the analytic set $\{(x,y) \in \mathbb{C}^2 \mid x^2=y^3\}$. Thus $(\Theta, \mathcal{R})$ is not a space with normal crossings. We must perform a sequence of quadratic transformations to resolve the singularity of Θ. We represent this resolution by the following sequence of symbols.

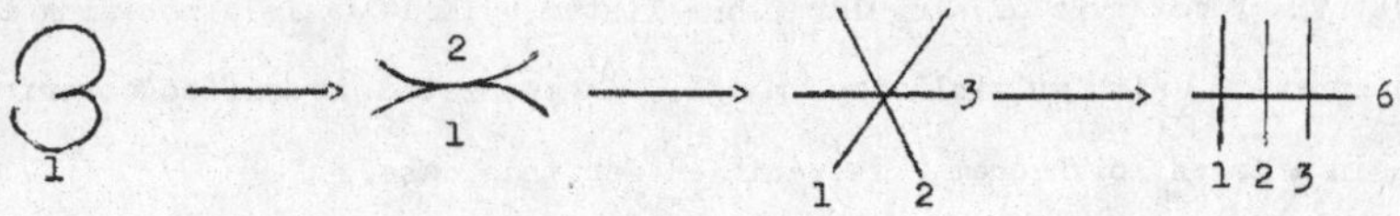

These symbols have the following meaning. The first symbol represents a rational curve with a cusp; it has multiplicity one. We blow up the cusp to get two non-singular rational curves which are tangent at one point. The cusp is replaced by a rational curve with multiplicity 2; The curve with the cusp has as proper transform a non-singular rational curve with multiplicity 1. Next we blow up the point of tangency to get three non-singular rational curves intersecting non-tangentially in the same point. The curve with multiplicity three is the result of blowing up the point of tangency. Finally we blow up the point of intersection to get a non-singular rational curve of multiplicity 6 intersecting the three other rational curves normally (in three different points). This space is a space with normal crossings $(C, \mathcal{E})$ where $C = C_1 \cup C_2 \cup C_3 \cup C_6$. The curves C_k have multiplicity k. We must construct the branched coverings (which we shall call V_k) of C_k described in §2 and fit them together to form $\nu(\Theta, \mathcal{R}) = \nu(C, \mathcal{E})$. V_6 is a 6-sheeted covering of $C_6 = \mathbb{P}^1$ branched over 3 points. Over the first point there is one branch point of V_6 of order 6, over the next point there are 2 branch points of order 3, and over the last point there are 3 branch points of order 2. First we separate the branch points of order 3 (they are pinched together in V_6) and then we separate those of order 2. This gives a non-singular ramified covering V_6' of C_6. Recall the Riemann-Hurwitz formula

$$g = \tfrac{1}{2} \sum_{\rho} (e_\rho - 1) - n + 1$$

for a branched covering R of the spehre $\mathbb{P}^1$ where g is the genus of R, the sum is over all branch points ρ in R, e_ρ is the branching order at point ρ, and n is the number of sheets in the covering. With $R = V_6'$ we get $g(V_6') = 1$, thus V_6' is a torus. Hence $\tilde{V}_6$ is a torus with 6 disks cut out of it. $\tilde{V}_1$ is clearly a sphere with a disk cut out. $\tilde{V}_2$ is a union of two disjoint spheres each with a disk cut out. Finally $\tilde{V}_3$ is a union of three disjoint spheres each with a disk cut out (V_3 is a union of 3 spheres pinched together at one point). Thus $\tilde{V}_6 + \tilde{V}_1 + \tilde{V}_2 + \tilde{V}_3 = \nu(\Theta, \mathfrak{R})$ is a torus.

(iv) Finally let us consider one of the most complicated cases. In Kodaira's notation this is fibre II^* which is represented by the following graph

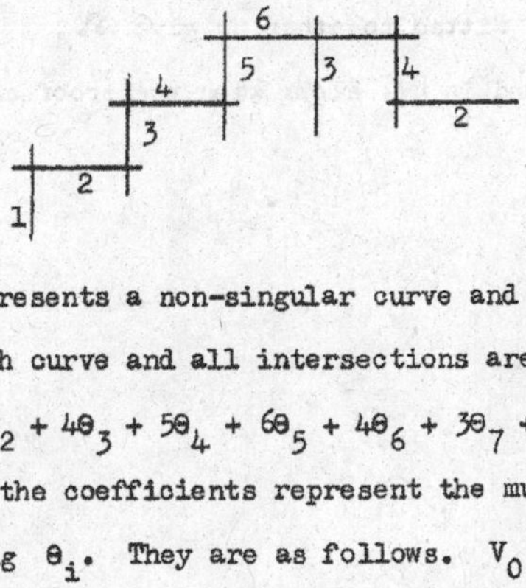

where each line represents a non-singular curve and the numbers indicate the multiplicity of each curve and all intersections are normal. Thus $II^* = \Theta_0 + 2\Theta_1 + 3\Theta_2 + 4\Theta_3 + 5\Theta_4 + 6\Theta_5 + 4\Theta_6 + 3\Theta_7 + 2\Theta_8$ where each Θ_i is a rational curve and the coefficients represent the multiplicities. We let V_i denote the variety covering Θ_i. They are as follows. V_0 is a copy of Θ_0. V_1 is a two sheeted covering of Θ_1 ramified at two points, each point of order 2. V_2 is a 3-sheeted covering of Θ_2, branched at two points each with branching order 3. V_3 and V_4 are done similarly, V_5 is a 6-sheeted covering of Θ_5. It has one branch point of order 6, three of order 2 which are then pinched together, and two of order 3 which are then pinched together. V_6 is a union of two 2-sheeted coverings of Θ_6, each branched at two points and then these coverings are pinched together at two points. At these pinch points one branch point from each of the two 2-sheeted coverings comes together. V_7 is a union of three copies of Θ_7, all pinched together at one point. Finally V_8 is a union of two copies of Θ_8, pinched together at one point. On V_5 separate the pinch points to get a variety V_5'. The Riemann-Hurwitz formula yields $g(V_5') = \frac{1}{2}(5+3.1+2.2) - 6+1 = 1$, hence V_5' is a torus. Thus $\tilde{V}_5$ is a torus with six disks cut out. Similarly one can check that V_1, V_2, V_3, V_4 are spheres, V_6' (V_6 with pinch points separated) is a union of two spheres, V_7' (separate pinch points of V_7) is a union of three spheres, and V_8' (similarly defined) is a union of two spheres. Glueing these manifolds together after cutting out disks around the branch points yields $\nu(II^*, \mathfrak{R})$. This is a torus as it is just a connected sum of a torus and several spheres.

(v) These examples yield the following information. If C is one of the singular fibres of an elliptic surface and M is an arbitrary compact complex manifold then $\nu(C \times M, \mathfrak{R}) = T^2 \times M$ where $\mathfrak{R}$ is chosen appropriately and T^2 is a 2-torus. For example let II^* be the curve just discussed in (iv). Then $\mathfrak{R}$ is the structure sheaf which can be defined on $II^* \times M$ in an obvious way locally and these local pieces can then be fitted together to give $\mathfrak{R}$.

Some more examples can be found in [3] along with the proof of the general result (Theorem 2).

Bibliography

[1] Hironaka, H. Resolution of singularities of an algebraic variety over a field of characteristic zero: I-II, Ann. of Math. 79, 109-326 (1964)

[2] Kodaira, K. On compact analytic surfaces, II, Ann. of Math. 77, 563-626 (1963)

[3] Morrow, J. Non-singular deformations of singular spaces, to appear.

SINGULARITIES IN RELATIVISTIC QUANTUM MECHANICS

D. Olive

1. Introduction

I shall discuss a branch of theoretical physics in which the notions of singularity theory prove useful. The subject concerns the elementary particles which are the smallest constituents of matter. They are so small and can move so fast that any theoretical description must embody the principles of quantum mechanics and special relativity. There are two main theoretical frameworks which do this, quantum field theory and S matrix theory. I shall talk about the latter because it deals more directly with the particles themselves. As we shall see, an important notion will be to think of an overall scattering process involving particles as occurring via the scattering and rescattering of intermediate particles. Singularity theory will help us in the description of the kinematics of these intermediate processes.

The subject is much larger than the space I have available and I shall just outline a few of the main ideas and developments. I shall use two series of references, one with capital letters referring to books which greatly amplify the material and one with numbers referring to original papers in some cases and in other cases to the most recent of a series.

2. The Scattering (or S-) matrix (1, D, E)

What one observes in experiments in sub-nuclear physics are isolated particles moving in straight lines with constant velocities. This is usually specified by giving the momentum $\underline{p}$ and the energy p^0 of the particle. Together these form a relativistic four-vector p^μ such that

$$p^2 \overset{\text{def}}{\equiv} p^\mu p_\mu \overset{\text{def}}{\equiv} p^{02} - \underline{p}^2 = m^2 \qquad (2.1)$$

where m is a constant called the rest mass of the particle. Points satisfying this equation lie on a hyperboloid in p^μ space.

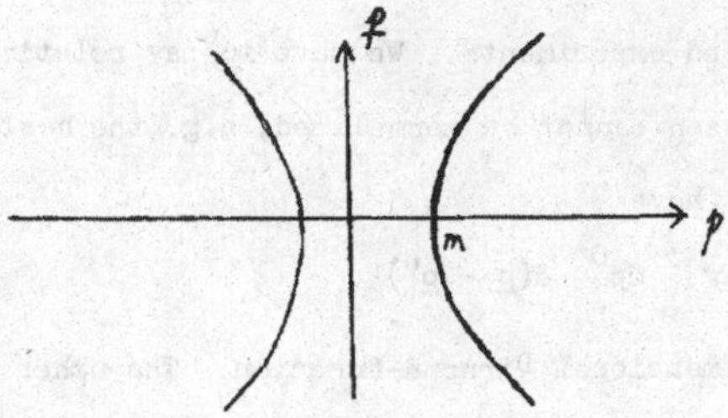

A particle must have positive energy and so lie on the positive sheet, so

$$p^0 = \sqrt{\underline{p}^2 + m^2} \qquad (2.2)$$

It then lies on the "mass shell". Owing to quantum mechanics the other sheet is also important.

Experiments consist of measuring momenta of particles before and after collisions set up in the laboratory. Let us call them $p_1, p_2, \ldots p_n$ and $p_{n+1} \ldots p_{n+m}$ respectively. In present day experiments the number of particles before the collision, n, is 2 but we shall consider any value of n . Notice that $n \neq m$ necessarily i.e. the number of particles need not be conserved. This is because according to relativity energy and mass can be converted (c.f. equation 2.2). Total energy and momentum are still conserved :

$$p_1 + p_2 + \ldots \qquad p_n = p_{n+1} + \ldots \qquad p_{n+m} \qquad (2.3)$$

(This is really a set of four separate equations since, as usual, we have suppressed the Lorentz index μ.)

Quantum mechanics comes in in the following way :
The initial and final states (before and after the collision) are described by vectors in an abstract space according to Dirac's ket notation

$$|p_1 \ldots p_n > \quad \text{and} \quad |p_{n+1} \quad \ldots \quad p_{n+m} >$$

and the probability amplitude for the scattering process is the matrix element of an operator S (called the S-matrix with S for scattering)

$$< p_{n+1} \ldots p_{n+m} |S| p_1 \ldots p_n > \qquad (2.4)$$

The square of the modulus of this probability amplitude is the (relative) probability for the process, and it is this which is measured by accumulating the results of many repeated experiments. We have to say relative probability because the states we have chosen cannot be normalised, e.g. the best we can do for a 1 particle state is to have

$$< p \mid p' > \;=\; (2\pi)^3\, 2p^0\;\; \delta(\underline{p} - \underline{p}') \tag{2.5}$$

with $\delta(\underline{p})$ a three dimensional Dirac δ-function. The other factors are for convenience.

The conservation of probability, i.e. the fact that unity must be the total probability for some outcome to a scattering experiment, leads to the fact that S is a unitary matrix

$$SS^+ \;=\; S^+S \;=\; 1 \tag{2.6}$$

In terms of the S-matrix elements (2.4) these equations are fairly complicated (and this is important). The index summation implied in the matrix product in (2.6) involves a sum over the number of particles in the intermediate states and also an integration over the momentum of each intermediate particle.

$$\int \frac{d^3p}{(2\pi)^3\, 2p^0} \;=\; \frac{1}{(2\pi)^3} \int d^4p\;\; \theta(p^0)\; \delta(p^2 - m^2) \tag{2.7}$$

where $\theta(p^0) = 1$ if $p^0 > 0$ and zero otherwise. This is constructed to give one when (2.5) is integrated.

Owing to energy and momentum conservation (2.3), this summation and integration is of finite range since it is restricted by the energies and momenta of the outside particles. Already we see that we have an equation involving sums of multiple integrals and as we shall see, this is one place where the singularity analysis will be helpful.

There are really even more terms in the sum than we have mentioned since the S-matrix elements themselves break up into separate parts describing the different possibilities of subsets of particles colliding or missing each other altogether. (This presupposes that the interparticle forces are shortrange like the nuclear forces we have in mind rather than longrange like the Coulomb forces.)

This can best be represented by introducing a "bubble notation" which is a sort of pictorial representation of the scattering process. The S-matrix element (2.4) is now written

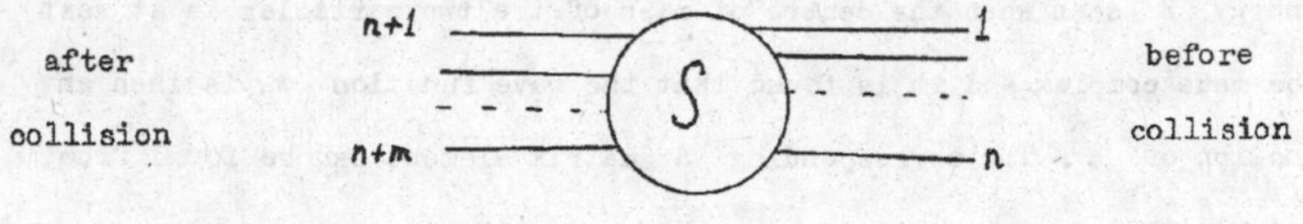

and the "cluster decomposition" just mentioned can be illustrated by :

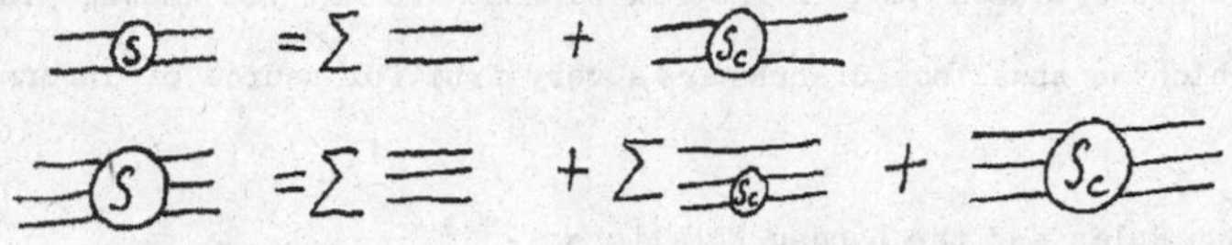

A straight through line — represents the fact that the particle concerned is not deflected and that the corresponding S-matrix element is simply (2.5). S_c is a new matrix (the "connected part" of the S matrix) which cannot be decomposed any further in this way and so is free of momentum conservation δ-functions like $\delta(\underline{p} - \underline{p}')$. In all these collisions total energy and momentum must be conserved (2.3) and this is represented by the fact that S_c contains as a factor just one four dimensional δ function guaranteeing overall energy and momentum conservation:

$$< S_c > \;=\; -i(2\pi)^4\, \delta\Big(\sum_{1}^{n} p_i - \sum_{n+1}^{n+m} p_i\Big) < A > .$$

The matrix elements of this new operator A are the fundamental quantities of the theory since they will turn out to be functions analytic in the sense of complex variable theory, when regarded as functions of the energies and momenta of the particles.

These matrix elements are defined for values of $p_1, \ldots, p_{n+m}$ satisfying the conservation (2.3) and mass shell (2.1) constraints. It is easy to prove that these points lie on a manifold except for exceptional points occurring when all vectors are parallel. These exceptional points can only occur if mass is conserved in the process.

The original indication of this analyticity came from the Schrödinger equation which describes the scattering of two slowly moving particles

$$\nabla^2\psi - V\psi = E\psi .$$

The energy E seen when the centre of mass of the two particles is at rest can easily be made complex and it is found that the wave function ψ is then an analytic function of E. The corresponding A matrix element can be found from ψ and is also analytic.

In the relativistic case such an argument is not possible and indeed the precise analytic properties of the A matrix element are not yet known. The Feynman rules which we shall now discuss are a very fruitful source of information.

3. The Feynman Rules and the Landau Equations

The general relativistic equations of motion corresponding to the Schrödinger equation are not known and may not even exist since all guesses up to now inevitably suffer mathematical sicknesses. So no rigorous solution can be constructed, yet it is possible to derive a formal solution for the guesses, called perturbation theory, which is a power series in terms of a parameter g, called a coupling constant, which measures the strength of the forces between the particles.

Contributions to the perturbation series for the A matrix elements can be specified by the "Feynman rules" (2, A, B, F, H, J, K) which tell us to consider all possible connected diagrams which enable the process to take place via virtual internal particles e.g.

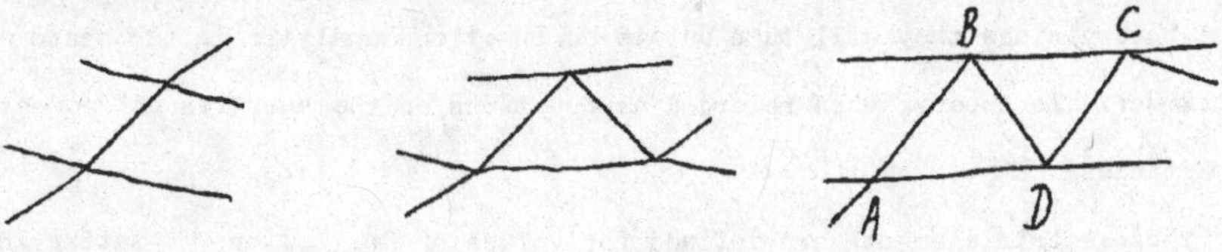

The perturbation series is a sum over all Feynman integrals where the Feynman integral corresponding to a particular Feynman diagram is made up of a product of factors read off the diagram by assigning

(i) g_r to each vertex with r lines attached

(ii) $\frac{1}{q^2 - m^2 + i\epsilon}$ to each internal line

(iii) $\int \frac{i\,d^4 k}{(2\pi)^4}$ to each independent loop. (3.1)

(iv) $1/n_F$ where n_F is a symmetry number

[These rules are appropriate for a simple theory with just one kind of spinless particle.]

The vertex constants g_r are the coupling constants just mentioned and are usually zero for $r \geq k$ (an integer).

q is the four momentum of a particle thought to move along a particular internal line of the graph. Since energy and momentum is to be conserved at each vertex q can be expressed as a linear combination of the momenta of the external particles, and of the loop momenta associated with an independent set of loops. The loop momenta are integrated over all real values and, as a result, the integral is independent of the particular way in which the independent loops are chosen. For example in the third diagram drawn above different possible choices of two independent loops are : ABD, BCD; ADB, BCD; ABD, ABCD; etc.

Notice that whereas energy and momentum is conserved at each vertex, the internal particles can be "off the mass shell" or virtual. Indeed the integrand almost has a pole singularity when an internal particle is on the mass shell. This is just avoided by the $i\epsilon$ (with ϵ small and positive) which pushes the pole off the real axis in a specific way and so away from the contour. Eventually we shall take the limit $\epsilon \to 0$.

For example, corresponding to the graph

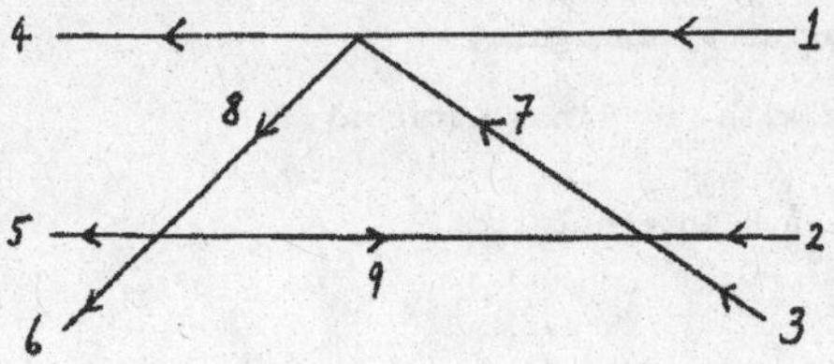

the Feynman integral is $\frac{i}{(2\pi)^4} \int \frac{d^4k \quad (g_4)^3}{(q_7^2 - m^2 + i\epsilon)(q_8^2 - m^2 + i\epsilon)(q_9^2 - m^2 + i\epsilon)}$

where
$$q_9 = k$$
$$q_7 = k + p_2 + p_3$$
$$q_8 = k + p_5 + p_6$$

The Feynman integrals usually diverge but they can be reinterpreted (by "renormalisation theory" (3, A, B, F, H, J, K)) to give finite results. Unfortunately the resultant series, the sum over all diagrams, is believed to diverge. In quantum electrodynamics, the first few terms make predictions, calculated in terms of the mass and charge of the electron, which are verified experimentally to an amazing accuracy. The basic problem nowadays is the theory of "strong interactions" or nuclear forces. Here "g" is large and the first few terms do not make good quantitative predictions, so Feynman diagrams are useless for calculations. Nevertheless these diagrams and their corresponding integrals provide a ready "laboratory" for seeking general properties of relativistic quantum theories. For example, as we shall see, it is possible to work out the positions of singularities (points of non analyticity) and the corresponding discontinuities in the A-matrix elements. The results do not depend on the details of the "force" assumed but only on the masses of the particles involved. This suggests the results are indeed very general and independent of perturbation theory.

We shall discuss mainly the singularities occurring when the momenta of the external particles are real. Logically one must understand these first before analytically continuing away from real values. When this is done new singularities may occur but these are more difficult to analyse mathematically and will not have such a direct physical interpretation. The interesting possibilities occur when we consider the scattering of several particles.

As Landau (4, E) first found, the real singularities occur on certain surfaces in the space of the external momenta. The surfaces are specified by implicit equations called the "Landau equations". The leading singularity for a particular Feynman graph is specified by equations involving real numbers α , there being one such parameter for each internal line of the graph :

$$\left.\begin{array}{ll} \text{For each internal line} & q \text{ is real} \\ & q^2 = m^2 \\ & \alpha > 0 \\ \text{For each loop} & \sum_k \alpha q = 0 \end{array}\right\} \qquad (3.2)$$

This latter equation defines the α's as coefficients of linear dependence, and $\sum_k$ denotes a sum over the lines of the chosen loop with loop momentum k , with the momenta q being measured in the sense round the loop.

In addition there are "lower order singularities" which are leading singularities for graphs obtained by contracting internal lines and identifying the vertices at the ends of the contracted lines. (These contracted graphs may have vertices with more than k lines attached.)

There need not be a solution to a particular set of equations.

We can illustrate with the graph immediately above. The leading Landau equations are

$$q_7^2 = m^2 \quad q_8^2 = m^2 \quad q_9^2 = m^2 \qquad \alpha_7 > 0,\ \alpha_8 > 0,\ \alpha_9 > 0$$

$$\alpha_7\, q_7 + \alpha_8\, q_8 + \alpha_9\, q_9 = 0 .$$

Since q_7, q_8 and q_9 are linearly dependent we can eliminate the α's by observing that the Gram determinant $\det\,(q_i\, q_j)$ of these vectors must vanish. The resulting equations can conveniently be expressed in terms of the variables

$$s_{23} = (p_2 + p_3)^2 \qquad s_{56} = (p_5 + p_6)^2 \qquad s_{14} = (p_1 - p_4)^2$$

If we fix $s_{14} < 0$ the resulting curve in the s_{23}, s_{56} plane is a hyperbola, one branch of which is drawn below

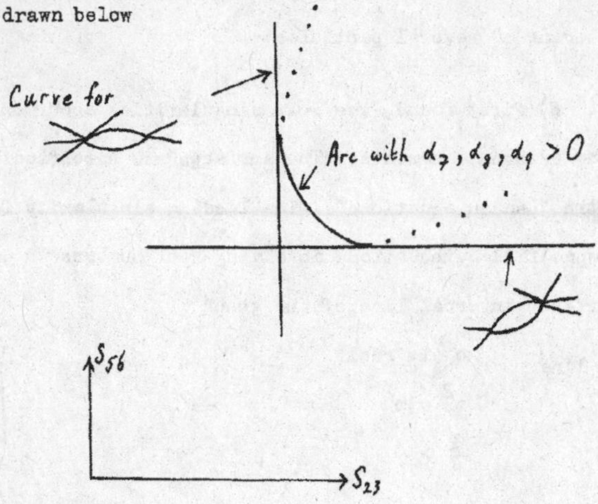

The singular arc must have α_7, α_8, and α_9 all positive and is only a segment of one branch of this hyperbola. It is bounded by points of tangency with the lower order curves obtained by contracting out lines 7 and 8 as indicated. This feature is very general and is called the "hierarchical structure" (5). The fact that the singular arc is an analytic manifold of codimension 1 is also general.

Now I want to explain that these Landau curves have a very nice physical interpretation (6).

First note that all internal particles are physical in the sense that they have real four momenta on the mass shell. In special relativity, if a particle moves with constant momentum, as these particles do between collisions, the four vector displacement in space time, Δx say, between two collisions must be proportional to the four momentum of the particle. The Landau equations suggest that the coefficients α provide this coefficient of proportionality. That is, for each line

$$\Delta x = \alpha q \qquad (3.3)$$

Then the condition $\sum_k{}' \alpha q = 0$ reads

$$\sum_k \Delta x = 0$$

which can now be understood as meaning that the space time separation between two collisions represented by vertices of the graph reads the same whatever way we trace between them along the lines of the graph. Looking in the frame of reference in which a particular internal particle is at rest, the time component of equation (3.3) gives, measuring in the sense $q^0 > 0$:

$$\Delta\tau = \alpha m \qquad \text{i.e.} \qquad \alpha = \Delta\tau/m$$

where $\Delta\tau = \{\text{sign}\,(\Delta t)\}\sqrt{\Delta t^2 - \Delta x^2}$ is the "proper time" for the flight of a positive energy particle between two collisions. The $\alpha > 0$ condition occurring in (3.2) can now be regarded as a sort of causality condition since it is saying that the positive energy particle must move forwards in time if it is to be responsible for a singularity in the S-matrix, i.e. we only see effects detected after emission.

We can sum up by saying that singularities occur only at those values of the momenta of the external particles which permit there to be an intermediate classical scattering process involving particles of negligible size. So classical (non quantum) physics corresponds to the singularities and the effect of quantum mechanics is to allow an analytic probability amplitude between the singularities.

So our A matrix elements are analytic near real values of the four momenta except for singularities at points which as we shall see later lie on analytic manifolds of codimension 1. At any point of one of these manifolds there is a normal and we can define a variable measured along this normal. The complex plane of this variable η near the singular point can be drawn

The singular point $\eta = 0$ is denoted by a cross and we have attached a branch cut. The original Feynman amplitude was defined on the real axis to either side of this branch point. Let us call the functions $F_>$ and $F_<$ in $\eta > 0$ and $\eta < 0$ respectively. The question arises whether $F_>$ and $F_<$ are analytically related and if so whether the path of analytic continuation avoids $\eta = 0$ by moving into the upper or lower half plane. It turns out that $F_>$ and $F_<$ are analytically related and that if the sense of η is defined by a simple rule, the path of continuation deviates into the upper half plane, and we call this the "natural distortion" (7).

One can also ask for the discontinuity across such a singularity, that is the difference between the values of the A matrix element found by going from $\eta < 0$ to $n > 0$ via paths avoiding $\eta = 0$ in the natural way and in the opposite way. The rule found for evaluating this is called the Cutkosky rule (8). For each line with $\alpha > 0$ for the singularity in question one replaces

$$\frac{1}{q^2 - m^2 + i\epsilon} \rightarrow -2\pi i\, \delta(q^2 - m^2)$$

in the Feynman integral.

Notice that for lower order singularities the propagator is unchanged for the lines which have been contracted. Thus for the Feynman integral

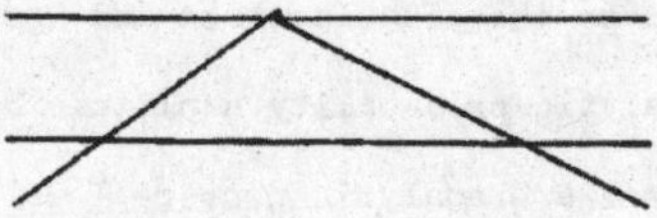

the discontinuity across the singularity associated with the contracted diagram

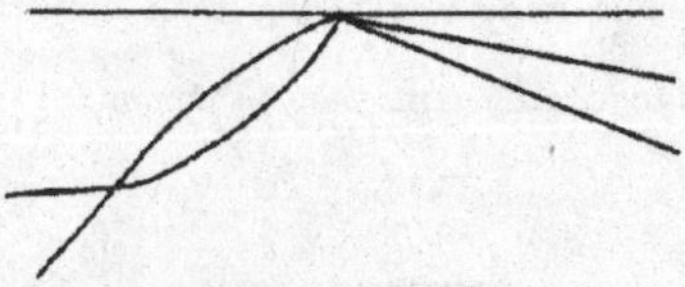

is

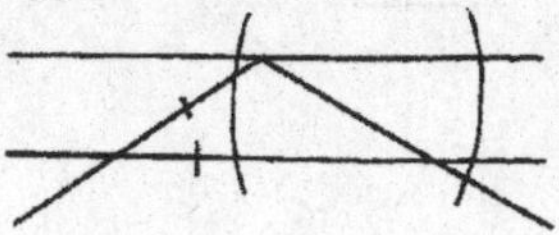

where this is to be interpreted by the Feynman rules (3.1) with the understanding that the lines —+— correspond to $-2\pi i\ \delta(q^2 - m^2)$ while the line inside the bracket has the usual propagator. The point is that the quantity inside the bracket is just the contribution to the amplitude

from the Feynman diagram

When one sums over all diagrams one expects to get

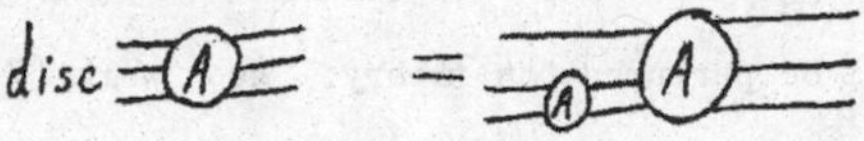

since we shall get all Feynman contributions to the individual bubbles on the right hand side.

Actually in this example, a technical point arises and the previous equation is not quite correct but we shall ignore this and assert the general result (valid with a few qualifications I do not wish to discuss) : the formula for any discontinuity of the amplitude A across a Landau curve can be expressed as a product of mass shell δ functions and mass shell A matrix amplitudes for the vertices. Notice the coupling constants have disappeared. This is what I meant earlier by saying the result was model independent.

The significance (C) of this result is that according to the analyticity properties one can use Cauchy's theorem to express the A matrix elements as integrals over their discontinuities. By the result above these discontinuities can be expressed as products of other A matrix elements. Hence, putting the two results together we find that the A matrix elements satisfy an infinite set of coupled non-linear integral equations. Although these equations are so complicated no one has ever been able to write them all down they have the virtue that in principle the integrations are well defined which was not the case for the Feynman integrals. It has been pointed out that in principle the equations could be solved by successive approximations and plausible arguments suggest that the result is renormalised (reinterpreted) pertubation theory. At one time it was hoped that alternative methods of solution could be developed in order to find the A matrix elements relevant to the strong interactions (C). It was later realised that boundary conditions in the form of asymptotic behaviour need also be added and that the solution of such a complicated scheme was wishful thinking (D).

Nowadays the subject of singularities has lost its fundamental importance and has become just an interesting thing in itself.

One question that arises is whether the properties mentioned above can be derived independently of the Feynman integrals and so put on a better footing. The likelihood of this is suggested by the fact that the final properties can be stated in a way independent of perturbation theory. Much work (7, 9) has been devoted to rederiving these properties from the unitarity equations mentioned earlier. I shall not describe this work but just say that the first step in the programme, the understanding of the analytic properties of unitarity integrals, their possible singularities and corresponding discontinuities will partly motivate my discussion of the next section, when I look at analytic properties of multiple integrals.

4. Singularities of Multiple Integrals (E, 7)

First consider the simple case of one integration variable k and one other "external" variable p .

$$I(p) = \int_{\Gamma} f(k, p)\, dk \qquad (4.1)$$

As long as there are no singularities in the k plane lying on the contour Γ we have

$$\frac{dI}{dp} = \int_{\Gamma} \frac{\partial f}{\partial p}\, dk$$

existing so that F(p) is analytic. The singularities of f(k, p) are given by, say

$$s(k, p) = 0 .$$

As p varies, the position of these singularities may vary in the k plane and move towards the contour

In this situation we can distort the contour away by Cauchy's theorem, and repeat the argument. If now two singularities approach the same point of the contour from opposite sides of the contour it is no longer possible to distort the contour away

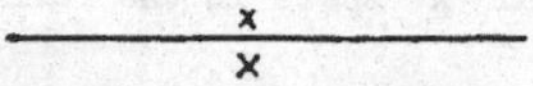

In this case we say the contour is "pinched" and can no longer prove that I is analytic for the value of p at which the pinch occurs. In this situation $\frac{\partial s}{\partial k} = 0$ since there is a double root to the equation. However, two singularities could coincide without causing a pinch

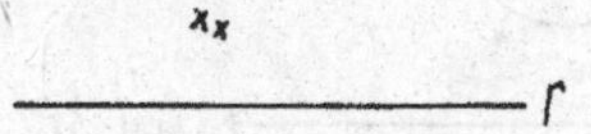

In the general case with many variables k, p and many singularity surfaces

$$s_i\,(k,\, p) \;=\; 0 \qquad\qquad i = 1 \ldots I\,. \qquad\qquad (4.2)$$

it is difficult to visualise the situation and tell whether or not a pinch occurs. In general topology is needed, but in the simple case that concerns us, p real and Γ a contour along the real axis, it is not necessary since there is a simple way of specifying the orientation of the contour with respect to the singularity (7).

Consider a particular s_i and a value of p such that not all its k-derivatives vanish. Then we can define a variable η_i defined along the normal to s_i by

$$d\eta_i \;=\; \Sigma\, \frac{\partial s_i}{\partial k}\; dk$$

We shall now suppose that the singular points of the integrand lie at points satisfying

$$s_i(k,\, p) + i\epsilon \;=\; 0 \quad .$$

The appearance of the $i\epsilon$, with ϵ small and positive, means that the singular points no longer lie on the real axis. This is the situation we have already seen for the Feynman integrals and enables a real contour of integration to be free of singularities. In the η_i plane we have the picture

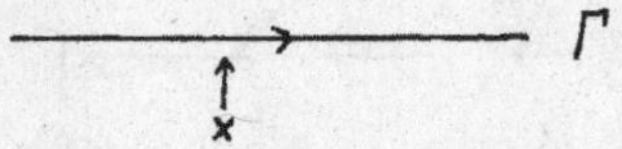

The singular point lies at $\eta_i = -i\epsilon$ and is denoted by a cross. As $\epsilon \to 0$ it moves up to the real axis.

We can picture this in the real k space in the following way. Draw the points satisfying

$$s(k, p) = 0$$

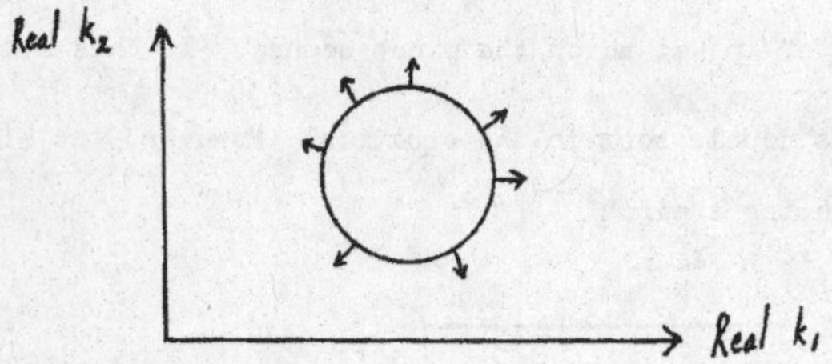

and attach normal arrows pointing in the direction $d\eta > 0$. Then we understand that the singular points are displaced from the real points on the curve drawn in such a way that as $\epsilon \to 0$ the singular points approach the real points in a direction given by the arrows times i . With this picture we can understand how pinches develop in two dimensional integrals. Consider the situation with two singular surfaces :

$$s_1(k_1, k_2, p) = 0 \qquad\qquad s_2(k_1, k_2, p) = 0$$

with arrow systems as indicated in the picture .

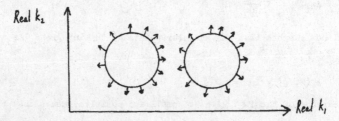

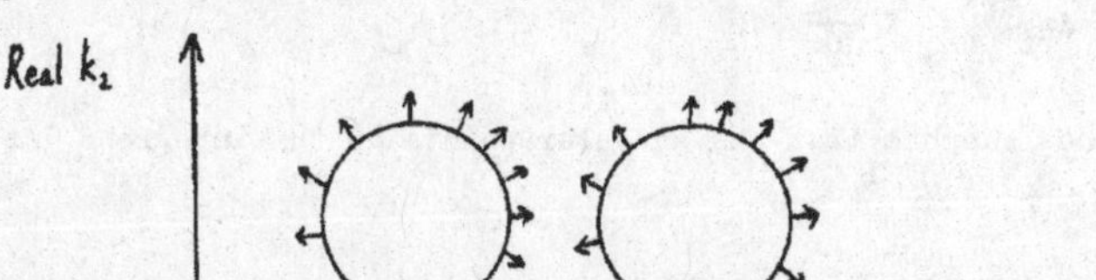

As p varies these shapes will vary in the Real k_1 Real k_2 plane. When they touch there is trouble because in the limit $\epsilon \to 0$ singularities approach the point of tangency in opposite directions from opposite sides of the contour (which is the real k_1 real k_2 plane). So the contour is "pinched". This situation occurs if the normals to the two surfaces are linearly dependent at a point of intersection

$$s_1(p, k) = 0 = s_2(k, p)$$

$$\alpha_1 \frac{\partial s_1}{\partial k_i} + \alpha_2 \frac{\partial s_2}{\partial k_i} = 0 \qquad i = 1, 2 .$$

For a pinch to occur we must have, in addition, that the arrows disagree, and this means

$$\alpha_1 > 0 , \qquad \alpha_2 > 0$$

(or they could be both negative, but it is conventional to write it this way).

A similar difficulty arises if one of the surfaces collapses to a point acnode

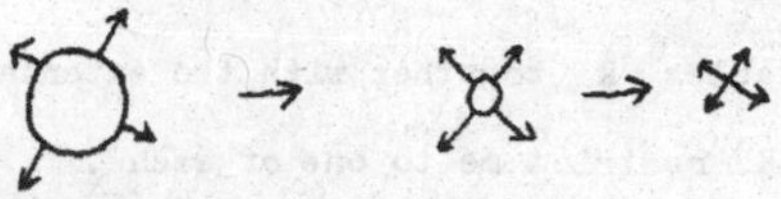

since again the arrows disagree and the contour is pinched. This can occur independently of the other surface and occurs if

$$s_1 = 0 \qquad \frac{\partial s_1}{\partial k} = 0$$

These examples lead us to guess the general condition that if

$$I(p) = \int_{\text{real}} dk \; f(p, k)$$

and f is singular at points satisfying

$$s_i(p, k) + i\epsilon = 0 \qquad i = 1 \ldots I$$

then the only real points p at which I can be singular must satisfy equations (with k real):

$$s_i(p, k) = 0$$

$$\sum_{i=1}^{I} \alpha_i \frac{\partial s_i}{\partial k_j} = 0 \qquad j = 1 \ldots J \qquad (4.3)$$

$$\alpha_i > 0$$

or else a similar set of equations involving only a subset of the possible surfaces s_i.

This can be proved without any visualisation (7) but we shall give no further proof and just remark that this result applied to Feynman integrals gives the Landau rules (3.2).

For then $s_i = q_i^2 - m^2$

and $\frac{\partial s_i}{\partial k_j} = \begin{cases} \pm 2 q_i & \text{if loop j contains line i} \\ 0 & \text{otherwise .} \end{cases}$

I shall now illustrate another way of visualising the result. Instead of considering the real space of integration variables alone, I shall consider the real space of the integration variables k together with the external variables p. The dimensionality of the paper will restrict me to one of each .

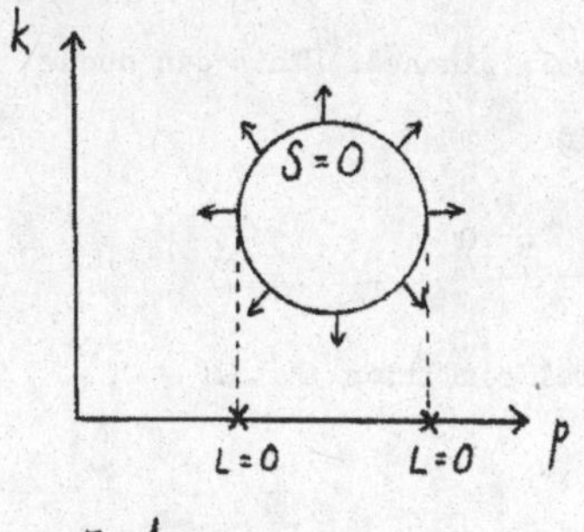

Ex.1

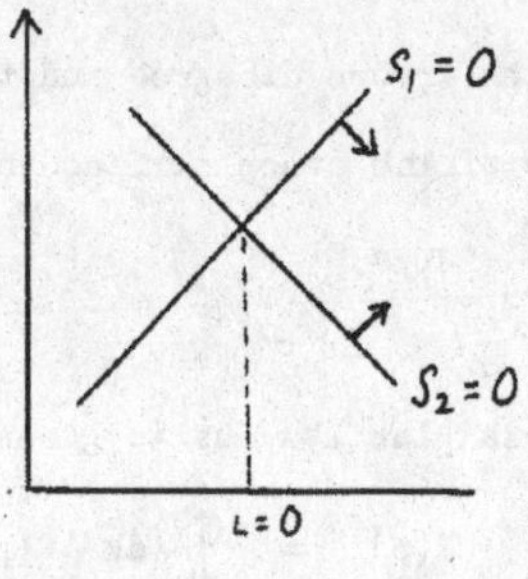

Ex.2.

For given p the integration contour is a vertical line. I have attached arrows. The k component of their direction corresponds to the previous arrows. In Ex. 1 with one singularity surface $s = 0$, there is a difficulty when the normals point in the p direction. Then the contour is pinched because to either side the arrows have projections onto opposite k directions. At these points there is a possible pinch.

In Ex. 2 we see that there is also a pinch at the intersection of s_1 and s_2 since the arrows have disagreeing k components. If one arrow is reversed there would be no pinch.

Ex. 1 illustrates what happens in general (Ex. 2 is rather exceptional since the number of intersecting singularities is just one more than the number of integration variables). Consider the points of intersection of the singularity surfaces in (p, k) space and suppose these form a manifold. Then consider the projection of this manifold into p space. (The vertical projection in our figures.) The pinch points are the critical points of the projection (Σ^1 if the α's form a one dimensional space as we shall always suppose). The projections of the critical points (the "apparent contour") are the Landau curves. This is Pham's way of looking at it, (10) suggested by Thom.

It is possible to deduce the "natural distortion" of the integral at a Landau singularity from this picture. At a critical point consider the linear combination of arrows (normals) which point into p space. This direction in p space is that of the natural distortion for it is in this direction times i, that the Landau curve moves in the limit $\epsilon \to 0$.

The condition for singularity above is not sufficient. Using the pinch criteria and the rule for the natural distortion above, it is possible to deduce (7) a formula for the discontinuity of the integral across one of its real singularities and we can check whether or not this discontinuity is zero. Roughly the result is that

$$\text{disc} \int dk \; f(p, k) = \int dk \prod_i \text{disc}_{s_i} f(p, k)$$

but see reference (7) for a more precise statment.

Since $$\text{disc} \frac{1}{s_i + i\epsilon} = -2\pi i \; \delta(s_i)$$

we obtain the Cutkosky rule previously mentioned for the Feynman integrals.

5. Manifold Properties

There is a gap in the argument up to now since I have not checked

(a) that the critical points form a manifold, and

(b) that their projections (the Landau curves) form a manifold.

These properties are important to understand and there is room to doubt them since there are knownexamples of cusps and acnodes on Landau curves (as we shall discuss later). In view of the purely geometrical way in which we understood the Landau curves at the end of the last section, this is now a problem in singularity theory. According to Pham (10), the properties (a) and (b) are valid if two "transversality conditions" (Ta) and (Tb) are satisfied. These are

$$\text{(Ta)}\quad \left.\begin{aligned} &\sum_j v_j \sum_i \alpha^i \frac{\partial^2 s^i}{\partial k_j \partial y} + \sum_i \beta^i \frac{\partial s^i}{\partial y} = 0;\ y = \text{all } k,\ p \text{ in turn} \\ &\sum_j v_j \frac{\partial s^i}{\partial k_j} = 0 \quad \text{all } i \end{aligned}\right\} + \text{Eq}\,(4.3) \Rightarrow v_j = 0$$

$$\text{(Tb)}\quad \left.\begin{aligned} &\sum_j v_j \sum_i \alpha^i \frac{\partial^2 s^i}{\partial k_j \partial y} + \sum_i \beta^i \frac{\partial s^i}{\partial y} = 0;\ y = \text{all } k \text{ in turn} \\ &\sum_j v_j \frac{\partial s^i}{\partial k_j} = 0 \quad \text{all } i \end{aligned}\right\} + \text{Eq}\,(4.3) \Rightarrow v_j = 0$$

The gist of these conditions is that certain coefficients v_j entering them can only satisfy the equations if they are zero. The two conditions differ only in that y runs over all k and p variables in (Ta) but only over the k-variables in (Tb). In fact it is easy to see that

$$(\text{Tb}) \Rightarrow (\text{Ta})$$

but this is a special feature of the simplifying assumption we have made that the α's defined by (4.3) form a 1 dimensional set (the corank 1 situation). The transversality conditions for higher coranks are slightly more complicated.

Pham proves the consequences of these transversality conditions at some length (11) but I should like to mention that in the course of preparing these lectures I found an alternative method which seems to me to be shorter and simpler. The basic idea is consider the enlarged space of variables α, k and p, and ask whether the points satisfying equations (4.3) form a manifold there. The resultant condition is that the equations in (Ta) should imply that $v_j = 0$ and $\beta_i = 0$. However if $v_j = 0$, then the β_i are automatically zero as a consequence of our assumption that the points satisfying $s_i = 0$ lie on a manifold.

Thus (Ta) guarantees that the points satisfying (4.3) form a manifold in (α, k, p) space and to obtain properties (a) and (b) we must study the projections of this manifold onto (k, p) space and p space respectively. The desired results follow from the following lemma whose proof I leave as an exercise.

Lemma

Suppose the equations $\mathcal{S}_i(u, v) = 0$ define a manifold M in (u, v) space and consider vectors W satisfying

$$\Sigma W_j \frac{\partial \mathcal{S}_i}{\partial v_j} = 0 \qquad \text{all } i .$$

These vectors correspond to elements of the subspace σ of the tangent space to M at (u, v) such that σ projects to zero under the projection onto u space.

Then if $\dim \sigma$ is constant in a neighbourhood of (u, v) in M, then the projection of a smaller such neighbourhood is a manifold of dimension $\dim \sigma$ less than that of M.

In the application of the lemma $\dim \sigma$ turns out to be 1 as a consequence of our corank 1 assumption.

Before applying (Ta) and (Tb) to Landau curves let us first see that there is a simple Hessian condition which implies (Tb) and hence (Ta).

In the first equation occurring in (Tb) choose $y = k_\ell$, multiply by v_ℓ and sum over ℓ. By the second equation

$$\sum_j v_j \frac{\partial s_i}{\partial k_j} = 0 \tag{5.1}$$

we find

$$\sum_{j,\ell} v_j v_\ell \sum_i \alpha_i \frac{\partial^2 s_i}{\partial k_j \partial k_\ell} = 0 \tag{5.2}$$

If these equations imply $v_j = 0$, then Tb and Ta follow and hence results (a) and (b).

Let us now consider leading Landau curves with $\alpha_i > 0$. The intersection of $s_i = 0$ is a manifold in (k, p) space if

$$\sum_p \alpha q = 0 \qquad \sum_k \alpha q = 0$$

imply that $\alpha_i = 0$. According to the physical interpretation we have discussed, the first equation implies that all vertices of the graph with attached external particles, must coincide in space time. Since the $\alpha_i > 0$, we easily deduce that all vertices must coincide, and hence all $\alpha_i = 0$, since they are the proper times.

Having checked that we are indeed dealing with a manifold we now turn to the Hessian conditions (5.1) and (5.2).

Now q_i is a certain linear combination of k_j's + another linear combination of the p's. Then let W_i denote the same linear combination of v_j's as the q_i are of the k_j's, so that equations (5.1) and (5.2) take the form

$$W^i \cdot q^i = 0 \qquad \text{each line } i \text{ of the graph} \tag{5.3}$$

$$\sum_i \alpha_i W_i^2 = 0 \tag{5.4}$$

All we need to show is that these equations imply that all W_i and hence all v_j vanish. Since each α_i is positive it looks as if (5.4) implies that each W_i^2 and hence each W_i vanishes but unfortunately we must remember that the W_i are Lorentz four vectors (with the Lorentz index suppressed) and so W_i can be real and W_i^2 positive or negative (see equation (2.1)).

However since $s_i = 0$, $q_i^2 = m^2$ and we can choose a frame of reference in which

$$q_i = (m, 0, 0, 0)$$

Then (5.3) implies that $W_i^2 \leq 0$ with equality only if W_i itself vanishes. Now (5.4) tells us that each W_i^2 and hence each W_i vanishes. From this we argue back to the desired statements (a) and (b) above, (12).

Pham's work goes further than I have space to describe. The fact that the Hessian quadratic form (5.4) is negative definite enables him to formulate the precise local type of the projection we have considered. It is possible to put the external particles on the mass shell, to consider cases with corank larger than 1 and to consider lower order curves when some α's are zero.

There are problems remaining :

(1) The discussion of effective intersections (places where higher order curves touch lower order curves as mentioned above). This needs a mathematical theory of singularities of composed maps.

(2) A discussion of arcs with some α's negative. All the arguments above have depended on all α's being ≥ 0. This corresponds to all internal particles moving forwards in time. As discussed earlier we want to talk about unitarity integrals which come from the equation

$$S S^{+} = 1 \quad .$$

Since S describes particles moving forwards in time $S^{+} = S^{-1}$ describes particles moving backwards in time. Hence in unitarity integrals we have particles moving both forwards and backwards and time and hence both positive and negative α's occurring.

6. Results concerning complex momenta

It is of course also important to understand the nature of the singularities occurring for complex values of the momemta. The generalisation of the approach discussed so far involves homology theory (G, I), and has not yielded clean results of any generality.

It is possible to make progress by introducing new integration variables α_i, one for each of the M internal lines, closely related to the previous α_i, and to carry out the integrations over the L loop momenta obtaining a new form for the Feynman integral proportional to

$$\int_0^1 \frac{d\alpha_1 \ldots d\alpha_N \, \delta(1 - \Sigma\alpha_i)}{C(\alpha)^2 \, \{Q(\alpha,p) - \Sigma\alpha_i m_i^2 + i\epsilon\}^{M-2L}}$$

where $C(\alpha)$ and $Q(\alpha, p)$ are functions which can be read off the Feynman graph by simple rules. This procedure was originally due to Feynman (2) and is helpful in calculations. The "α-space" integral written above can also be used to make an analytic continuation in p away from the physical values, without distorting the α contours away from the real axis. Suppose that for a given diagram D the complex region of analyticity in p so found is $\mathcal{D}_D$. In general it is prohibitively difficult to evaluate $\mathcal{D}_D$, but Symanzik (13) was able to show, playing upon the physical interpretation of the α's as proper times, that if one diagram D is more complicated than another D^1 in a well defined way, then

$$\mathcal{D}_D \supset \mathcal{D}_{D^1}$$

Hence, by looking at certain simple diagrams regions of analyticity can be found which are applicable to all Feynman diagrams. This is called the "majorisation programme" and there is a very recent account (L).

Soon after Symanzik's work (13), Landau curves (4) were discovered (1958) and it was hoped to derive a larger region of analyticity (5), applicable to amplitudes describing two particle scattering, called the Mandelstam representation (14, C). The idea was to allow distortions of the α contour so long as it was not pinched, and prove that the Landau curves were absent from a larger domain. The arguments used assumed that Landau surfaces were manifolds but this was soon found to be wrong (1961) when the following diagram was analysed (15).

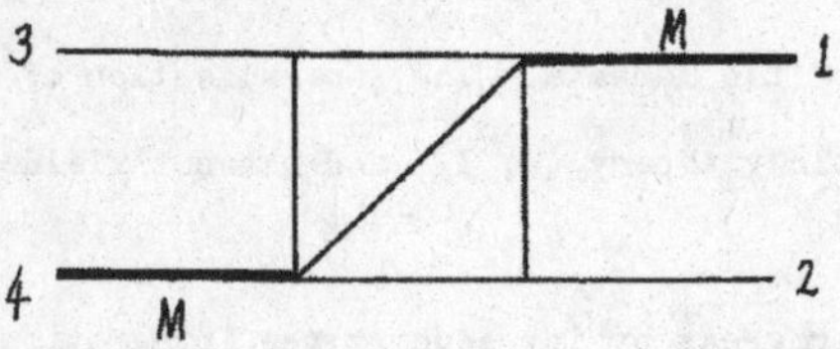

All particles have unit mass except for the two heavy lines which have mass M. In terms of the "Mandelstam variables"

$$s = (p_1 + p_2)^2 \qquad\qquad t = (p_2 - p_4)^2$$

the leading Landau curve in the real s, t plane takes the form as M increases

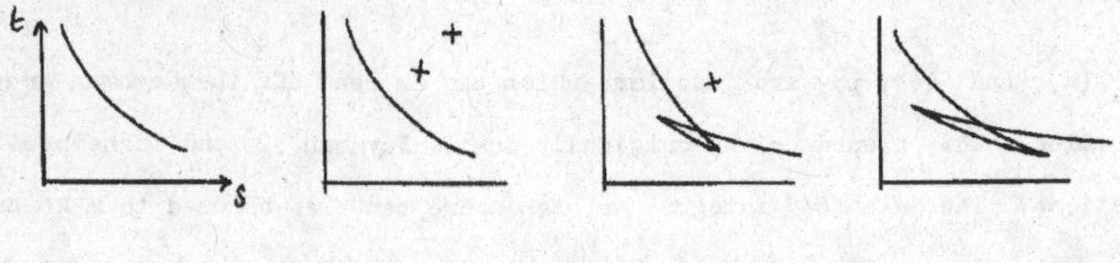

The crosses denote acnodes. This formation is very familiar to those who have read the work of Thom. It stopped further progress in this particular study of Feynman graphs. This example does not disagree with Pham's results discussed earlier since the singular points occur in a region where s and t are real but where the corresponding momenta p_i are complex.

Finally I should like to mention the main work done today on the analytic properties of Feynman integrals. This is due to a group at Princeton (16) and concerns the topological properties of the complete Riemann sheet structure of specific Feynman integrals regarded as functions not only of the momenta of the external particles but also of the masses of the external and internal particles.

REFERENCES

[1] W. Heisenberg, Die beobachtbären Grössen in der Theorie der Elementarteilchen. Zeitschrift für Physik 120 (1942), 513-538.

[2] R.P. Feynman, Space-time approach to quantum electrodynamics. Phys. Rev. 76 (1949), 769-789.

Mathematical formulation of the quantum theory of electromagnetic interaction. Phys. Rev. 80, (1950) 440-457.

[3] F.J. Dyson, The S-matrix quantum electrodynamics. Phys. Rev. 75 (1949) 1736-1755.

[4] L.D. Landau, On analytic properties of vertex parts in quantum field theory, Nuclear Phys. 13 (1959) 181-192.

[5] P.V. Landshoff J.C. Polkinghorne and J.C. Taylor, A proof of the Mandestam representation in pertubation theory. Il Nuovo Cimento 19 (1961) 939-952.

[6] S. Coleman and R.E. Norton, Singularities in the Physical Region. Il Nuovo Cimento 38 (1965) 438-442.

[7] M.J.W. Bloxham, D.I. Olive and J.C. Polkinghorne, S-matrix singularity structure in the physical region. I Properties of multiple integrals. J.Math. Phys: 10, (1969) 494-502.

[8] R.E. Cutkosky, Singularities and discontinuities of Feynman amplitudes. J. Math. Phys. 1 (1960) 429-433.

[9] M.J.W. Bloxham, D.I. Olive and J.C. Polkinghorne, S-matrix singularity structure in the physical region II Unitarity Integrals. J. Math. Phys. 10 (1969) 545-552.

III General discussion of simple Landau singularities. J. Math. Phys. 10 (1969), 553-561.

J. Coster and H.P. Stapp, Physical Region Discontinuity Equations for Many-Particle Scattering Amplitudes I. J. Math. Phys. 10 (1969) 371-396.

II. J. Math. Phys. 11 (1970) 1441-1463.

[10] F. Pham, Singularites des processus de diffusion multiple. Ann. Inst. Henri Poincare 6A (1967) 89-204.

[11] F. Pham, in Symposia on Theoretical Physics vol.7 (Plenum, New York)

[12] The manifold properties were also proved by:

C. Chandler and H.P. Stapp, Macroscopic causality conditions and properties of scattering amplitudes. J. Math. Phys. 10 (1969) 826-859.

C. Chandler, Causality in S-matrix theory. Phys. Rev. 174, (1968) 1749-1758.

[13] K. Symanzik, Dispersion relations and Vertex Properties in Perturbation Theory. Prog. Theor. Phys. 20 (1958), 690-702.

[14] S. Mandelstam, Determination of the pion-nucleon scattering amplitude from dispersion relations and unitarity. General Theory. Phys. Rev. 112 (1958) 1344-1360.

[15] R.J. Eden, P.V. Landshoff, J.C. Polkinghorne and J.C. Taylor, Acnodes and cusps on Landau curves. J. Math. Phys. 2 (1961) 656-663.

[16] G. Ponzano, T. Regge, E.R. Speer and M.J. Westwater, The monodromy rings of a class of self-energy graphs. Commun Math. Phys. 15 (1969), 83-132.

T. Regge, p.433 in "Battelle Rencontres 1967" ed C.M. De Witt and J.A.Wheeler (W.A. Benjamin, New York 1968)

T. Regge, in "Nobel Symposium 8 : Elementary Particles Theory". ed Nils Svartholm (Interscience, New York, 1969).

Reprints of (4), (8) and (14) appear in (C)

Reprints of (2) and (3) appear in (K)

REFERENCE BOOKS

[A] J.D. Bjorken and S.D. Drell, Relativistic quantum fields. McGraw-Hill, 1965.

[B] N.N. Bogoliubov and D.V. Shirkov, Introduction to the Theory of Quantised Fields, Interscience, New York, 1959.

[C] G.F. Chew, S-Matrix Theory of Strong Interactions. W.A. Benjamin, New York, 1961.

[D] G.F. Chew, The Analytic S-Matrix. W.A. Benjamin, New York, 1966.

[E] R.J. Eden, P.V. Landshoff, D.I. Olive, & J.C. Polkinghorne, The Analytic S-Matrix, Cambridge 1966.

[F] R.P. Feynman, Quantum Electrodynamics. W.A. Benjamin, New York, 1962.

[G] R.C. Hwa and V.L. Teplitz, Homology and Feynman Integrals. W.A. Benjamin, New York, 1966.

[H] J.M. Jauch and F. Rohrlich, The Theory of Photons and Electrons . (Addison-Wesley, Cambridge, Mass., 1955.

[I] F. Pham, Introduction a l'Etude Topologique des Singularites de Landau. Gauthier-Villars, Paris, 1967.

[J] S. Schweber, An Introduction to Relativistic Quantum Field Theory. Harper and Row, New York, 1961 .

[K] J. Schwinger, Quantum Electrodynamics. Dover, 1958.

[L] I.T. Todorov, Analytic Properties of Feynman Graphs in Quantum Field Theory. Bulgarian Academy of Sciences, 1966. (To appear in English)

LANDAU SINGULARITIES AND A PROBLEM IN GRAPH THEORY

M.J. Westwater

§1. The Problem

Graphs enter into the theory of the S-matrix of particle physics in two ways (in principle quite different) :

a) as a notation for the terms of the formal power series expansion of the S-matrix in the coupling constant if the attempt is made to calculate the S-matrix in the framework of quantum field theory - Feynman graphs.

b) as a notation for the singularities which one may expect on general grounds to find in the connected parts of S-matrix elements - Landau graphs. (Recall from David Olive's lectures [1] that the connected amplitudes are expected to be analytic functions in the physical region except for values of the momenta for which the scattering can take place by means of a succession of point scatterings in spacetime, and that the Landau graph is the abstract graph which is realized in spacetime by this classical scattering process).

In the case of Feynman graphs the question as to which graphs are allowed is determined by the structure of the Lagrangian. In the case of Landau graphs the question is not so easy to answer. Although one may write down Landau equations

$$q_i^2 - z_i = 0 \qquad 1 \leqslant i \leqslant N \qquad \text{(L1)}$$

$$\sum_{i \in \text{loop}} \alpha_i q_i = 0 \qquad \text{for all loops} \qquad \text{(L2)}$$

for any graph G they do not necessarily define a subvariety of codimension 1 in the space of external momenta.

The purpose of this talk is to outline some partial results obtained by Gene Speer and myself on this question in the course of studying the structure of Feynman amplitudes as analytic functions [2].

From the point of view of David Olive's talks the natural form of the question would be: Given a causally oriented graph G (one with an orientation which induces a partial ordering of its vertices) when do the equations (L1), (L2) have a solution for which $z_i = m^2$ for all i, and the α_i are uniquely defined up to a factor by (L2) and may be taken to be all positive? Here the external momenta p (which enter into (L1), (L2) implicitly through the momentum conservation equations corresponding to the vertices of G) are supposed to lie in the physical region, and at each vertex the stability condition that there are at least two incoming and two outgoing lines is supposed to be satisfied by G.

However (corresponding to our different viewpoint which will not be developed in detail here) the question considered in [2] is put somewhat differently. Inspection of the proofs shows that the only essential difference is that we do not require that all the z_i be equal, and that we pay no attention to the stability condition. [Also for our results we require that G have at most 5 external vertices; this is because we need to be able to choose the external momenta so that the sums $p_1, .., p_n$ of external momenta entering at the external vertices satisfy only the linear relation $p_1 + \dots p_n = 0$, and no others]. A graph G for which (L1), (L2) have solutions of this kind is called <u>normal</u> in [2]. It is shown that the property of being normal depends only on the structure of G as an unoriented graph. The problem of determining which G are normal is the graph theoretic problem of the title.

§2. Examples

With the help of the representation of (L1), (L2) by means of dual diagrams [3] it is easily seen that the graphs of Figs 1 and 2 are <u>not</u> normal.

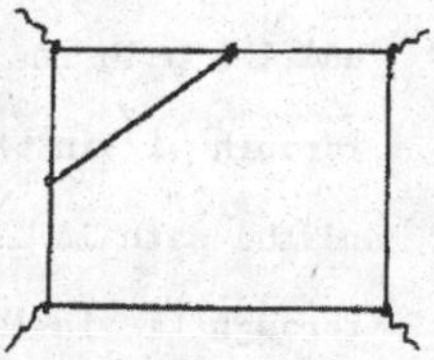

Fig 1

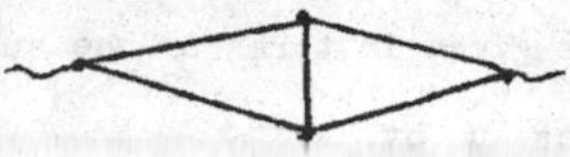

Fig 2

§3. Results

The basic result of [2] is that G is normal iff a certain rational map $f(G) : \alpha \to g(\alpha)$ of the real projective space $\mathbb{P}^{N-1}$ into an affine space has Jacobian of rank $N - 1$ for some α. From this result some necessary and some sufficient conditions of a graph theoretic nature for G to be normal are derived. The sufficient conditions require G to be reducible to a single loop graph by means of certain operations. It is possible that these conditions are also necessary - we did not find a counterexample, and the nature of the allowed operations fits in well with the general classes of non-normal graphs discussed in [4].

The map $f(G)$ is the map sending α to the matrix of the linear transformation expressing internal in terms of external momenta obtained by solving the linear equations (L2) for the internal momenta (this is always possible if $\alpha_i > 0$ for all i). It is given explicitly as follows :

Define the Symanzik polynomial $d(\alpha)$ by

$$d(\alpha) = \sum \prod_{i \notin T} \alpha_i$$

where the sum is taken over all (maximal) trees in G. Thus $d(\alpha)$ is a polynomial of degree ℓ, the number of independent loops in G. Next define for each internal line i of G and each pair u,v of distinct external vertices of G the polynomial $g_{iuv}(\alpha)$ by

$$g_{iuv}(\alpha) = \sum \epsilon_{T,iuv} \prod_{i \notin T} \alpha_i$$

where the sum is taken over maximal trees in G as before and

$$\epsilon_{T,iuv} = +1 \quad \text{if } i \in T \text{ and the path in } T \text{ from } u \text{ to } v \text{ passes through } i \text{ in the positive sense}$$

$$= -1 \quad \text{if } i \in T \text{ and the path in } T \text{ from } u \text{ to } v \text{ passes through } i \text{ in the negative sense}$$

$$= 0 \quad \text{otherwise} .$$

Then the momentum q_i in line i is given in terms of the sums p_u of external momenta entering the external vertices u by

$$q_i(\alpha,p) = \sum_{u \neq v} \frac{1}{d} g_{iuv} p_u \qquad (1)$$

(Here v is any fixed external vertex. The corresponding sum of external momenta p_v has been eliminated via $p_v = -\sum_{u \neq v} p_u$. Of course, the formulae obtained by making different choices for v are equivalent). Thus the map $f(G)$ is given by

$$\alpha \rightarrow \left\{ \frac{1}{d} g_{iuv} \right\} \qquad (2)$$

The proof of the basic result is obtained by writing (L1), (L2) in the equivalent form

$$\frac{\partial D}{\partial \alpha_i} = 0 \qquad 1 \leqslant i \leqslant N ,$$

where $D = \sum_i \alpha_i (q_i(\alpha,p)^2 - z_i)$, and obtaining for the Hessian form associated with D

$$H(\beta) = \sum_{i,j} \frac{\partial^2 D}{\partial \alpha_i \partial \alpha_j} \beta_i \beta_j$$

the identity $$H(\beta) = -\sum_i \alpha_i \left(\sum_j \frac{\partial q_i}{\partial \alpha_j} \beta_j \right)^2 . \qquad (3)$$

(3) is very close to the expression for the Hessian which appears in Pham's analysis of Landau singularities but the two Hessians are not identical because we do not keep the masses fixed.

An immediate corollary of the basic result is that for G to be normal it is necessary that

$$\ell(n_E - 1) \geqslant N - 1 .$$

For the statement of the sufficient conditions (some of which require nontrivial results in graph theory for their proofs) we refer to [2] .

REFERENCES

[1] D. Olive, Singularities in Relativistic Quantum Mechanics. This volume pp. 244-269.

[2] E.R. Speer & M.J. Westwater, Generic Feynman Amplitudes, Vol. XIV No.1, 1971, 1-55, Ann. Inst. H. Poincare.

[3] R.J. Eden et.al. The Analytic S-matrix. Cambridge University Press, 1966.

[4] L.B. Okun & A.P. Rudik, On a method of finding singularities of Feynman graphs. Nucl. Phys. 15, (1960) 261-288.

THE DEFINITION AND OCCURRENCE OF SINGULARITIES IN GENERAL RELATIVITY

S.W. Hawking

In the General Theory of Relativity space-time is represented by a pair (M, g), where M is a four-dimensional manifold and g_{ab} is a metric of Lorentz signature (+ - - -) on M which obeys the Einstein field equations

$$R_{ab} - \frac{1}{2} g_{ab} R = 8\pi G T_{ab}$$

where R_{ab} is the Ricci tensor of the metric g_{ab}, G is the gravitational constant and T_{ab} is the energy-momentum tensor of the matter fields. Intuitively, one would like to define a space-time singularity as a point of M at which the metric is undefined or is not suitably differentiable. The trouble with this is that the space-time manifold is not defined a priori. Thus one could simply remove the singular points from M and say that the resulting manifold represented the whole of space-time. In fact if one regards the space-time manifold as the set of points at which measurement can be made, one must leave out the singular points. The problem of defining a space-time singularity then becomes a question of detecting whether points have been left out. For this a suitable definition of completeness is needed. If g_{ab} were a positive definite metric tensor one could define a topological distance function d(x, y) as the lower bound of the lengths of curves from x to y. One can then say that a space is <u>metrically complete</u> (m-complete) if every Cauchy sequence with respect to d converges. This is equivalent to the requirement that every curve should be extendable to arbitrary length. Unfortunately this definition of completeness cannot be used in general relativity because the metric is not positive-definite and so there are curves of zero length. However, there is another kind of completeness, b-completeness, which is equivalent to m-completeness where there is a positive-definite metric tensor on M but which can be defined whenever there is an affine connection on M. This notion of completeness seems to have been suggested originally by Ehresmann [1], and has recently been reformulated in an elegant manner by Schmidt [2].

Let p be a point of M and let $\gamma(t)$ be a curve passing through p. Choose a basis of the tangent space T_p and parallelly propagate it along $\gamma(t)$ so as to obtain a basis for the tangent spaces at all the other points of $\gamma(t)$. One can then define a positive definite metric m_{ab} along $\gamma(t)$ such that in this metric the parallelly propagated basis is orthonormal. Using this positive definite metric one can measure the length of $\gamma(t)$. Of course this length depends upon the original choice of the basis at p but whether or not the length is infinite is independent of the choice of basis. One says that M is b-complete if every inextendable curve in M has infinite length in this sense.

Another formulation of b-completeness has been given by Schmidt in terms of the bundle of linear frames L(M) over M. Let $q \in L(M)$ then q represents a point $x = \pi(q)$ in M and a basis $E_a (a = 1, 2, 3, 4)$ of T_x. The connection in L(M) defined by the Lorentz metric tensor g_{ab} defines a splitting of the tangent space T_q into a vertical subspace U_q which is tangent to the fibre $\pi^{-1}(x)$ and a horizontal subspace H_q. In H_q there are four canonically defined vectors Y_a which are the horizontal lifts of the vectors E_a, i.e. $\pi_*(Y_a) = E_a$. The general linear group acts in the fibre of L(M). If one chooses a basis for the Lie algebra of the general linear group, this action defines sixteen vertical vector fields $Z_\alpha (\alpha = 1..16)$. The vectors Y_a and Z_α together form a basis for T_q. One can then define a positive definite metric tensor m_{ij} on L(M) such that this basis is orthonormal with respect to m_{ij}. The metric m_{ij} depends on the choice of basis in the Lie algebra of the general linear group, but it turns out that this dependence is unimportant; different metrics defined from different bases give the same result. Using this positive definite metric tensor on L(M) one can ask whether L(M) is complete in the ordinary metric sense; it turns out that L(M) is m-complete if and only if M is b-complete. This procedure also enables one to construct boundaries for L(M) and M which can be thought of as representing the singular points which have been left out. One forms the metric space completion $\tilde{L}(M)$ of $L(M)$, i.e. the points of $\tilde{L}(M)$ are equivalence classes of Cauchy sequences in $L(M)$. $\tilde{L}(M)$ consists of a part which is homeomorphic to $L(M)$ and a set of boundary points Δ. The action of the general linear group in $L(M)$ maps equivalence classes of Cauchy sequences into equivalence classes of Cauchy sequences. Thus the action of the general linear group can be extended to $\tilde{L}(M)$ and one can define $\tilde{M}$ to be the quotient of $\tilde{L}(M)$ by the general linear group. $\tilde{M}$ consists of a part homeomorphic to M and a set of boundary points ∂. Although ∂ is a topological space it is unfortunately in general not any kind of manifold. It is this which makes it so hard to study the nature of singularities in General Relativity since one does not have enough structure defined at the singular points.

In order to prove theorems about the occurrence of singularities in General Relativity one uses the fact that the Einstein equations give one a relation between the Ricci tensor R_{ab} and the energy momentum tensor T_{ab} of the matter. In general the form of T_{ab} will be very complicated as there may be many kinds of matter present but there are certain inequalities which it is reasonable to expect T_{ab} to satisfy. The one I shall use is $T_{ab} K^a K^b \geq \frac{1}{2} g_{ab} K^a K^b T^c_c$ for any timelike or null vector K^a, i.e. any vector such that $K^a K^b g_{ab} \geq 0$. For a fluid with density μ and pressure p this will be satisfied if and only if $\mu \geq 0$ and $\mu + 3p \geq 0$. These requirements should be satisfied by any physically reasonable fluid. The Einstein equations then imply that $R_{ab} K^a K^b \geq 0$.

The significance of this is that it is the condition for irrotational congruences of timelike or null geodesics to contain focal points. By the standard variation of arc length one can show that a timelike or null geodesic $\gamma(t)$ between two points p and q is not the longest timelike curve between p and q if the congruence of timelike geodesics which pass through p has a focal point on $\gamma(t)$ between p and q. The method then of proving the existence of a space-time singularity is basically to prove that if M were b-complete (and hence geodesically complete) one would find points p and q such that every timelike or null geodesic joining p and q contains a focal point.

This establishes a contradiction only if one can prove that there must be a longest timelike or null geodesic between p and q . This will not be true in general. However, it is the case for all pairs of points p and q belonging to an open set N which satisfies the global hyperbolicity condition of Leray [3]. This requires that the space of all timelike or null paths between p and q should be compact in the C^0 topology. It can be shown that this condition is equivalent to the requirement that the intersection of the future of p with the past of q should be compact and should not contain any almost closed timelike curves. It is also equivalent to the existence of a Cauchy surface for the sub-manifold N, [4] i.e. an imbedded space-like three manifold which intersects every inextendable timelike or null curve in N once and once only.

Using these concepts one can prove a number of theorems which seem to indicate that practically every reasonable model of space time contains singularities [5]. As an example I shall quote theorem which is not necessarily applicable to our universe but which is fairly simple to present [6].

Space time is not b-complete if :

1) $R_{ab} K^a K^b \geq 0$ for every timelike or null vector K^a .
2) There exists an imbedded space-like compact three-manifold S (i.e. the universe is spatially closed).
3) The unit normals to S are converging everywhere on S (i.e. the trace of the second fundamental form of S is negative).

REFERENCES

[1] C. Ehresmann, Les connexions infinitiomales dans un espace fibre differentiable. In Colloque de Topologie (espaces fibres), Bruxelles 1950, pp. 29-50 (Publ. Masson, Paris 1957).

[2] B.G. Schmidt, A new definition of singularities in general relativity, J.G.R. (Journal of General Relativity) 1971, 3.

[3] H.J. Seifert, Z. Natur forsch. 22a, 1356, 1967.

[4] R.P. Geroch, Domain of Dependence. In J. Math. Phys. 11, 437, 1970.

[5] S.W. Hawking and R. Penrose, The singularities of gravitational collapse and cosmology. In Proc. Roy. Soc. A 314, 1970. pp.529-548.

[6] S.W. Hawking, The occurrence of singularities in cosmology. III. Causality and singularities. In Proc. Roy. Soc. A 300, 187, 1967.

AUTHOR ADDRESSES

G. Glaeser, Département de Mathématique, Université de Strasbourg, Rue René Descartes, 67 - Strasbourg, France.

A. Haefliger, Institut de Mathématiques, 16 bd d'Yvoy, Genève, Switzerland.

H.A. Hamm, D - 34 Göttingen, Mathematisches Institut, Bunsenstrasse, Germany.

S.W. Hawking, University of Cambridge, Department of Pure Mathematics, 16 Mill Lane, Cambridge.

F. Hirzebruch, Mathematisches Institut der Universität Bonn, 53 Bonn, Germany.

S.A. Khabbaz, Department of Mathematics and Astronomy, College of Arts and Scienc Lehigh University, Bethlehem, PA.18015, U. S. A.

N.H. Kuiper, Math. Instituut der Universiteit van Amsterdam, Amsterdam C, Hollan

H.I. Levine, Department of Mathematics, Brandeis University, Waltham, Ma. 02154, U. S. A.

J.A. Little, Department of Mathematics, University of Michigan, Ann Arbor, Michigan, U. S. A.

J. Martinet, Département de Mathématique, Université de Strasbourg, Rue René Descartes, 67 - Strasbourg, France.

J.A. Morrow, Department of Mathematics, University of Washington, Seattle, Washington, U. S. A.

D.I. Olive, Department of Applied Mathematics, University of Cambridge, Silver Street, Cambridge.

F. Pham, Département de Mathématiques, Faculte des Sciences, Parc Valrose, Nice, France.

W.F. Pohl, Department of Mathematics, University of Minnesota, Minneapolis, Minn. 55455, U. S. A.

I.R. Porteous, Department of Pure Mathematics, The University, P.O. Box 147, Liverpool, L69 3BX, England.

G. Stengle, Department of Mathematics and Astronomy, College of Arts and Scien Lehigh University, Bethlehem, PA 18015, U. S. A.

D. Sullivan, Massachusetts Institute of Technology, Department of Mathematics, Cambridge, Mass. 02139, U. S. A.

C.J. Titus, Department of Mathematics, University of Michigan, Ann Arbor, Michigan, U. S. A.

J.B. Wagoner, Department of Mathematics, University of California, Berkeley, California 94720, U. S. A.

C.T.C. Wall, Department of Pure Mathematics, The University, P.O. Box 147, Liverpool, L69 3BX, England.

M.J. Westwater, Department of Mathematics, University of Washington, Seattle, Washington 98105, U. S. A.

Lecture Notes in Mathematics

Bisher erschienen/Already published

Vol. 1: J. Wermer, Seminar über Funktionen-Algebren. IV, 30 Seiten. 1964. DM 3,80

Vol. 2: A. Borel, Cohomologie des espaces localement compacts d'après. J. Leray. IV, 93 pages. 1964. DM 9,–

Vol. 3: J. F. Adams, Stable Homotopy Theory. Third edition. IV, 78 pages. 1969. DM 8,–

Vol. 4: M. Arkowitz and C. R. Curjel, Groups of Homotopy Classes. 2nd. revised edition. IV, 36 pages. 1967. DM 4,80

Vol. 5: J.-P. Serre, Cohomologie Galoisienne. Troisième édition. VIII, 214 pages. 1965. DM 18,–

Vol. 6: H. Hermes, Term Logic with Choise Operator. III, 55 pages. 1970. DM 6,–

Vol. 7: Ph. Tondeur, Introduction to Lie Groups and Transformation Groups. Second edition. VIII, 176 pages. 1969. DM 14,–

Vol. 8: G. Fichera, Linear Elliptic Differential Systems and Eigenvalue Problems. IV, 176 pages. 1965. DM 13,50

Vol. 9: P. L. Ivănescu, Pseudo-Boolean Programming and Applications. IV, 50 pages. 1965. DM 4,80

Vol. 10: H. Lüneburg, Die Suzukigruppen und ihre Geometrien. VI, 111 Seiten. 1965. DM 8,–

Vol. 11: J.-P. Serre, Algèbre Locale. Multiplicités. Rédigé par P. Gabriel. Seconde édition. VIII, 192 pages. 1965. DM 12,–

Vol. 12: A. Dold, Halbexakte Homotopiefunktoren. II, 157 Seiten. 1966. DM 12,–

Vol. 13: E. Thomas, Seminar on Fiber Spaces. IV, 45 pages. 1966. DM 4,80

Vol. 14: H. Werner, Vorlesung über Approximationstheorie. IV, 184 Seiten und 12 Seiten Anhang. 1966. DM 14,–

Vol. 15: F. Oort, Commutative Group Schemes. VI, 133 pages. 1966. DM 9,80

Vol. 16: J. Pfanzagl and W. Pierlo, Compact Systems of Sets. IV, 48 pages. 1966. DM 5,80

Vol. 17: C. Müller, Spherical Harmonics. IV, 46 pages. 1966. DM 5,–

Vol. 18: H.-B. Brinkmann und D. Puppe, Kategorien und Funktoren. XII, 107 Seiten, 1966. DM 8,–

Vol. 19: G. Stolzenberg, Volumes, Limits and Extensions of Analytic Varieties. IV, 45 pages. 1966. DM 5,40

Vol. 20: R. Hartshorne, Residues and Duality. VIII, 423 pages. 1966. DM 20,–

Vol. 21: Seminar on Complex Multiplication. By A. Borel, S. Chowla, C. S. Herz, K. Iwasawa, J.-P. Serre. IV, 102 pages. 1966. DM 8,–

Vol. 22: H. Bauer, Harmonische Räume und ihre Potentialtheorie. IV, 175 Seiten. 1966. DM 14,–

Vol. 23: P. L. Ivănescu and S. Rudeanu, Pseudo-Boolean Methods for Bivalent Programming. 120 pages. 1966. DM 10,–

Vol. 24: J. Lambek, Completions of Categories. IV, 69 pages. 1966. DM 6,80

Vol. 25: R. Narasimhan, Introduction to the Theory of Analytic Spaces. IV, 143 pages. 1966. DM 10,–

Vol. 26: P.-A. Meyer, Processus de Markov. IV, 190 pages. 1967. DM 15,–

Vol. 27: H. P. Künzi und S. T. Tan, Lineare Optimierung großer Systeme. VI, 121 Seiten. 1966. DM 12,–

Vol. 28: P. E. Conner and E. E. Floyd, The Relation of Cobordism to K-Theories. VIII, 112 pages. 1966. DM 9,80

Vol. 29: K. Chandrasekharan, Einführung in die Analytische Zahlentheorie. VI, 199 Seiten. 1966. DM 16,80

Vol. 30: A. Frölicher and W. Bucher, Calculus in Vector Spaces without Norm. X, 146 pages. 1966. DM 12,–

Vol. 31: Symposium on Probability Methods in Analysis. Chairman. D. A. Kappos. IV, 329 pages. 1967. DM 20,–

Vol. 32: M. André, Méthode Simpliciale en Algèbre Homologique et Algèbre Commutative. IV, 122 pages. 1967. DM 12,–

Vol. 33: G. I. Targonski, Seminar on Functional Operators and Equations. IV, 110 pages. 1967. DM 10,–

Vol. 34: G. E. Bredon, Equivariant Cohomology Theories. VI, 64 pages. 1967. DM 6,80

Vol. 35: N. P. Bhatia and G. P. Szegö, Dynamical Systems. Stability Theory and Applications. VI, 416 pages. 1967. DM 24,–

Vol. 36: A. Borel, Topics in the Homology Theory of Fibre Bundles. VI, 95 pages. 1967. DM 9,–

Vol. 37: R. B. Jensen, Modelle der Mengenlehre. X, 176 Seiten. 1967. DM 14,–

Vol. 38: R. Berger, R. Kiehl, E. Kunz und H.-J. Nastold, Differentialrechnung in der analytischen Geometrie IV, 134 Seiten. 1967 DM 12,–

Vol. 39: Séminaire de Probabilités I. II, 189 pages. 1967. DM 14,–

Vol. 40: J. Tits, Tabellen zu den einfachen Lie Gruppen und ihren Darstellungen. VI, 53 Seiten. 1967. DM 6.80

Vol. 41: A. Grothendieck, Local Cohomology. VI, 106 pages. 1967. DM 10,–

Vol. 42: J. F. Berglund and K. H. Hofmann, Compact Semitopological Semigroups and Weakly Almost Periodic Functions. VI, 160 pages. 1967. DM 12,–

Vol. 43: D. G. Quillen, Homotopical Algebra. VI, 157 pages. 1967. DM 14,–

Vol. 44: K. Urbanik, Lectures on Prediction Theory. IV, 50 pages. 1967. DM 5,80

Vol. 45: A. Wilansky, Topics in Functional Analysis. VI, 102 pages. 1967. DM 9,60

Vol. 46: P. E. Conner, Seminar on Periodic Maps. IV, 116 pages. 1967. DM 10,60

Vol. 47: Reports of the Midwest Category Seminar I. IV, 181 pages. 1967. DM 14,80

Vol. 48: G. de Rham, S. Maumary et M. A. Kervaire, Torsion et Type Simple d'Homotopie. IV, 101 pages. 1967. DM 9,60

Vol. 49: C. Faith, Lectures on Injective Modules and Quotient Rings. XVI, 140 pages. 1967. DM 12,80

Vol. 50: L. Zalcman, Analytic Capacity and Rational Approximation. VI, 155 pages. 1968. DM 13.20

Vol. 51: Séminaire de Probabilités II. IV, 199 pages. 1968. DM 14,–

Vol. 52: D. J. Simms, Lie Groups and Quantum Mechanics. IV, 90 pages. 1968. DM 8,–

Vol. 53: J. Cerf, Sur les difféomorphismes de la sphère de dimension trois (Γ_4= O). XII, 133 pages. 1968. DM 12,–

Vol. 54: G. Shimura, Automorphic Functions and Number Theory. VI, 69 pages. 1968. DM 8,–

Vol. 55: D. Gromoll, W. Klingenberg und W. Meyer, Riemannsche Geometrie im Großen. VI, 287 Seiten. 1968. DM 20,–

Vol. 56: K. Floret und J. Wloka, Einführung in die Theorie der lokalkonvexen Räume. VIII, 194 Seiten. 1968. DM 16,–

Vol. 57: F. Hirzebruch und K. H. Mayer, O (n)-Mannigfaltigkeiten, exotische Sphären und Singularitäten. IV, 132 Seiten. 1968. DM 10,80

Vol. 58: Kuramochi Boundaries of Riemann Surfaces. IV, 102 pages. 1968. DM 9,60

Vol. 59: K. Jänich, Differenzierbare G-Mannigfaltigkeiten. VI, 89 Seiten. 1968. DM 8,–

Vol. 60: Seminar on Differential Equations and Dynamical Systems. Edited by G. S. Jones. VI, 106 pages. 1968. DM 9,60

Vol. 61: Reports of the Midwest Category Seminar II. IV, 91 pages. 1968. DM 9,60

Vol. 62: Harish-Chandra, Automorphic Forms on Semisimple Lie Groups X, 138 pages. 1968. DM 14,–

Vol. 63: F. Albrecht, Topics in Control Theory. IV, 65 pages. 1968. DM 6,80

Vol. 64: H. Berens, Interpolationsmethoden zur Behandlung von Approximationsprozessen auf Banachräumen. VI, 90 Seiten. 1968. DM 8,–

Vol. 65: D. Kölzow, Differentiation von Maßen. XII, 102 Seiten. 1968. DM 8,–

Vol. 66: D. Ferus, Totale Absolutkrümmung in Differentialgeometrie und -topologie. VI, 85 Seiten. 1968. DM 8,–

Vol. 67: F. Kamber and P. Tondeur, Flat Manifolds. IV, 53 pages. 1968. DM 5,80

Vol. 68: N. Boboc et P. Mustată, Espaces harmoniques associés aux opérateurs différentiels linéaires du second ordre de type elliptique. VI, 95 pages. 1968. DM 8,60

Vol. 69: Seminar über Potentialtheorie. Herausgegeben von H. Bauer. VI, 180 Seiten. 1968. DM 14,80

Vol. 70: Proceedings of the Summer School in Logic. Edited by M. H. Löb. IV, 331 pages. 1968. DM 20,–

Vol. 71: Séminaire Pierre Lelong (Analyse), Année 1967 – 1968. VI, 190 pages. 1968. DM 14,–

Vol. 72: The Syntax and Semantics of Infinitary Languages. Edited by J. Barwise. IV, 268 pages. 1968. DM 18,–

Vol. 73: P. E. Conner, Lectures on the Action of a Finite Group. IV, 123 pages. 1968. DM 10,–

Vol. 74: A. Fröhlich, Formal Groups. IV, 140 pages. 1968. DM 12,–

Vol. 75: G. Lumer, Algèbres de fonctions et espaces de Hardy. VI, 80 pages. 1968. DM 8,–

Vol. 76: R. G. Swan, Algebraic K-Theory. IV, 262 pages. 1968. DM 18,–

Vol. 77: P.-A. Meyer, Processus de Markov: la frontière de Martin. IV, 123 pages. 1968. DM 10,–

Vol. 78: H. Herrlich, Topologische Reflexionen und Coreflexionen. XVI, 166 Seiten. 1968. DM 12,–

Vol. 79: A. Grothendieck, Catégories Cofibrées Additives et Complexe Cotangent Relatif. IV, 167 pages. 1968. DM 12,–

Vol. 80: Seminar on Triples and Categorical Homology Theory. Edited by B. Eckmann. IV, 398 pages. 1969. DM 20,–

Vol. 81: J.-P. Eckmann et M. Guenin, Méthodes Algébriques en Mécanique Statistique. VI, 131 pages. 1969. DM 12,–

Vol. 82: J. Wloka, Grundräume und verallgemeinerte Funktionen. VIII, 131 Seiten. 1969. DM 12,–

Vol. 83: O. Zariski, An Introduction to the Theory of Algebraic Surfaces. IV, 100 pages. 1969. DM 8,–

Vol. 84: H. Lüneburg, Transitive Erweiterungen endlicher Permutationsgruppen. IV, 119 Seiten. 1969. DM 10,–

Vol. 85: P. Cartier et D. Foata, Problèmes combinatoires de commutation et réarrangements. IV, 88 pages. 1969. DM 8,–

Vol. 86: Category Theory, Homology Theory and their Applications I. Edited by P. Hilton. VI, 216 pages. 1969. DM 16,–

Vol. 87: M. Tierney, Categorical Constructions in Stable Homotopy Theory. IV, 65 pages. 1969. DM 6,–

Vol. 88: Séminaire de Probabilités III. IV, 229 pages. 1969. DM 18,–

Vol. 89: Probability and Information Theory. Edited by M. Behara, K. Krickeberg and J. Wolfowitz. IV, 256 pages. 1969. DM 18,–

Vol. 90: N. P. Bhatia and O. Hajek, Local Semi-Dynamical Systems. II, 157 pages. 1969. DM 14,–

Vol. 91: N. N. Janenko, Die Zwischenschrittmethode zur Lösung mehrdimensionaler Probleme der mathematischen Physik. VIII, 194 Seiten. 1969. DM 16,80

Vol. 92: Category Theory, Homology Theory and their Applications II. Edited by P. Hilton. V, 308 pages. 1969. DM 20,–

Vol. 93: K. R. Parthasarathy, Multipliers on Locally Compact Groups. III, 54 pages. 1969. DM 5,60

Vol. 94: M. Machover and J. Hirschfeld, Lectures on Non-Standard Analysis. VI, 79 pages. 1969. DM 6,–

Vol. 95: A. S. Troelstra, Principles of Intuitionism. II, 111 pages. 1969. DM 10,–

Vol. 96: H.-B. Brinkmann und D. Puppe, Abelsche und exakte Kategorien, Korrespondenzen. V, 141 Seiten. 1969. DM 10,–

Vol. 97: S. O. Chase and M. E. Sweedler, Hopf Algebras and Galois theory. II, 133 pages. 1969. DM 10,–

Vol. 98: M. Heins, Hardy Classes on Riemann Surfaces. III, 106 pages. 1969. DM 10,–

Vol. 99: Category Theory, Homology Theory and their Applications III. Edited by P. Hilton. IV, 489 pages. 1969. DM 24,–

Vol. 100: M. Artin and B. Mazur, Etale Homotopy. II, 196 Seiten. 1969. DM 12,–

Vol. 101: G. P. Szegö et G. Treccani, Semigruppi di Trasformazioni Multivoche. VI, 177 pages. 1969. DM 14,–

Vol. 102: F. Stummel, Rand- und Eigenwertaufgaben in Sobolewschen Räumen. VIII, 386 Seiten. 1969. DM 20,–

Vol. 103: Lectures in Modern Analysis and Applications I. Edited by C. T. Taam. VII, 162 pages. 1969. DM 12,–

Vol. 104: G. H. Pimbley, Jr., Eigenfunction Branches of Nonlinear Operators and their Bifurcations. II, 128 pages. 1969. DM 10,–

Vol. 105: R. Larsen, The Multiplier Problem. VII, 284 pages. 1969. DM 18,–

Vol. 106: Reports of the Midwest Category Seminar III. Edited by S. Mac Lane. III, 247 pages. 1969. DM 16,–

Vol. 107: A. Peyerimhoff, Lectures on Summability. III, 111 pages. 1969. DM 8,–

Vol. 108: Algebraic K-Theory and its Geometric Applications. Edited by R. M.F. Moss and C. B. Thomas. IV, 86 pages. 1969. DM 6,–

Vol. 109: Conference on the Numerical Solution of Differential Equations. Edited by J. Ll. Morris. VI, 275 pages. 1969. DM 18,–

Vol. 110: The Many Facets of Graph Theory. Edited by G. Chartrand and S. F. Kapoor. VIII, 290 pages. 1969. DM 18,–

Vol. 111: K. H. Mayer, Relationen zwischen charakteristischen Zahlen. III, 99 Seiten. 1969. DM 8,–

Vol. 112: Colloquium on Methods of Optimization. Edited by N. N. Moiseev. IV, 293 pages. 1970. DM 18,–

Vol. 113: R. Wille, Kongruenzklassengeometrien. III, 99 Seiten. 1970. DM 8,–

Vol. 114: H. Jacquet and R. P. Langlands, Automorphic Forms on GL (2). VII, 548 pages. 1970. DM 24,–

Vol. 115: K. H. Roggenkamp and V. Huber-Dyson, Lattices over Orders I. XIX, 290 pages. 1970. DM 18,–

Vol. 116: Sèminaire Pierre Lelong (Analyse) Année 1969. IV, 195 pages. 1970. DM 14,–

Vol. 117: Y. Meyer, Nombres de Pisot, Nombres de Salem et Analyse Harmonique. 63 pages. 1970. DM 6.–

Vol. 118: Proceedings of the 15th Scandinavian Congress, Oslo 1968. Edited by K. E. Aubert and W. Ljunggren. IV, 162 pages. 1970. DM 12,–

Vol. 119: M. Raynaud, Faisceaux amples sur les schémas en groupes et les espaces homogènes. III, 219 pages. 1970. DM 14,–

Vol. 120: D. Siefkes, Büchi's Monadic Second Order Successor Arithmetic. XII, 130 Seiten. 1970. DM 12,–

Vol. 121: H. S. Bear, Lectures on Gleason Parts. III, 47 pages. 1970. DM 6,–

Vol. 122: H. Zieschang, E. Vogt und H.-D. Coldewey, Flächen und ebene diskontinuierliche Gruppen. VIII, 203 Seiten. 1970. DM 16,–

Vol. 123: A. V. Jategaonkar, Left Principal Ideal Rings. VI, 145 pages. 1970. DM 12,–

Vol. 124: Séminare de Probabilités IV. Edited by P. A. Meyer. IV, 282 pages. 1970. DM 20,–

Vol. 125: Symposium on Automatic Demonstration. V, 310 pages. 1970. DM 20,–

Vol. 126: P. Schapira, Théorie des Hyperfonctions. XI, 157 pages. 1970. DM 14,–

Vol. 127: I. Stewart, Lie Algebras. IV, 97 pages. 1970. DM 10,–

Vol. 128: M. Takesaki, Tomita's Theory of Modular Hilbert Algebras and its Applications. II, 123 pages. 1970. DM 10,–

Vol. 129: K. H. Hofmann, The Duality of Compact Semigroups and C*- Bigebras. XII, 142 pages. 1970. DM 14,–

Vol. 130: F. Lorenz, Quadratische Formen über Körpern. II, 77 Seiten. 1970. DM 8,–

Vol. 131: A Borel et al., Seminar on Algebraic Groups and Related Finite Groups. VII, 321 pages. 1970. DM 22,–

Vol. 132: Symposium on Optimization. III, 348 pages. 1970. DM 22,–

Vol. 133: F. Topsøe, Topology and Measure. XIV, 79 pages. 1970. DM 8,–

Vol. 134: L. Smith, Lectures on the Eilenberg-Moore Spectral Sequence. VII, 142 pages. 1970. DM 14,–

Vol. 135: W. Stoll, Value Distribution of Holomorphic Maps into Compact Complex Manifolds. II, 267 pages. 1970. DM 18,–

Vol. 136: M. Karoubi et al., Séminaire Heidelberg-Saarbrücken-Strasbuorg sur la K-Théorie. IV, 264 pages. 1970. DM 18,–

Vol. 137: Reports of the Midwest Category Seminar IV. Edited by S. MacLane. III, 139 pages. 1970. DM 12,–

Vol. 138: D. Foata et M. Schützenberger, Théorie Géométrique des Polynômes Eulériens. V, 94 pages. 1970. DM 10,–

Vol. 139: A. Badrikian, Séminaire sur les Fonctions Aléatoires Linéaires et les Mesures Cylindriques. VII, 221 pages. 1970. DM 18,–

Vol. 140: Lectures in Modern Analysis and Applications II. Edited by C. T. Taam. VI, 119 pages. 1970. DM 10,–

Vol. 141: G. Jameson, Ordered Linear Spaces. XV, 194 pages. 1970. DM 16,–

Vol. 142: K. W. Roggenkamp, Lattices over Orders II. V, 388 pages. 1970. DM 22,–

Vol. 143: K. W. Gruenberg, Cohomological Topics in Group Theory. XIV, 275 pages. 1970. DM 20,–

Vol. 144: Seminar on Differential Equations and Dynamical Systems II. Edited by J. A. Yorke. VIII, 268 pages. 1970. DM 20,–

Vol. 145: E. J. Dubuc, Kan Extensions in Enriched Category Theory. XVI, 173 pages. 1970. DM 16,–

Vol. 146: A. B. Altman and S. Kleiman, Introduction to Grothendieck Duality Theory. II, 192 pages. 1970. DM 18,–

Vol. 147: D. E. Dobbs, Cech Cohomological Dimensions for Commutative Rings. VI, 176 pages. 1970. DM 16,–

Vol. 148: R. Azencott, Espaces de Poisson des Groupes Localement Compacts. IX, 141 pages. 1970. DM 14,–

Vol. 149: R. G. Swan and E. G. Evans, K-Theory of Finite Groups and Orders. IV, 237 pages. 1970. DM 20,–

Vol. 150: Heyer, Dualität lokalkompakter Gruppen. XIII, 372 Seiten. 1970. DM 20,–

Vol. 151: M. Demazure et A. Grothendieck, Schémas en Groupes I. (SGA 3). XV, 562 pages. 1970. DM 24,–

Vol. 152: M. Demazure et A. Grothendieck, Schémas en Groupes II. (SGA 3). IX, 654 pages. 1970. DM 24,–

Vol. 153: M. Demazure et A. Grothendieck, Schémas en Groupes III. (SGA 3). VIII, 529 pages. 1970. DM 24,–

Vol. 154: A. Lascoux et M. Berger, Variétés Kähleriennes Compactes. VII, 83 pages. 1970. DM 8,–

Vol. 155: Several Complex Variables I, Maryland 1970. Edited by J. Horváth. IV, 214 pages. 1970. DM 18,–

Vol. 156: R. Hartshorne, Ample Subvarieties of Algebraic Varieties. XIV, 256 pages. 1970. DM 20,–

Vol. 157: T. tom Dieck, K. H. Kamps und D. Puppe, Homotopietheorie. VI, 265 Seiten. 1970. DM 20,–

Vol. 158: T. G. Ostrom, Finite Translation Planes. IV. 112 pages. 1970. DM 10,–

Vol. 159: R. Ansorge und R. Hass. Konvergenz von Differenzenverfahren für lineare und nichtlineare Anfangswertaufgaben. VIII, 145 Seiten. 1970. DM 14,–

'ol. 160: L. Sucheston, Constributions to Ergodic Theory and Probaility. VII, 277 pages. 1970. DM 20,–

'ol. 161: J. Stasheff, H-Spaces from a Homotopy Point of View. , 95 pages. 1970. DM 10,–

ol. 162: Harish-Chandra and van Dijk, Harmonic Analysis on Reducve p-adic Groups. IV, 125 pages. 1970. DM 12,–

ol. 163: P. Deligne, Equations Différentielles à Points Singuliers eguliers. III, 133 pages. 1970. DM 12,–

ol. 164: J. P. Ferrier, Seminaire sur les Algebres Complètes. II, 69 paes. 1970. DM 8,–

ol. 165: J. M. Cohen, Stable Homotopy. V, 194 pages. 1970. DM 16,–

ol. 166: A. J. Silberger, PGL_2 over the p-adics: its Representations, pherical Functions, and Fourier Analysis. VII, 202 pages. 1970. M 18,–

l. 167: Lavrentiev, Romanov and Vasiliev, Multidimensional Inverse oblems for Differential Equations. V, 59 pages. 1970. DM 10,–

l. 168: F. P. Peterson, The Steenrod Algebra and its Applications: conference to Celebrate N. E. Steenrod's Sixtieth Birthday. VII, 7 pages. 1970. DM 22,–

l. 169: M. Raynaud, Anneaux Locaux Henséliens. V, 129 pages. 1970. 12,–

. 170: Lectures in Modern Analysis and Applications III. Edited by T. Taam. VI, 213 pages. 1970. DM 18,–

. 171: Set-Valued Mappings, Selections and Topological Properties 2^X. Edited by W. M. Fleischman. X, 110 pages. 1970. DM 12,–

. 172: Y.-T. Siu and G. Trautmann, Gap-Sheaves and Extension Coherent Analytic Subsheaves. V, 172 pages. 1971. DM 16,–

. 173: J. N. Mordeson and B. Vinograde, Structure of Arbitrary ely Inseparable Extension Fields. IV, 138 pages. 1970. DM 14,–

. 174: B. Iversen, Linear Determinants with Applications to the ard Scheme of a Family of Algebraic Curves. VI, 69 pages. 1970. 8,–

. 175: M. Brelot, On Topologies and Boundaries in Potential Theory. 76 pages. 1971. DM 18,–

. 176: H. Popp, Fundamentalgruppen algebraischer Mannigfaltigen. IV, 154 Seiten. 1970. DM 16,–

. 177: J. Lambek, Torsion Theories, Additive Semantics and Rings Quotients. VI, 94 pages. 1971. DM 12,–

178: Th. Bröcker und T. tom Dieck, Kobordismentheorie. XVI, Seiten. 1970. DM 18,–

179: Seminaire Bourbaki – vol. 1968/69. Exposés 347-363. IV. pages. 1971. DM 22,–

180: Séminaire Bourbaki – vol. 1969/70. Exposés 364-381. IV, pages. 1971. DM 22,–

181: F. DeMeyer and E. Ingraham, Separable Algebras over nmutative Rings. V, 157 pages. 1971. DM 16.–

182: L. D. Baumert. Cyclic Difference Sets. VI, 166 pages. 1971. 16,–

183: Analytic Theory of Differential Equations. Edited by P. F. Hsieh A. W. J. Stoddart. VI, 225 pages. 1971. DM 20,–

Vol. 184: Symposium on Several Complex Variables, Park City, Utah, 1970. Edited by R. M. Brooks. V, 234 pages. 1971. DM 20,–

Vol. 185: Several Complex Variables II, Maryland 1970. Edited by J. Horváth. III, 287 pages. 1971. DM 24,–

Vol. 186: Recent Trends in Graph Theory. Edited by M. Capobianco/ J. B. Frechen/M. Krolik. VI, 219 pages. 1971. DM 18.–

Vol. 187: H. S. Shapiro, Topics in Approximation Theory. VIII, 275 pages. 1971. DM 22,–

Vol. 188: Symposium on Semantics of Algorithmic Languages. Edited by E. Engeler. VI, 372 pages. 1971. DM 26,–

Vol. 189: A. Weil, Dirichlet Series and Automorphic Forms. V, 164 pages. 1971. DM 16,–

Vol. 190: Martingales. A Report on a Meeting at Oberwolfach, May 17-23, 1970. Edited by H. Dinges. V, 75 pages. 1971. DM 12,–

Vol. 191: Séminaire de Probabilités V. Edited by P. A. Meyer. IV, 372 pages. 1971. DM 26,–

Vol. 192: Proceedings of Liverpool Singularities – Symposium I. Edited by C. T. C. Wall. V, 319 pages. 1971. DM 24,–

Vol. 193: Symposium on the Theory of Numerical Analysis. Edited by J. Ll. Morris. VI, 152 pages. 1971. DM 16,–

Vol. 194: M. Berger, P. Gauduchon et E. Mazet. Le Spectre d'une Variété Riemannienne. VII, 251 pages. 1971. DM 22,–

Vol. 195: Reports of the Midwest Category Seminar V. Edited by J.W. Gray and S. Mac Lane. III, 255 pages. 1971. DM 22,–

Vol. 196: H-spaces – Neuchâtel (Suisse)- Août 1970. Edited by F. Sigrist, V, 156 pages. 1971. DM 16,–

Vol. 197: Manifolds – Amsterdam 1970. Edited by N. H. Kuiper. V, 231 pages. 1971. DM 20,–

Vol. 198: M. Hervé, Analytic and Plurisubharmonic Functions in Finite and Infinite Dimensional Spaces. VI, 90 pages. 1971. DM 16.–

Vol. 199: Ch. J. Mozzochi, On the Pointwise Convergence of Fourier Series. VII, 87 pages. 1971. DM 16,–

Vol. 200: U. Neri, Singular Integrals. VII, 272 pages. 1971. DM 22,–

Vol. 201: J. H. van Lint, Coding Theory. VII, 136 pages. 1971. DM 16,–

Vol. 202: J. Benedetto, Harmonic Analysis on Totally Disconnected Sets. VIII, 261 pages. 1971. DM 22,–

Vol. 203: D. Knutson, Algebraic Spaces. VI, 261 pages. 1971. DM 22,–

Vol. 204: A. Zygmund, Intégrales Singulières. IV, 53 pages. 1971. DM 12,–

Vol. 205: Séminaire Pierre Lelong (Analyse) Année 1970. VI, 243 pages. 1971. DM 20,–

Vol. 206: Symposium on Differential Equations and Dynamical Systems. Edited by D. Chillingworth. XI, 173 pages. 1971. DM 16,–

Vol. 207: L. Bernstein, The Jacobi-Perron Algorithm – Its Theory and Application. IV, 161 pages. 1971. DM 16,–

Vol. 208: A. Grothendieck and J. P. Murre, The Tame Fundamental Group of a Formal Neighbourhood of a Divisor with Normal Crossings on a Scheme. VIII, 133 pages. 1971. DM 16,–

Vol. 209: Proceedings of Liverpool Singularities Symposium II. Edited by C. T. C. Wall. V, 280 pages. 1971. DM 22,–